Contents

Reactive Science

Reactive Science

Science

For GCSE

Jean Martin

Bryan Milner

CAMBRIDGE

Series editors: Jean Martin, Bryan Milner
Subject editors: Jean Martin, Bryan Milner, John Mills, Ray Oliver
Biology authors: Jenny Burden, Chris Christofi, Geraint Evans, Jean Martin
Chemistry authors: Peter Evans, Helen Norris, Ray Oliver
Physics authors: Michael Chyriwsky, Anne Cohen, Bryan Milner, Alan Yate
Consultants: Nigel Heslop, Martyn Keeley, Helen Norris
Authors who have worked on this revision: Sam Ellis, Jean Martin, Nicky Thomas

Much of the material in this book was previously published as the Science Foundations series

PUBLISHED BY THE PRESS SYNDICATE OF THE UNIVERSITY OF CAMBRIDGE
The Pitt Building, Trumpington Street, Cambridge, United Kingdom

CAMBRIDGE UNIVERSITY PRESS
The Edinburgh Building, Cambridge CB2 2RU, UK
40 West 20th Street, New York, NY 10011-4211, USA
477 Williamstown Road, Port Melbourne, VIC 3207, Australia
Ruiz de Alarcón 13, 28014 Madrid, Spain
Dock House, The Waterfront, Cape Town 8001, South Africa

http://www.cambridge.org

© Cambridge University Press 1997, 2001, 2004

First published 2004

Printed in the United Kingdom at the University Press, Cambridge

Typeface Photina   *System* QuarkXPress®

A catalogue record for this book is available from the British Library

ISBN 0 521 60920 8   paperback ✓

Cover and book design by Hardlines Ltd
Page layout by Hardlines Ltd

Physics

Acknowledgements

12, Action Plus; 16, David Scharf/SPL; 18, Oxford Scientific (www.osf.co.uk); 24, Jerry Mason/SPL; 25, 26, 35, 121*br*, 142, 166, 298, Biophoto Associates; 29*t*, Secchi-Lecaque/Roussel-UCLAF/CNRI/SPL; 29*b*, Michael Brooke; 30, Barry Dowsett/SPL; 46, 157, Nigel Cattlin/Holt Studios International; 48, Andrew Syred/SPL; 70, 483, 494, Mike Wyndham Picture Collection; 79, 136, Graham Portlock; 82, *Report of the Royal College of Physicians, 1962 – Smoking and Health*/ courtesy of the Royal College of Physicians, London; 85, Bryan & Cherry Alexander; 93, Stephen Dalton/NHPA; 94, Anthony Bannister/NHPA; 96, Bill Wood/NHPA; 97, Marty Stouffer/Animals Animals/Oxford Scientific (www.osf.co.uk); 102, Getty Images/Jay Freis; 104, Geoff Dore/Bruce Coleman Ltd; 115, Agripicture Images; 117, 362, 376*t*, Ecoscene; 120*c* Paul Glendell/Environmental Images; 120*r*, 121*bl*, Nick Hawkes/Ecoscene; 146, John Walmsley; 150, Bill Longcore/SPL; 151*t* Ralph Reinhold/Animals Animals/Oxford Scientific (www.osf.co.uk); 151*c, b*, Gerard Lacz/NHPA; 159*bl*, 161*br*, Kevin Schafer/NHPA; 158, 159*br*, A S Gould; 160, *213 bl*, Photo Images/Werner Reith; 161*tl*, Novosti/SPL; 161*tr*, Derek Bromhall/Oxford Scientific (www.osf.co.uk); 167, David Fox/Oxford Scientific (www.osf.co.uk); 168, Thomas Dressler/Planet Earth Pictures; 169, Mary Evans Picture Library; 170*t* Garry Watson/SPL; 170*b* Science Source/SPL; 178, 181, 184, 185; 188, 190*c*, 192, 203, 205, *207t, 213tr, br*, 224, 258, 259, 260, 290, 299, 318, 393, 482, 492*t* Andrew Lambert; 188, Copper in Architecture; 190, TRH Pictures/MOD; 209, Ben Osborne/Oxford Scientific (www.osf.co.uk); 210, courtesy of British Cement Association; 227*tl*, Adrian Davies/Bruce Coleman Ltd; 227*tr*, Erik Shaffer/Ecoscene; 229, *233tr*, 276, Geoscience Features Picture Library; 230, Ludek Pesek/Science Photo Library; 233*tl*, Science Photo Library; 239, Jim Sugar/CORBIS; 370, Kevin King/Ecoscene; 371, 461, NASA/SPL; 376*b*, Anthony Cooper/Ecoscene; 402, IBM UK Labs, Hursley; 428, Takashi Takahura/SPL; 436, Sutton Motorsport Images; 441, NASA/Image Select; 442, Mike Hewitt/Action Plus; 451, Luke Dodd/SPL; 458, 462, NOAO/Science Photo Library; 460, NASA/ courtesy of the Lunar and Planetary Institute; 492*b* courtesy of HR Wallingford Ltd; 516, Chris Priest/SPL

NHPA = Natural History Photographic Agency
SPL   = Science Photo Library

We have made every effort to trace copyright holders, but if we have inadvertently overlooked any we will be pleased to make the necessary arrangements at the earliest opportunity.

How to use this book

An introduction for students and teachers

Reactive Science includes material from biology, chemistry and physics. Each subject is colour-coded and broken down into a number of sections for ease of use.

The number of pages devoted to each topic will vary depending on the complexity of the concepts covered. Coverage is not restricted to double-page spreads.

There are three main types of material:

- Ideas from your previous studies of Science at Key Stage 3;

- Scientific ideas that all Key Stage 4 students are expected to know, whether they are entered for the Foundation Tier or the Higher Tier of GCSE Science tests and examinations.

- Scientific ideas that only candidates entered for the Higher Tier GCSE tests and examinations need to know.

Ideas from your previous science studies at Key Stage 3

You need to understand these ideas before you start on the new science for Key Stage 4. But you will <u>not</u> be assessed <u>directly</u> on these Key Stage 3 ideas in GCSE Science tests and examinations.

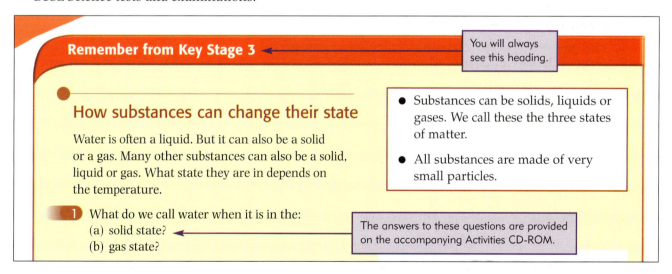

Remember from Key Stage 3

You will always see this heading.

How substances can change their state

Water is often a liquid. But it can also be a solid or a gas. Many other substances can also be a solid, liquid or gas. What state they are in depends on the temperature.

1 What do we call water when it is in the:
  (a) solid state?
  (b) gas state?

The answers to these questions are provided on the accompanying Activities CD-ROM.

- Substances can be solids, liquids or gases. We call these the three states of matter.

- All substances are made of very small particles.

Each time you are introduced to a new idea you will usually be asked a question. This is so you can make sure that you really understand the ideas.

Science that all Key Stage 4 students need to know

Most of the material in the book is of this type. It does not have any special border or heading.

> Each concept is introduced by a section of text providing a clear explanation and questions to help understanding.

How light is reflected

A mirror that is flat is called a <u>plane</u> mirror.
The diagrams show how a plane mirror **reflects** light.

Look at the angle with the mirror:

- of the beam of light that comes from the lamp;
- of the beam of light after it is reflected.

1 Look at the diagrams.
What can you say about the angle for beams that arrive at, and reflect from, the mirror?

Light is reflected from a plane mirror at the same <u>angle</u> as it strikes the mirror.

> The answers to these questions are provided on the accompanying Activities CD-ROM.

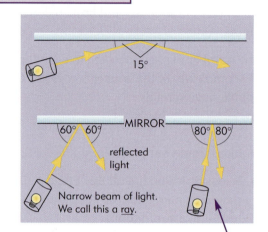

15°

MIRROR
60° 60°   80° 80°

reflected light

Narrow beam of light.
We call this a <u>ray</u>.

> Artwork, diagrams and photographs are placed near the relevant text to support the explanation of concepts.

Science that only Higher Tier students need to remember

In order to provide clear progression of topic coverage, Higher Tier material is placed immediately after the Foundation Tier material that it builds on.

Higher

> Higher Tier material is clearly differentiated by using the 'Higher' heading and a coloured background.

More about anaerobic respiration

In anaerobic respiration, glucose isn't broken down completely so less energy is released than in aerobic respiration.

The lactic acid produced is one cause of muscle fatigue. Your muscles get tired and ache, and don't work as well as usual.

11 In the first few minutes of exercise, is it carbon dioxide or lactic acid that makes you pant? Explain your answer.

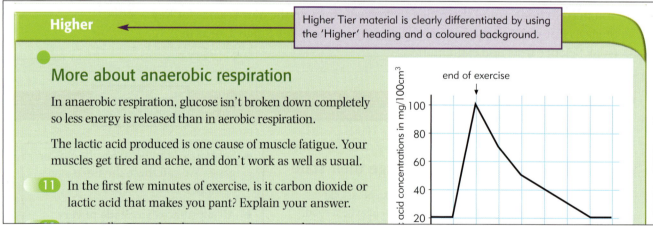

The questions in these sections are like many of the questions you will meet in Higher Tier tests and examinations. You have to use ideas from the topic to explain something new. You are not expected to remember the answers to these questions.

A note about Ideas and Evidence

All science specifications must now assess candidates' understanding of what the National Curriculum calls *Ideas and Evidence*. Those parts of this book which include material about this aspect of Science are indicated on the contents pages like this:

I&E Humans against microorganisms   32

1 Humans as organisms

What are our bodies built from?

What are cells like?

Your body is made of billions of cells. Larger organisms like humans have more cells than smaller organisms like ants. Some organisms have only one cell.

Most animal cells have the same basic parts:

- a **nucleus** which controls everything that happens in the cell;
- **cytoplasm** where most of the cell's chemical reactions happen;
- a **cell membrane** to control which substances pass in and out of the cell. It also holds the cell together.

1. Make a table to show the <u>three</u> main cell parts and what each part does.

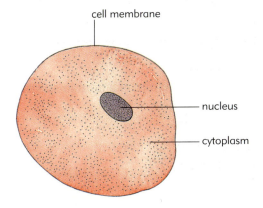

Parts of an animal cell.

Some cells look different

Cells may be different shapes and sizes but they still have a nucleus, cytoplasm and cell membrane. They may also have other parts so that they can do their jobs.

We say the cells are **specialised** to do their job.

For example, red blood cells carry oxygen around the body. They are full of a substance that can combine with oxygen, but which also releases oxygen again.

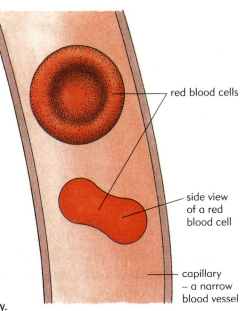

Red blood cells carry oxygen around the body.

2 How is a nerve cell specialised to do its job?

3 How is a red blood cell specialised to do its job?

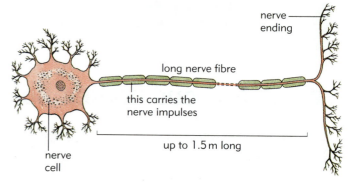

Nerve cells can carry signals (nerve impulses) between the brain and other parts of the body.

Sperm cells have a tail

Look at the diagrams.

Sperm cells are placed in a woman's vagina. From there they swim up the uterus and along the egg tube (**oviduct**). If they reach an egg cell in the oviduct they can fertilise it.

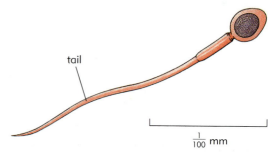

Sperm cells swim to fertilise an egg cell.

4 Copy the diagram of a human sperm cell. Add these labels:
(a) nucleus
(b) cytoplasm
(c) cell membrane.

5 What part of a sperm cell helps it reach the egg?

Cells that line the oviducts have hairs

Each oviduct is a tube which carries eggs from the ovary to the womb. Egg cells are released from an ovary and travel down an oviduct.

Each oviduct is lined with special cells. These cells have tiny hairs which can move forwards and backwards.

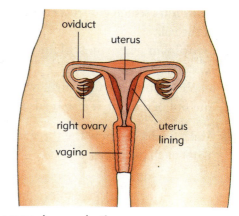

A woman's reproductive organs.

6 Make a large copy of Cell X in the oviduct diagram. Label the nucleus, cytoplasm and cell membrane.

7 Why do the cells lining an oviduct have tiny hairs on their surface?

You should be able to match specialised cells to the jobs that they do in tissues, organs or the whole organism when you are given information about the structure of cells.

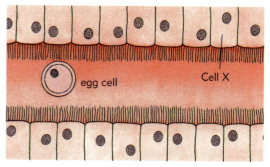

Inside an oviduct.

Digesting our food

Remember from Key Stage 3

- All the food you eat is your diet.
- Your diet includes carbohydrates, proteins and fats.
- Your digestive system breaks down food.
- Your gullet, stomach, liver, pancreas, small intestine and large intestine are organs in your digestive system.
- Digested food passes into your bloodstream.

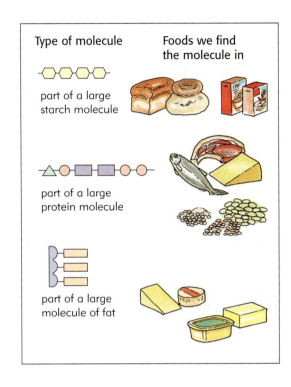

Type of molecule

Foods we find the molecule in

part of a large starch molecule

part of a large protein molecule

part of a large molecule of fat

Why do we need to digest our food?

Most of the food we eat is made up of fairly large molecules. Starch, protein and fat are all made up of large molecules.

1 Name two foods that contain large starch molecules.

2 What sort of large molecules does butter contain?

Our bodies cannot use large molecules like starch. The large molecules cannot pass through the lining of our small intestine. This is because the large food molecules cannot dissolve. We say they are **insoluble**.

Large food molecules must be broken down into smaller ones that can dissolve. These molecules are **soluble** and can pass into the blood.

3 What does the word 'soluble' mean?

How do we break down large molecules?

Our bodies break down large food molecules into smaller ones. This breakdown is called **digestion**.

These small molecules dissolve and pass through the lining and into the bloodstream in the wall of the small intestine. The diagrams show you the small molecules that are made by digestion.

4 Write sentences to show the molecules that we make when we digest starch, protein and fat.

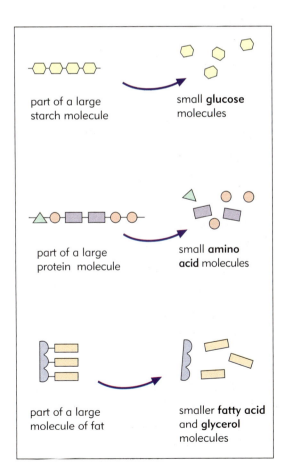

part of a large starch molecule

small **glucose** molecules

part of a large protein molecule

small **amino acid** molecules

part of a large molecule of fat

smaller **fatty acid** and **glycerol** molecules

5 Look at the picture of starch digestion. Then copy and complete the picture for the breakdown of protein.

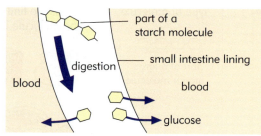

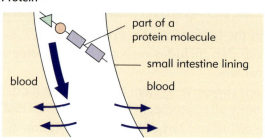

Only small, soluble molecules can pass through the intestine lining into the blood.

Is all food digested?

Vitamins do not need to be digested.

We cannot digest fibre. It makes up most of the undigested waste that we call **faeces**. This leaves the body via the anus.

6 Do you think vitamins are large or small molecules?

7 Do you think fibre molecules are large or small?

Enzymes digest our food

Our digestive systems contain **glands**. These glands produce substances called **enzymes**. Enzymes are **catalysts**. Catalysts make chemical reactions happen quickly and easily. Digestive enzymes help us to break down food more easily. Our bodies make lots of different digestive enzymes. Each enzyme breaks down a particular food. When an enzyme has broken down one food molecule, it can then break down another molecule of the same kind. It can do this over and over again. It makes the reaction happen without being used up.

8 Look at the diagram. Write a short paragraph to explain how enzymes break down fat.

9 Why can a small amount of enzyme break down a large amount of food?

How you digest fat.

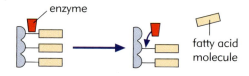

The enzyme snips off a fatty acid molecule.

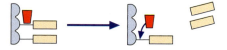

Then it snips off another two.

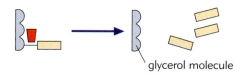

The same enzyme molecule can do this over and over again to more fat molecules.

It makes the reaction happen without being used up. We say this is a catalyst.

Different foods need different enzymes

- Enzymes which break down fats are called **lipases**.

- Enzymes which break down starch are called **amylases**.

- Enzymes which break down proteins are called **proteases**.

Our bodies make several different enzymes in each of these groups.

10 You can show starch digestion like this:

starch $\xrightarrow{\text{amylase}}$ glucose

Draw similar diagrams for the digestion of proteins and fats.

11 Make a copy of the diagram.
Add more stages to show the part of the starch molecule being completely digested.

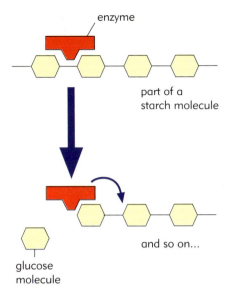

How a starch molecule is digested.

Your digestive glands don't just make enzymes

Protease enzymes in your stomach work best in acid conditions. So glands in your stomach lining produce hydrochloric acid. When the food from the stomach passes into the small intestine, it is still acidic.

The enzymes in your small intestine need alkaline conditions. Salts in **bile** neutralise the acid so these enzymes can work properly.

Bile also breaks fat down into tiny droplets of oil. We say that bile **emulsifies** the fats. This increases the surface area for lipase enzymes to act on.

12 Where is bile
(a) made?
(b) stored?

13 Describe the jobs that bile does.

14 What does the word 'emulsifies' mean?

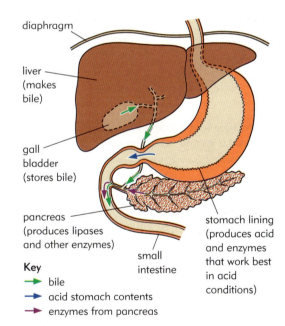

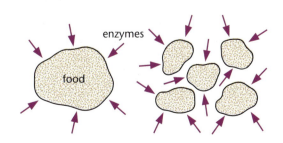

Enzymes break down food faster when the surface area of the food is large.

Your digestive system – what happens where?

In your mouth and gullet

You start to digest your food as soon as you put it into your mouth. Your salivary glands produce **saliva**. Saliva contains **amylase**.

Your gullet (or **oesophagus**) carries food from your mouth to your stomach. Muscles in the wall of your gullet squeeze the food along. It is a bit like pushing toothpaste all the way from the bottom of a nearly empty tube.

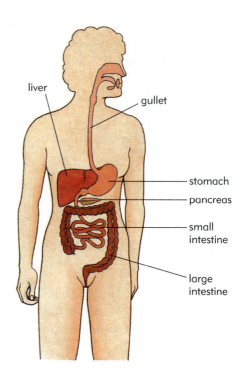

The human digestive system breaks down and absorbs food.

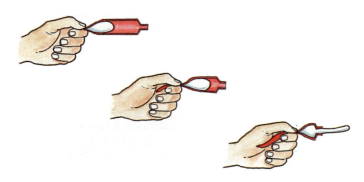

Pushing toothpaste from the bottom of a tube.
Food is moved along the gullet in the same way.

1 What sort of food does saliva digest?

2 How else does saliva help you to digest food?
   (Hint: think of eating a dry cracker!)

3 How does chewing food help digestion?

4 Describe how your gullet moves food to your stomach.

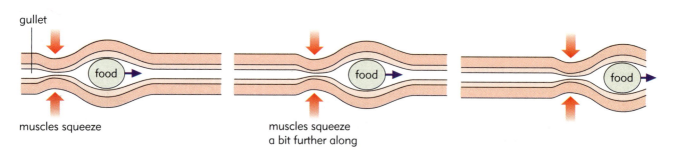

Muscles in the gullet wall squeeze the food along, section by section.

Next stop: your stomach

Your stomach is a bag of muscle tissue.
It churns up food for about three hours.

Your stomach has a lining of glandular tissue. This makes hydrochloric acid which kills most of the bacteria in food. The stomach lining also produces enzymes. These work best in acid conditions.

5 What two things does the glandular tissue of the stomach produce?

6 What part of your food starts being digested in the stomach?

7 Write down two reasons why your stomach produces hydrochloric acid.

Into the small intestine

Partly digested food leaves the stomach a little at a time. It goes into the small intestine. More enzymes are added from the pancreas and the small intestine lining.

8 Which parts of food are digested in the small intestine?

9 How do you think food is moved along the small intestine?

Think of the intestines as a tube passing through your body. Food moves through this tube, but it doesn't really enter your body until it is absorbed.

10 About how long is the human digestive system from the mouth to the anus?

11 Why do you think it has to be this long?

After digestion, what next?

After digestion the food molecules are small and soluble. They pass into the bloodstream in the wall of the small intestine. The molecules of digested food are absorbed into the bloodstream, so we call this **absorption**.

12 Write down seven substances that we absorb into our blood.

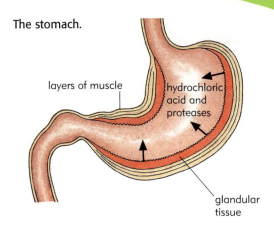

The stomach.

layers of muscle

hydrochloric acid and proteases

glandular tissue

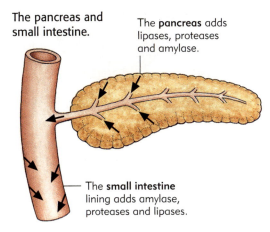

The pancreas and small intestine.

The **pancreas** adds lipases, proteases and amylase.

The **small intestine** lining adds amylase, proteases and lipases.

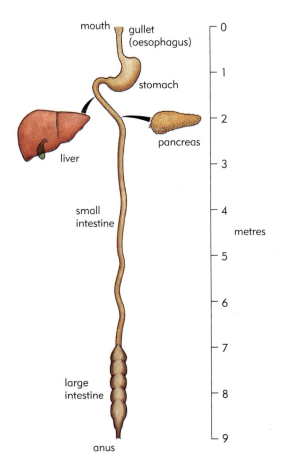

mouth

gullet (oesophagus)

stomach

liver

pancreas

small intestine

large intestine

anus

0
1
2
3
4
5
6
7
8
9

metres

What's the small intestine like?

The way that the small intestine is built makes it very good at absorbing food. It is long and the lining is very folded. This gives it a big surface area.

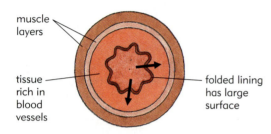

key ➡ digested food passes into the blood

muscle layers

tissue rich in blood vessels

folded lining has large surface

Section through the small intestine.

13 What features of the small intestine make it so good at absorbing digested food?

14 Why does the small intestine need a good blood supply?

Remember

In your digestive system:

is digested to

● starch ⟶ glucose

● protein ⟶ amino acids

● fat ⟶ fatty acids and glycerol

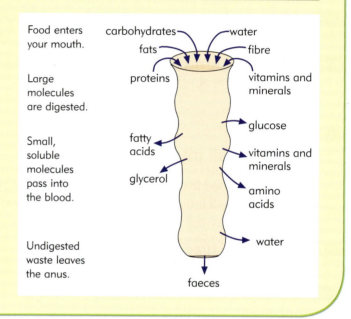

Food enters your mouth.

Large molecules are digested.

Small, soluble molecules pass into the blood.

Undigested waste leaves the anus.

carbohydrates ⟶ water
fats ⟶ fibre
proteins ⟶ vitamins and minerals
⟶ glucose
fatty acids ⟶ vitamins and minerals
glycerol ⟶ amino acids
⟶ water
faeces

Water and waste

As well as solid food, our intestines have a lot of water in them.

15 Where does this water come from?

Our bodies must absorb water. If this doesn't happen our bodies will be short of water and we will be **dehydrated**. Also we will suffer from diarrhoea as too much water will leave the body in the faeces. Our faeces contain the indigestible parts of our food. They leave the body through the anus.

16 What part of your digestive system absorbs water?

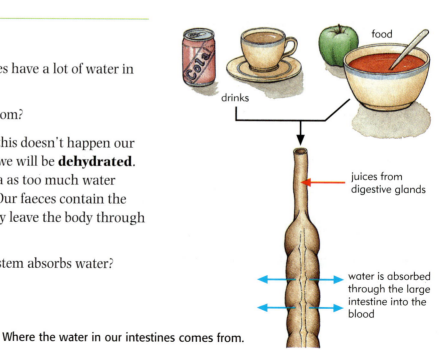

food

drinks

juices from digestive glands

water is absorbed through the large intestine into the blood

Where the water in our intestines comes from.

Energy from food

The food that you eat is the fuel that provides you with energy. When children rush about, some people say that they 'have a lot of energy'. One of the things you need energy for is to move about. You also need energy for warmth, and for growth and repair.

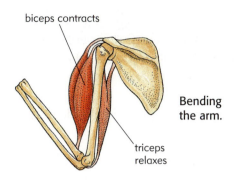

Bending the arm.

Energy for movement

When you move, your muscles contract. This means they get shorter and fatter. The diagram shows what happens when you bend and straighten your arm.

Muscles are made of muscle cells. Muscle cells need energy to contract. Some of the energy in muscles is released as heat.

1 Which muscle contracts to:
   (a) bend your arm?
   (b) straighten your arm?

2 What happens to the triceps when the biceps contracts?

3 Why do you get hot when you run?

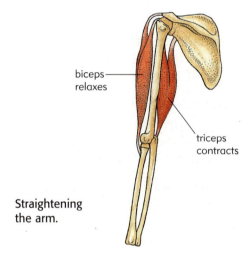

Straightening the arm.

Energy for warmth

Your normal body temperature is 37 °C. In the UK, the air around us is usually colder than this. The heat energy released in your cells is important for keeping you warm.

4 (a) What happens if your body temperature drops by 2 °C?
   (b) What do we call this condition?

5 How much further does the body temperature need to fall before a person goes into a coma?

We die if our body temperature falls too low. Most winters in Britain, 300 to 400 old people die of **hypothermia**.

6 Why is it hard for old people to keep warm?

When we are cold we shiver. This is because our muscle cells contract.

7 Why do you think you shiver when you're cold?

Body temperature (°C)	How does the body behave?
37	normal behaviour
35	shivering, body movements and speech become slow, drowsiness, start of hypothermia
30	goes into coma
28	breathing stops

Energy for growth and repair

Cells are mainly built up of **proteins**. We get our proteins from food. These proteins are broken down (**digested**) in our bodies into **amino acids**. Our cells then build the amino acids back up again into different proteins. The proteins are different because your cells join the amino acids together in different orders. The cells need energy to do this.

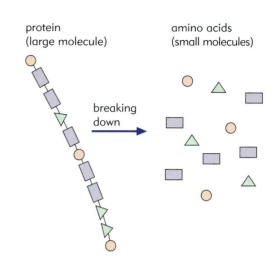

protein (large molecule) amino acids (small molecules)

breaking down

8 The diagram shows the breakdown of one protein. Draw a diagram to show the same amino acids built up into a <u>different</u> protein.

9 You eat proteins in food like eggs or cheese. Your body uses this to make different proteins in your muscle and skin. How does it do this?

Releasing energy from food

All the cells of your body obtain energy by respiring. Cells normally use oxygen to respire. You get oxygen from the air so we call this <u>aerobic respiration</u>.

Remember from Key Stage 3

- Breathing is taking air in and out of your lungs.
- Respiration is the breakdown of food to release energy.
- Respiration happens in all living cells.

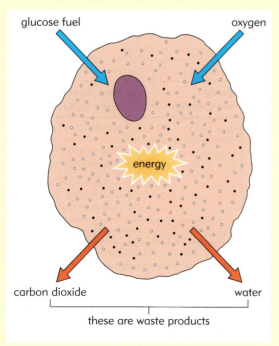

glucose fuel   oxygen

energy

carbon dioxide   water

these are waste products

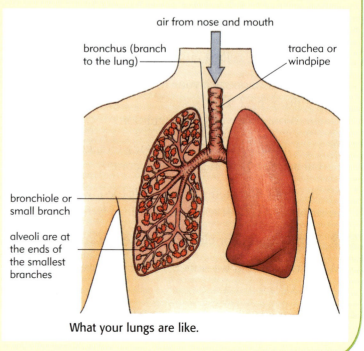

air from nose and mouth

bronchus (branch to the lung)   trachea or windpipe

bronchiole or small branch

alveoli are at the ends of the smallest branches

What your lungs are like.

Breathing

Breathing is taking air in and out of your lungs. To make air move in, your ribcage moves outwards and your diaphragm moves down. As you breathe out, the opposite happens.

When you breathe, you exchange stale air in your lungs for fresh air. We call this **ventilation**.

1 Describe the movements of your rib-cage and diaphragm when you:
(a) breathe in;
(b) breathe out.

Breathing in

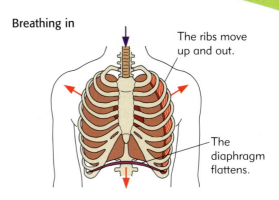

The ribs move up and out.

The diaphragm flattens.

Breathing out

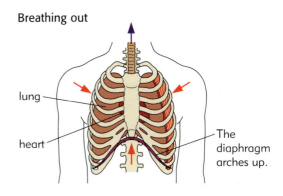

lung

heart

The diaphragm arches up.

We call the part of the body cavity
● above the diaphragm the thorax
● below the diaphragm the abdomen.

What happens to the air in your lungs?

The air that goes into your lungs ends up in millions of tiny **alveoli**. Gases move between the air in the alveoli and the blood in the capillaries around them. Each gas moves from where it is in high concentration to where it is in low concentration. We call this **diffusion**.

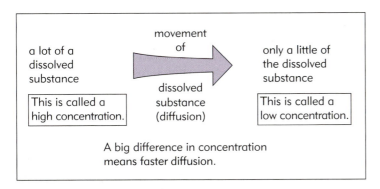

a lot of a dissolved substance

This is called a high concentration.

movement of

dissolved substance (diffusion)

only a little of the dissolved substance

This is called a low concentration.

A big difference in concentration means faster diffusion.

2 'Oxygen diffuses from where it is in high concentration to where it is in low concentration.' Explain this sentence in simpler language for a younger person.

So, in the alveoli:
● the oxygen your body needs diffuses from the air into your blood;
● waste carbon dioxide from your body diffuses from your blood into the air.

Oxygen travels in your blood to all the cells in your body.

3 Make a copy of the diagram. Draw and label arrows to show which gases diffuse into and out of the blood.

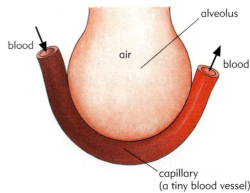
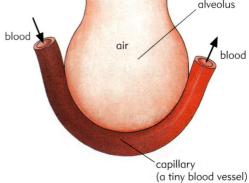

blood

alveolus

air

blood

capillary (a tiny blood vessel)

What happens inside an alveolus.

How can breathed-out air keep someone alive?

If you have a bad accident, you may stop breathing. You could die if you didn't get any oxygen for more than a few minutes.

Another person could save your life by breathing out into your lungs. This is called the 'kiss of life'.

The table shows the differences between the air you breathe in and the air you breathe out.

4 Describe these differences.

5 The kiss of life uses the air we breathe out. This can still help to keep someone alive. Explain why.

Gas	Air breathed in	Air breathed out
oxygen	21%	17%
carbon dioxide	almost zero	4%
nitrogen	79%	79%

What happens to the oxygen that you breathe in?

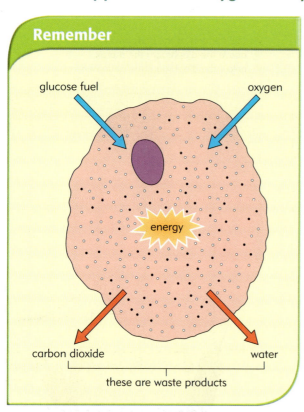

Remember

glucose fuel

oxygen

energy

carbon dioxide

water

these are waste products

Oxygen travels in your blood to all parts of your body. It diffuses from the blood in your capillaries to all the cells.

Think of glucose as a store of energy. Cells break down the glucose to release the energy. This is **respiration**.

Normally cells respire using oxygen from the air. When they use oxygen we call it **aerobic** respiration. Cells produce waste carbon dioxide and water when they respire in this way.

6 Write a word equation to show what happens in aerobic respiration.

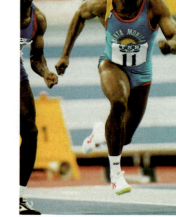

A sprinter pants and his heart beats faster during a race.

Look at the picture of the athlete. He is a sprinter so he only runs short races. During a race, his heart and lungs work harder so that his blood carries more oxygen to his muscle cells. They don't run out of oxygen.

7 During a short race the athlete's muscle cells get extra oxygen. Explain how this happens.

What happens if he doesn't get enough oxygen?

When the athlete trains, he does longer runs. At first his muscles get enough oxygen. Later his cells use oxygen faster than they can take it in. They have to respire without oxygen. We call this **anaerobic** respiration.

8 When do the athlete's muscle cells use anaerobic respiration?

9 Look at the word equation. Write down the name of the waste that is produced in anaerobic respiration.

glucose → lactic acid + energy

Lactic acid is a mild poison. It makes muscles ache and too tired to work. So cells can use anaerobic respiration for a short time only. Later (when the athlete is resting) they use extra oxygen to get rid of the lactic acid.

When athletes run long distances, they must run slowly enough to respire aerobically.

10 A marathon runner will not finish her race if she runs too fast. Explain why.

Remember

Oxygen diffuses from the air in the alveoli into the blood.

↓

The blood carries oxygen to all parts of the body.

↓

Oxygen diffuses from the blood to the cells.

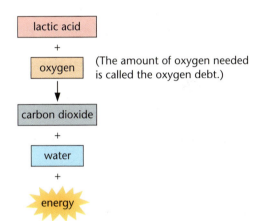

(The amount of oxygen needed is called the oxygen debt.)

Higher

More about anaerobic respiration

In anaerobic respiration, glucose isn't broken down completely so less energy is released than in aerobic respiration.

The lactic acid produced is one cause of muscle fatigue. Your muscles get tired and ache, and don't work as well as usual.

11 In the first few minutes of exercise, is it carbon dioxide or lactic acid that makes you pant? Explain your answer.

12 Your cells can only release energy by anaerobic respiration for a short time. Why is this?

13 (a) Make a table of differences between aerobic and anaerobic respiration.
(b) Why is less energy released in anaerobic than in aerobic respiration?

14 What is:
(a) muscle fatigue?
(b) oxygen debt?

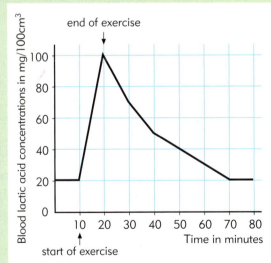

Lactic acid concentration in the blood of an athlete before, during and after exercise.

15 Look at the graph and then write down:
(a) the lactic acid concentration before exercise;
(b) the highest lactic acid concentration reached;
(c) how long it then takes for the lactic acid concentration to go back to what it was before exercise.

More about breathing

Movements of your ribs and diaphragm help to get air into your lungs. They do it by altering the volume of your thorax.

We call the movement of air into and out of your lungs **ventilation**.

> **Remember**
>
> Your lungs are in your thorax or chest.
> Breathing is taking air in and out of your lungs.

Looking at a model thorax

In this model, you can get air into the balloons by pulling the rubber sheet downwards.

sheet goes down
↓
volume in bell jar increases
↓
air pressure decreases
↓
air goes into balloons

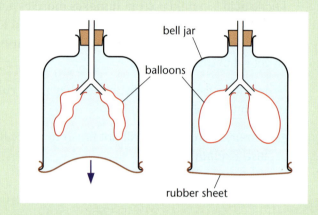

bell jar

balloons

rubber sheet

16 An important movement is <u>not</u> shown in the model thorax. Describe the missing movement.

Your thorax

Your diaphragm is a bit like a rubber sheet. It flattens when the muscles contract. At the same time, the muscles between your ribs contract and pull your rib-cage upwards and outwards. These movements of your diaphragm and rib-cage increase the volume inside your thorax. As the volume increases, the pressure decreases. To keep the pressures inside and outside your thorax the same, air goes into your lungs.

17 Your thorax is air-tight. Describe the only route for air to go into your lungs.

18 Draw flow charts, like the one for the model thorax, to show why:
(a) air goes into your lungs;
(b) air goes out of your lungs.

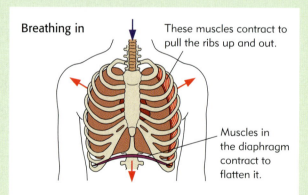

Breathing in

These muscles contract to pull the ribs up and out.

Muscles in the diaphragm contract to flatten it.

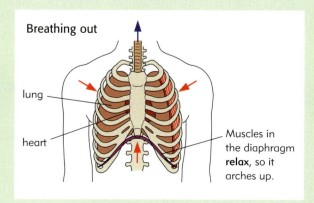

Breathing out

lung

heart

Muscles in the diaphragm **relax**, so it arches up.

Absorbing and transporting oxygen

'Folds' called **alveoli** increase the surface area of your lungs. Air reaches your alveoli through tiny tubes called **bronchioles**. The alveoli have a rich supply of capillaries. The walls of the alveoli and the capillaries are thin, so oxygen doesn't have far to diffuse. But it can only diffuse through cells in solution. So, the lining of the alveoli has to be moist.

Blood coming to your lungs contains only a little oxygen because the rest has been used in your body. The air in the alveoli has a higher concentration of oxygen than the blood, so oxygen diffuses easily from the air into the blood. The blood then carries the oxygen away to the rest of your body.

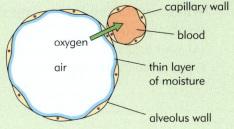

19 Describe, as fully as you can, how oxygen passes from the air in an alveolus into the blood.

20 Explain why each of the following features of an alveolus is important:
(a) the wall is thin;
(b) it has a rich capillary supply;
(c) the stale air is constantly being replaced with fresh;
(d) the lining is moist.

Using the oxygen

Respiration happens in all living cells. Like other chemical reactions in cells, it is controlled by enzymes. Most energy is released in structures called **mitochondria**.

21 Look at the diagram.
In what part of a cell are the mitochondria?

22 What are the raw materials for aerobic respiration?

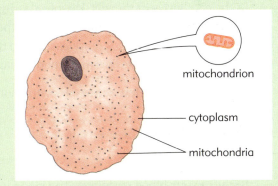

Another use of the energy from respiration

Substances often diffuse into or out of cells. This can only happen down a concentration gradient. But sometimes cells need to take in substances faster than they can diffuse in, or even against a concentration gradient. They use energy released in respiration to do this. So we call it **active transport**.

23 (a) Describe <u>two</u> differences between diffusion and active transport.
(b) Write down <u>four</u> uses for energy released in respiration.

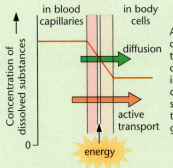

Active transport can speed up the movement of dissolved substances into and out of cells or move substances against the concentration gradient.

Exchange surfaces and diffusion

How your cells get all the things they need

Humans are large organisms made from many millions of cells. Your small intestine and your lungs have a large surface area so that they can absorb all the dissolved food and oxygen that your cells need.

Your small intestine is where you absorb most of your food. It has a thin moist lining so that dissolved substances can pass through easily. Tiny folds in the intestine, called **villi**, contain lots of blood capillaries.

1 Look at the photograph and the Remember box. Which structures provide a large surface area for absorption in
   (a) your lungs?
   (b) your small intestine?

The concentration of dissolved substances, such as glucose and amino acids, is higher inside the small intestine than in the capillaries in the villi. Dissolved substances always move from where there is a lot to where there is less. So glucose and amino acids move easily into the blood capillaries in the villi. This movement is called **diffusion**.

The greater the difference in concentration, the faster substances diffuse.

2 Copy the diagram of the villus.
   Complete the labels.

Because your blood circulates, the blood in the capillaries of your villi is constantly replaced. So the concentration of dissolved substances stays low.

3 If your blood didn't circulate, what would happen to the rate of diffusion of dissolved food from your small intestine?

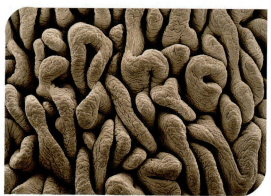

The intestine lining is folded. The surface area is increased even more by microscopic 'folds' called villi.

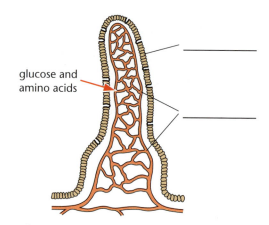

glucose and amino acids

A villus.

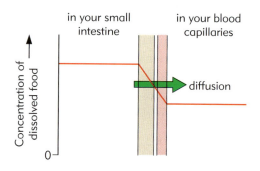

in your small intestine

in your blood capillaries

Concentration of dissolved food

diffusion

0

If your blood didn't circulate the concentration of dissolved food would quickly rise.

More about diffusion

Oxygen and carbon dioxide are exchanged in your alveoli. They are also exchanged between your blood and your body cells.

4 Look at the diagram.
(a) Is the concentration of carbon dioxide higher in your body cells or in your blood?
(b) In which direction does carbon dioxide diffuse?

Dissolved particles and particles in gases move in all directions. So they spread out.

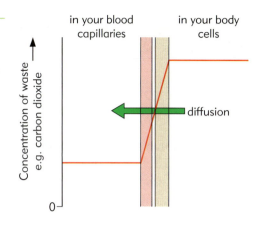

Look at the diagram.
There are more particles of carbon dioxide in the more concentrated solution. So more move from the cells into the blood than from the blood into the cells. We say that there is a **net movement** from the higher concentration in the cells to the lower concentration in the blood.

5 (a) Particles in a solution move in all directions. Explain why the diagram shows a net movement of carbon dioxide from right to left.
(b) A bigger difference in concentration makes diffusion faster. Why is this?

Low concentration of carbon dioxide	High concentration of carbon dioxide
Fewer carbon dioxide molecules	More carbon dioxide molecules
More molecules of other substances	Fewer molecules of other substances

Key

○ carbon dioxide

· other particles (mainly water)

← net movement of carbon dioxide in this direction (diffusion)

Higher

More about exchanges

Remember, to work well an exchange surface has:

- a large surface area;
- thin walls;
- a moist lining;
- a rich capillary supply.

6 Write down <u>one</u> reason for <u>each</u> of the features of a good exchange surface.

7 Alf smoked for 40 years. He had a bad 'smoker's cough' and easily got out of breath. His health got worse so he went to see his doctor. The doctor said that he had emphysema. She explained that coughing had damaged a lot of the alveoli in his lungs and reduced their surface area.
(a) Explain, as fully as you can, why Alf got out of breath easily.
(b) Alf's illness got worse. He couldn't walk very far and he had to breathe oxygen from a cylinder. Explain why.

Higher

Using what you know about exchanges

Plants and animals **exchange** substances with their surroundings. These exchanges happen in solution. Substances exchanged include gases such as oxygen and carbon dioxide, as well as nutrients such as sugars and mineral ions.

Exchanges also happen between cells and body fluids. To get in and out of cells, substances have to cross cell membranes.

8 (a) Write down <u>two</u> processes by which exchanges happen.
   (b) What is the difference between the two processes?

9 The liver fluke takes in oxygen and gets rid of carbon dioxide through its skin. The tadpole uses gills as well as its skin for this.
   (a) Explain why the liver fluke can get enough oxygen through its skin but the tadpole can't.
   (b) Design a tadpole gill. You can draw it or describe it. Explain your design.

Liver fluke.   Tadpole.

10 The villi in the small intestine of people with coeliac disease don't work properly. Children with the disease don't grow as quickly as children with normal villi. Why is this?

11 (a) The diagram shows a section through an alveolus. Explain how oxygen passes from the air in the alveolus into the blood.
   (b) Explain how oxygen passes from the blood to a body cell.

Remember

Substances pass in and out of cells by:

- Diffusion – this just happens.

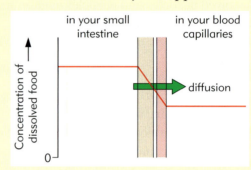

- Active transport – this can only happen if energy is supplied from respiration

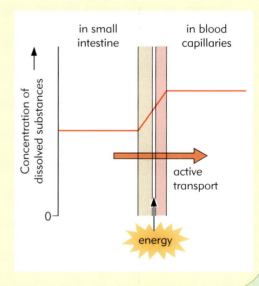

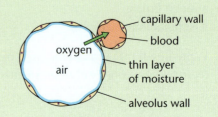

The heart – a pump for blood

Why doesn't your heart ache?

Your heart pumps blood around your body. To do this, it beats about 70 times a minute. Imagine squeezing a tennis ball 70 times. The muscles in your hand would soon become tired and begin to ache. The walls of the heart are made of special muscle called **cardiac** muscle. It does not get tired.

1. Paul's heart beats 70 times per minute. How many times does it beat in:
   (a) one hour?
   (b) one day?

Parts of the heart

As you can see from the diagram, the heart is made up of four **chambers**. The top two are thin walled chambers called **atria**. (If we are talking about only one of these we call it an **atrium**.) Below these are two larger, thick walled chambers. These are called **ventricles**.

2. Describe the route that blood takes through the heart from:
   (a) the head and body;
   (b) the lungs.

3. The left ventricle has a thicker wall than the right ventricle. Why do you think this is?

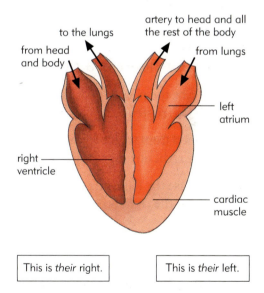

This is *their* right.   This is *their* left.

You are looking at another person's heart. Diagrams of the heart are always drawn this way round.

How does the heart work ?

The atria contract. They squeeze the blood into the ventricles. The ventricles then contract. They push the blood into the arteries.

The valves in the heart make sure that blood flows the right way.

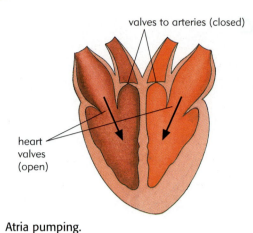

valves to arteries (closed)

heart
valves
(open)

Atria pumping.

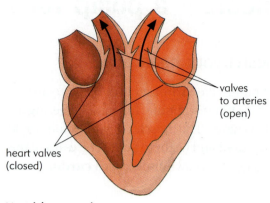

valves
to arteries
(open)

heart valves
(closed)

Ventricles pumping.

4 Describe the position of the heart valves and the valves to the arteries when:
(a) the atria are pumping;
(b) the ventricles are pumping.

Why your heart is a double pump

5 Describe the path of your blood round your body. Start with this sentence.

- Blood from your body, with little oxygen, goes into the right atrium of your heart.

Then use the following sentences. You will need to put them in the right order.

- The blood is pumped through an artery to the lungs.

- The blood goes into the left atrium of your heart.

- The blood goes into the right ventricle of your heart.

- The blood goes into the left ventricle of your heart.

- The blood picks up oxygen and gets rid of carbon dioxide.

- The blood is pumped through arteries to the rest of your body.

6 What feature of the heart ensures that only blood with very little oxygen goes to the lungs?

> ### Remember
>
> In your lungs, your blood:
> - collects oxygen;
> - gets rid of carbon dioxide.

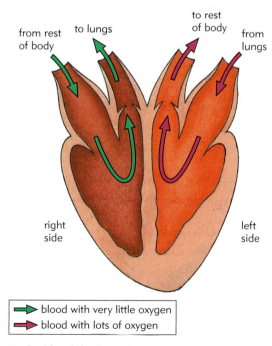

from rest of body | to lungs | to rest of body | from lungs

right side

left side

→ blood with very little oxygen
→ blood with lots of oxygen

Each side of the heart is a separate pump.

Know your blood vessels – you could save a life!

Imagine you are the first at the scene of a road accident. Two people are hurt. One has blood spurting from a cut while the other has blood oozing from a cut.

7 Which person should you treat first? Give a reason for your answer.

How to stop blood spurting from a cut

To stop blood spurting from a cut, you need to know how blood travels round the body.

8 Write down the structures that the blood passes through, from when it leaves the heart until it gets back to the heart.

9 Blood spurts from a cut artery. It oozes slowly from a cut vein. Why do you think this is?

10 To stop blood spurting from a cut artery you must press on the side of the cut nearer to the heart. Explain why.

Arteries and veins

Arteries have thick walls. This helps them withstand the high blood pressure caused by the pumping action of the heart. As the blood is forced into the arteries the walls stretch and then spring back. We feel this as a **pulse**.

11 (a) Where can you feel your arteries stretching?
(b) Why should you use your fingers to take someone's pulse and not your thumb?

Blood goes back to the heart in **veins**. They have thin walls and contain valves.

12 What job do these valves do?

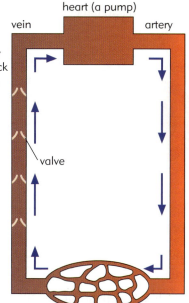

heart (a pump)

vein — This carries blood at low pressure back to the heart.

artery — This carries blood at high pressure away from the heart.

valve

There are lots of tiny blood vessels called capillaries in all organs.

If an artery is near the surface of the skin you can feel a pulse at each heart beat. You can feel it in your neck, your wrist and your thumb.

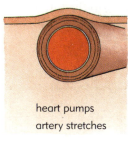

heart pumps
artery stretches

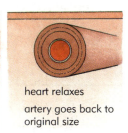

heart relaxes
artery goes back to original size

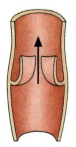

the blood flows forwards and pushes the valve open

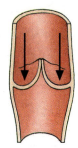

when the blood tries to flow backwards, the valve closes

The diagram shows a slice across (a cross section of) an artery, a vein and a capillary.

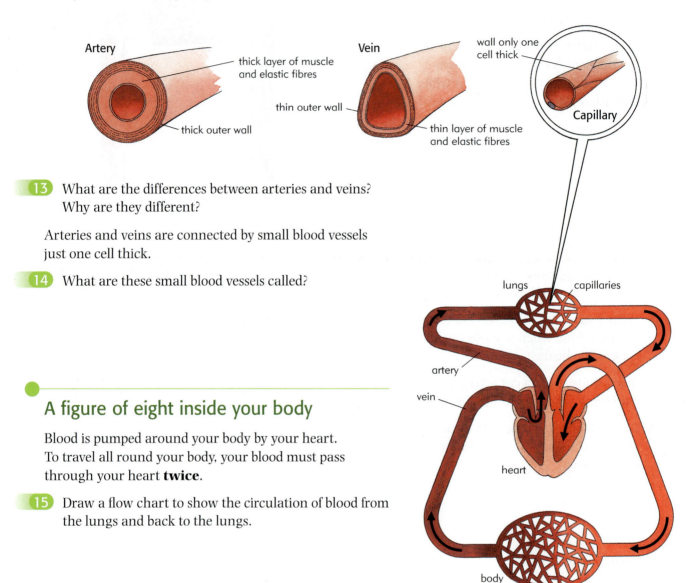

Artery

thick layer of muscle and elastic fibres

thick outer wall

Vein

thin outer wall

thin layer of muscle and elastic fibres

wall only one cell thick

Capillary

13 What are the differences between arteries and veins? Why are they different?

Arteries and veins are connected by small blood vessels just one cell thick.

14 What are these small blood vessels called?

lungs   capillaries

artery

vein

heart

body

A figure of eight inside your body

Blood is pumped around your body by your heart. To travel all round your body, your blood must pass through your heart **twice**.

15 Draw a flow chart to show the circulation of blood from the lungs and back to the lungs.

Your body's transport system

Blood is the body's transport system. It travels around the body, picking things up in some places and dropping them off in different places.

1. Look at the diagram.
   Where are the following waste substances made:
   (a) carbon dioxide?
   (b) urea?

2. Write down <u>one</u> substance that blood takes to all body cells.

What is transported where?

- In the lungs the blood drops off waste carbon dioxide and picks up oxygen.

- The blood picks up dissolved food in the small intestine.

- The blood supplies dissolved food and oxygen to all the body cells. Muscle cells need a lot of food and oxygen when you are working hard.

- The blood carries away the carbon dioxide these cells produce.

- Blood picks up a waste substance called **urea** in the liver. It drops it off in the kidneys.

3. (a) Write down the main substances that the blood picks up in each of the following parts:
   - the lungs;
   - the small intestine;
   - the muscles;
   - the liver.
   (b) Write down the main substances that the blood drops off in each of the following parts:
   - the lungs;
   - the muscles;
   - the kidneys.

How do things get into or out of your blood?

Inside the organs of your body, arteries divide to form **capillaries**. These are very narrow blood vessels with thin walls. There are lots and lots of capillaries. This means that every cell in the body is near to a capillary.

4. Make a big copy of the diagram to show gases changing places in the lungs. Label each arrow with the words 'oxygen' or 'carbon dioxide'.

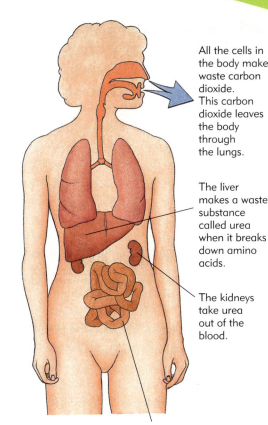

All the cells in the body make waste carbon dioxide. This carbon dioxide leaves the body through the lungs.

The liver makes a waste substance called urea when it breaks down amino acids.

The kidneys take urea out of the blood.

Dissolved food gets into the blood through the small intestine and it goes to all body cells.

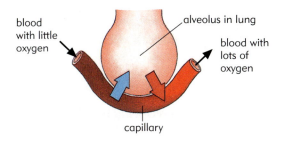

blood with little oxygen

alveolus in lung

blood with lots of oxygen

capillary

5 Copy the diagram of dissolved food getting into the blood from the small intestine. Complete the 'blood' labels with the words 'with little dissolved food' or 'with lots of dissolved food'.

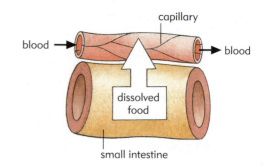

6 Look at the diagram. Then describe what happens. Start with this sentence:

- Blood from an artery goes into a capillary.

Then write down the following sentences in the right order:

- The liquid picks up carbon dioxide from the cells.

- Some liquid from the plasma leaks out of the capillaries. It washes over the body cells.

- The blood then flows on into a vein.

- The liquid then seeps back into the capillaries.

- It gives up oxygen and dissolved food to the cells.

7 Capillaries make it easy for blood to give dissolved food and oxygen to cells and pick up carbon dioxide from cells. Write down <u>two</u> reasons for this.

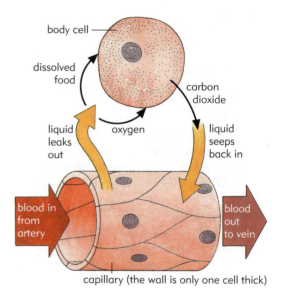

Replacing lost blood

Your body contains about 6 litres of blood.
If you lose more than 2 litres of this you could die.

8 How many cans of cola have the same volume as 6 litres of blood?

If you lost a lot of blood you might need a **blood transfusion**. You would be given blood from a donor. A blood donor can safely give about half a litre of blood at a time.

9 What fraction of their blood do blood donors give?

10 A person receiving blood usually needs blood from several donors. Why is this?

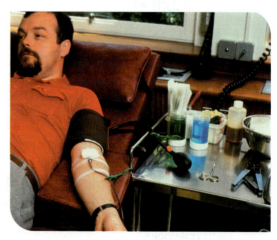

Each donor gives half a litre of blood.

Why is blood so important?

Blood is your body's transport system. It is made up of water, dissolved substances and blood cells. The water and dissolved substances together are called **plasma**. The pie chart shows how much of each there is.

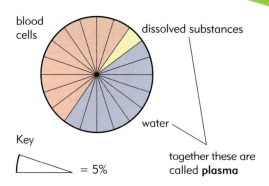

1. What percentage of blood is:
   (a) plasma?
   (b) water?
   (c) dissolved substances?
   (d) cells?

2. Look at the photograph. Write down the names of the two kinds of cell in blood.

3. Why does blood look red?

Plasma carries dissolved substances. We say they are **in solution**. The cells are solid but float about in the plasma. We say the cells are **in suspension**. Some substances that plasma transports around your body in solution are:

- carbon dioxide from body cells to the lungs;
- digested foods from the small intestine to the body cells;
- urea from the liver to the kidneys.

4. Write down <u>one</u> important job of your blood plasma.

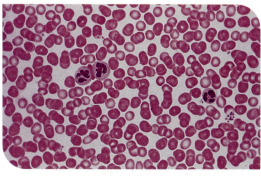

Blood seen through a microscope. White blood cells are stained purple so you can see them. Most of the cells are red blood cells. Blood also contains broken bits of cells called platelets.

What do different kinds of blood cells do?

Bacteria which get into your body can cause disease. White blood cells help to defend the body against disease. They can do this by destroying the bacteria.

When you cut yourself, **platelets** soon gather around the cut. The white cells which 'eat' bacteria also do this.

5. Write down <u>two</u> different ways that white blood cells can destroy bacteria.

6. Why do you think the platelets gather around a cut?

7. Make a table to show the types of blood cells and what each type does.

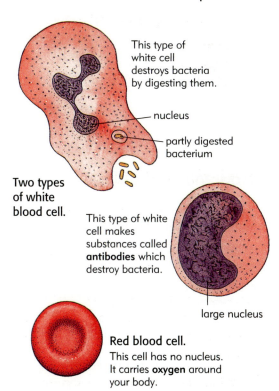

Two types of white blood cell.

This type of white cell destroys bacteria by digesting them.

nucleus

partly digested bacterium

This type of white cell makes substances called **antibodies** which destroy bacteria.

large nucleus

Red blood cell.
This cell has no nucleus. It carries **oxygen** around your body.

cell fragments with no nucleus

Platelets (these are bits of cells).
Platelets help blood to **clot** when we cut ourselves.

What is the job of red blood cells?

The job of the red blood cells is to carry oxygen around the body. They can do this because they are packed with a chemical called **haemoglobin**. Haemoglobin joins with oxygen to form **oxyhaemoglobin**.

haemoglobin + oxygen ⇌ oxyhaemoglobin

8 The **reversible reaction** sign (⇌) shows that oxyhaemoglobin splits easily.
What is formed when it splits?

Red blood cells are unusual in another way. They have no nucleus. They lose their nucleus before they go into the bloodstream. Without a nucleus the cell has:

- a larger surface area;
- as much haemoglobin inside as possible.

9 In terms of exchanging and carrying oxygen, explain the advantages of having:
(a) a large surface area;
(b) as much haemoglobin in the cell as possible.

Remember

Blood contains:
- red cells, white cells and platelets
- a liquid called plasma which has things dissolved in it.

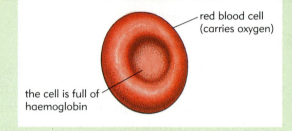

red blood cell (carries oxygen)

the cell is full of haemoglobin

10 You use the iron in your diet to make haemoglobin. If you don't get enough iron, you become anaemic. You have:
- fewer red blood cells;
- less haemoglobin in each red cell.

Explain why people who are anaemic
(a) look pale;
(b) often feel tired and cold.

How many red blood cells do we have?

There are about 5 million red blood cells in every cubic millimetre of blood. The number varies. Doctors can send people for blood counts to find out if they have enough. Technicians use a grid like the one on the right to help them to count the cells in a very tiny amount of blood.

11 (a) Count the red cells in one small square of the picture.
(b) About how many red cells are there altogether?
(c) About how many times more red blood cells are there than white blood cells?

12 What else is there in blood that you cannot see in the picture?

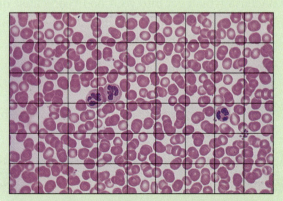

Technicians use a grid like this to count blood cells.

Higher

13 Look at the graph.
   (a) What is the relationship between the number of red cells and the height above sea level?
   (b) Suggest a reason for this relationship.
   (c) Athletes often train at altitude before an important event. Explain why this could help them do better.

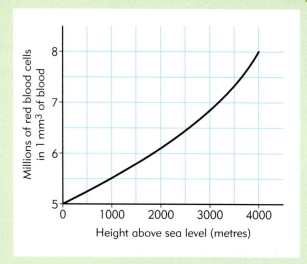

Where is oxygen exchanged?

Look at the diagram which shows what happens to red blood cells as they move round the body.

14 The sentences below are in the wrong order. Put them into the correct order to show what's happening in the diagram. Start with:

- Blood carries the red cells to the lungs.

- The red cells are now bright red.

- In the lungs haemoglobin joins up with oxygen.

- Oxygen goes to the cells.

- Blood carries the red cells to the organs.

- Oxyhaemoglobin is formed.

- The red cells are now dark red.

- In the organs oxyhaemoglobin splits up into haemoglobin and oxygen.

What happens to red cells in the lungs.

capillary
haemoglobin (dark red)
oxyhaemoglobin (bright red)
red blood cell
OXYGEN goes into the blood
to the lungs
from the lungs
through the heart
from the organs
to the organs
OXYGEN goes into the cells
haemoglobin (dark red)
capillary
oxyhaemoglobin (bright red)

What happens to red cells in other parts of the body.

Invading microorganisms

Small but dangerous

Very small living things are called **microorganisms**.

Microorganisms such as **bacteria** and **viruses** can get into your body. Some of them cause disease.
They can make you ill.

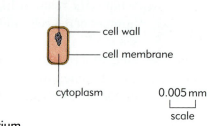

A bacterium.

Know your enemy

The cells of bacteria are even smaller than the cells of your body. The bacterium in the diagram is 1400 times larger than it is in life.

The plant and the animal cell are also 1400 times larger than in life.

1   Write down the parts which all three types of cell have.

All cells have **genes** inside them. These genes contain information which controls what happens in a cell.

The genes are contained in **chromosomes**.
Chromosomes are in the nucleus of animal and plant cells. The genes of a bacterium are not in a nucleus.

2   Write down <u>one</u> other difference between an animal cell and the cell of a bacterium.

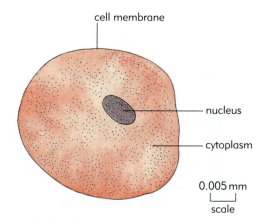

An animal cell.

Bacteria are everywhere

Like you, bacteria need food and water to survive.
Many of them need oxygen too. They grow best when it is warm. We find bacteria in places where there are all the things they need.

3   Write down <u>three</u> places where you would expect to find bacteria.

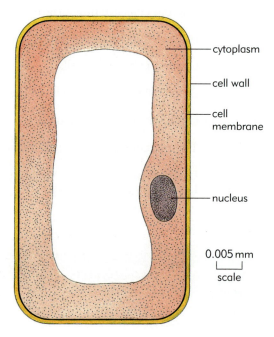

A plant cell.

You cannot escape

Your body is full of places where bacteria can live and reproduce.

4 Why is your body a good home for bacteria?

Millions of bacteria live between the cells of even the cleanest skin. Most of them do no harm, but some make you smell sweaty. Others can cause diseases such as sore throats and food poisoning.

5 Draw the shapes of <u>two</u> of the bacteria shown.

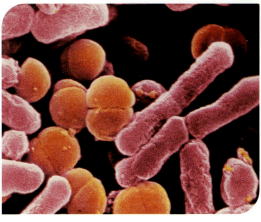

Bacteria that live in the human body.

What is the difference between bacteria and viruses?

Viruses are even smaller than bacteria. They are made of a few genes in a **coat** made of protein.

6 The diagram shows a bacterial cell infected by a virus. Which part of the virus goes into the cell?

Viruses cannot reproduce by themselves. They need to invade living cells and use the living cells to make more viruses. This damages the cells.

7 Why are viruses usually harmful?

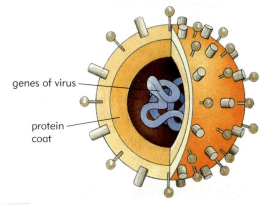

genes of virus

protein coat

A virus. This is about 300,000 times bigger than in real life.

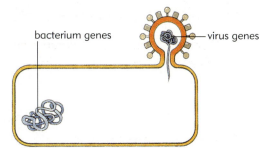

bacterium genes

virus genes

A virus infecting a bacterium.

Do microorganisms always make you ill?

Not all microorganisms make you ill. Many are easily destroyed by your white blood cells. Others are harder to destroy.

When the numbers of microorganisms are small, you do not notice any effect. But microorganisms can breed very quickly. As their numbers get larger and they produce more toxins, you begin to feel ill. This may take a few hours, days or even weeks.

Toxic

Microorganisms make toxins (poisons).

8 What kind of cell destroys microorganisms?

9 Jan's sister Carol caught scarlet fever. As soon as her mother knew Carol was infected, she kept the two girls apart. It was too late. Jan became ill a week later.

Why was it so long before Jan became ill?

10 What were the effects of the scarlet fever bacteria on the girls?

I am hot
I feel sick
I have a rash

Carol describes her symptoms.

How do microorganisms cause these symptoms?

The pictures show how microorganisms can affect your body.

11 Your body temperature rises when you have an infection. Give <u>one</u> reason why.

12 You may also have a rash or a headache. Write down <u>one</u> possible cause for each of these symptoms.

Your cells release more energy as heat. You can't control your body temperature by sweating.

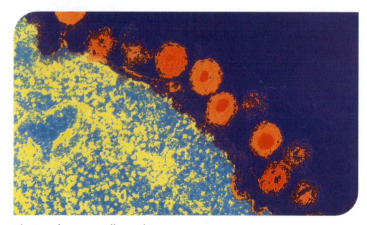

Viruses damage cells as they escape.

How quickly can bacteria breed?

Bacteria reproduce by dividing into two. At human body temperature, some bacteria can divide every 20 minutes.

13 Imagine you have eaten a pie with 50 bacteria in it. If the number of bacteria doubles every 20 minutes, how many bacteria will there be after one hour?

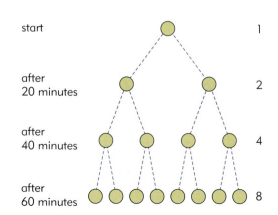

start   1

after 20 minutes   2

after 40 minutes   4

after 60 minutes   8

How can we prevent diseases spreading?

You should not keep food in a warm or dirty place. In such unhygienic conditions food can soon have enough bacteria in it to make you ill.

You can become ill with food poisoning within a few hours of eating infected food.

14 Look at the picture. Explain <u>two</u> things you can do to make sure you do not get food poisoning.

Tuberculosis (TB) is a lung infection. It is caused by bacteria. In overcrowded conditions the bacteria easily spread from one person to another.

15 People are more likely to get tuberculosis if they live in overcrowded conditions. Why is this?

> You need to be able to use evidence to explain how the conditions people live in and the way they behave affect the spread of disease.

Microorganisms cannot breed quickly in cold conditions. They are killed when it gets very hot.

The spread of infection

'Coughs and sneezes spread diseases'

This old rhyme is true.
However, diseases are also spread in other ways.

16 The pictures show ways that diseases are spread. Write a sentence about each one.

> **Remember**
>
> Microorganisms such as bacteria and viruses can cause disease.

How do microorganisms get inside your body?

Some microorganisms get inside your body through wounds. Cuts, injections and animal bites can all let microorganisms in. Others get in through the thinner skin lining the natural 'openings' of your body.

17 Make lists to show how microorganisms get into your body. Use these headings:
(a) Natural openings;
(b) Wounds.

18 You need to wash your hands before you touch food. Why is this?

19 Germicides kill bacteria. Doctors and nurses always wipe your skin with germicide before you have an injection. Why is this?

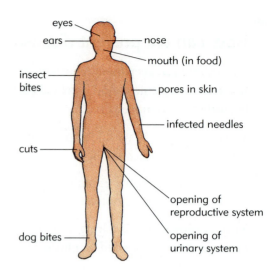

Where microorganisms can get into your body.

Humans against microorganisms

Skin – your first line of defence

Your body has ways of stopping microorganisms getting in. The cells of the outer layers of your skin are dead. They act like a barrier. However, there are some natural openings in the skin which can let bacteria in.
For example, you get spots when bacteria infect pores in the skin.

1 Name the other kind of natural opening in skin.

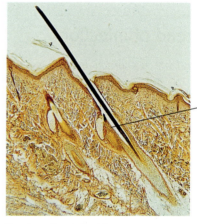

Each hair grows out of an opening called a hair follicle.

Section through skin.

What happens if your skin is broken?

Luckily your blood puts up another barrier when you cut yourself. Your blood dries up and goes hard. We say that it **clots**. Clotting starts as soon as the blood meets the air. It takes only a few minutes.

2 When blood clots, blood can no longer flow out through a cut. What cannot get in?

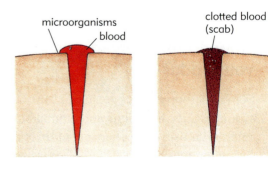

Fighting off the air-borne division

There are microorganisms in the air. These can get into the air passages to your lungs through your nose and mouth. Some of the microorganisms get trapped by the hairs and sticky mucus in your nose. Mucus also traps dust.

3 It is more healthy to breathe in through your nose than through your mouth. Why is this?

Mucus in your trachea and bronchi also traps microorganisms. This mucus is the **phlegm** which comes up into your mouth when you cough.

The cells in the air passsages have **cilia**. These are like little hairs. Smoking stops these cilia working, so if you smoke, microorganisms get into your lungs more easily.

4 Tuberculosis (TB) is caused by bacteria that infect your lungs. List the places where these bacteria could get trapped before they reach your lungs.

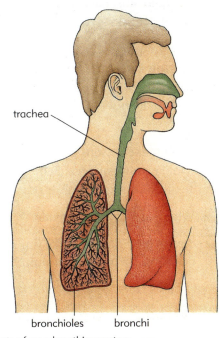

 Parts of your breathing system where mucus is made

A case of whooping cough

Sharon has had **whooping cough**. Bacteria cause whooping cough. They spread easily through the air from one person to another.

5 Describe Sharon's illness. Use the chart to help you.

6 How could Sharon have caught the disease? Write down your idea in a few sentences.

Within a few weeks she was better.
Her body had destroyed the bacteria.

Day	5	10	15	20
	runny nose, sneezing, feeling ill			feeling better
	fever			
		very bad coughing fits		still coughing

What happened to Sharon.

How did Sharon's body destroy the bacteria?

The white cells in Sharon's blood destroyed the bacteria.

7 Look at this diagram of a white blood cell. Write down <u>one</u> way white blood cells can destroy bacteria and other microorganisms.

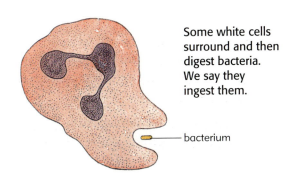

Some white cells surround and then digest bacteria. We say they ingest them.

bacterium

Some other white cells make **antibodies**. These destroy microorganisms. Each kind of antibody works against only one kind of microorganism.

When a new kind of bacterium or virus gets into your body, white cells have to start making a new kind of antibody. It takes time for the cells to make enough of the right kind of antibody. That is why it takes time to get better.

Bacteria make poisons called **toxins**. Other white blood cells make antitoxins which destroy toxins. Each kind of antitoxin works against only one kind of toxin.

8 Where are antitoxins made?

9 Toxins have time to make you ill before they are all destroyed. Why do you think this is?

Why didn't Daneya get whooping cough?

Daneya is Sharon's best friend. They spend most of their time together. But Daneya didn't get whooping cough. Two years ago she had an injection of a weak form of the bacteria which cause whooping cough. We say that she was **vaccinated** against the disease. Her white blood cells made antibodies against the bacteria. So Daneya is **immune**.

10 Look at the diagram. Explain what happens when bacteria from Sharon get into Daneya's blood.

11 Daneya's brother Adil had whooping cough last year. Will he catch it again? Explain your answer.

What if your defence system doesn't work?

Some medical treatments damage the white cells which destroy bacteria. **HIV** (human immunodeficiency virus) also destroys some kinds of white cells.

The HIV infects white blood cells. These cells then cannot do their job. They cannot make antibodies.

12 Illnesses which are normally quite mild can kill people with HIV who develop AIDS.
Why do you think this is?

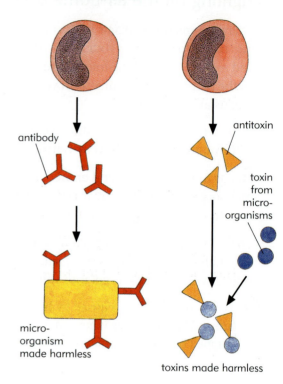

Other white cells make antibodies or antitoxins.

antibody

antitoxin

toxin from micro-organisms

micro-organism made harmless

toxins made harmless

Bacteria from Sharon get into Daneya's blood.

White blood cells recognise them. They have made the antibody before so they can quickly make it again.

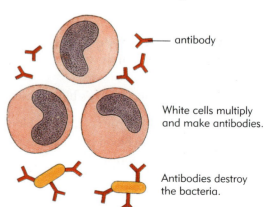

antibody

White cells multiply and make antibodies.

Antibodies destroy the bacteria.

Maintenance of life

How are plants built?

All plants are made from tiny parts called **cells**.
Cells are so small that you can only see them using
a microscope.

The cells in the drawing on the right are the first ones ever
seen. Robert Hooke drew them more than 300 years ago.

1 Why do you think he called them 'cells'?

2 Draw a line 1 centimetre (cm) long. About 400 plant
   cells will fit along this line. Now use a sharp pencil to
   mark off each millimetre along your line.
   How many cells will fit into a space of 1 mm?

We know now that all plants are made up of cells.

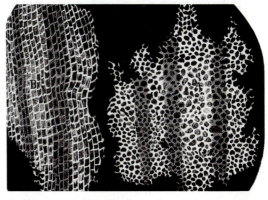

Cells from a cork oak plant.

These little spaces in a beehive are all the same
size and shape. They are called cells.

Do all cells look the same?

Some things are the same in all cells, but other things are
different. Look at the diagrams of the plant cells and the
animal cell.

3 Write down <u>three</u> parts you can see in both
   types of cell.

The **nucleus** controls what the cell does. The **cell
membrane** controls what passes in and out of a cell.
The cytoplasm is where most of the chemical reactions
take place.

4 Which part can you see in all three of the plant cells,
   but not in the animal cell?

5 Write down the <u>two</u> parts you can see in some, but not
   all, of the plant cells.

All plant cells have **cell walls** made of cellulose.
This makes the cells stronger and more rigid.
The spaces in plant cells called **vacuoles** are filled with
a watery fluid called **sap**.

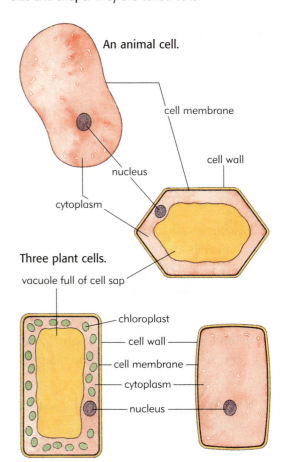

An animal cell.

cell membrane

cell wall

nucleus

cytoplasm

Three plant cells.

vacuole full of cell sap

chloroplast

cell wall

cell membrane

cytoplasm

nucleus

Why don't all plant cells have chloroplasts?

Chloroplasts contain **chlorophyll**. This is what gives plants their green colour. Chlorophyll absorbs light energy so that plants can make food. Chlorophyll is made only in the light. After a few weeks in the dark it disappears from the cells.

6 Look at the pictures of the lawn.
   (a) What happens to the grass under the tent?
   (b) Why does this happen?

7 Root cells do not have chloroplasts. Why not?

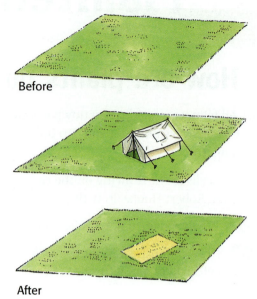

Before

After

8 This diagram shows a section cut through a stem of a plant. Only the outer layer is green. Why is it not green in the middle?

Potatoes are the underground stems of potato plants. They grow under the ground so they are not green usually, but they do go green in the light.

● The bad news: the green parts of potatoes are poisonous.

● The good news: you would have to eat a lot to make you ill.

9 How should you store potatoes so they don't go green?

A slice across a stem.

The cell for the job!

All plants are made from tiny parts called cells. But not all plant cells are the same. You can see this if you cut a slice through part of a plant.

10 You would need to look at the slice under a microscope. Why is this?

Different jobs in a plant are done by different kinds of cells. A group of cells with the same shape and job is called a **tissue**.

Only plant cells which have chloroplasts can use light energy to make food.

Look at the drawing of a slice across a leaf.

11 Write down <u>three</u> kinds of tissue shown in the leaf section.

12 What job does a leaf do?

13 Which of the tissues can make food?
Give a reason for your answer.

14 In which part of a leaf is most of the food made?

The leaves, stems and roots of plants are called **organs**.
Organs are made of more than one kind of tissue.

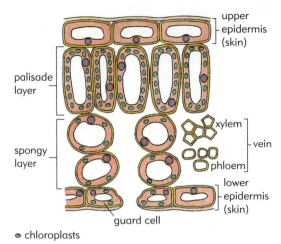

A slice across a leaf.

Why do leaves have veins?

Plants use veins to transport substances. Veins are made of two main kinds of tissue. The first is called **xylem** tissue. (You say this word zy-lem).

In xylem tissue there are rows of dead cells with the ends missing. They form a long tube like a drinking straw. Water travels up xylem tissue from the roots.

Look at the drawing of xylem tissue.

15 What else besides water travels from the roots to the stems and leaves through the xylem tissue?

16 Xylem tissue also does another job.
What do you think this job is?

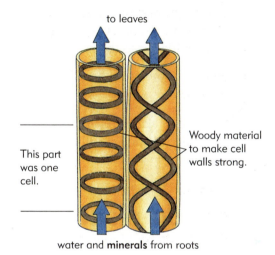

Xylem vessels are made of dead cells.

The second kind of tissue is called **phloem**.
(You say this word flo-em).

Look at the drawing of phloem tissue.

17 Write down <u>two</u> differences between xylem tissue and phloem tissue.

Phloem carries sugar from where it is made to other parts of the plant.

18 (a) Which part of a plant does phloem transport sugar from?
(b) Where does phloem carry sugar to?

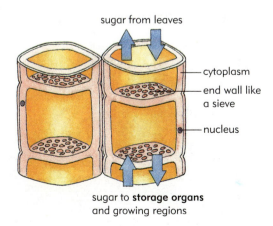

Phloem tissue is made of living cells.

Where are the transport tissues?

19 Copy the drawings of the slice of a root and a stem. Then colour the tissue which transports water in one colour. Use a different colour for the tissue which carries sugar.
Add a key to show what your colours mean.

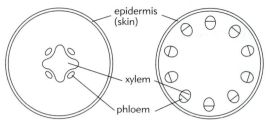

Slice of a root.   Slice of a stem.

What other kinds of plant cells are there?

Other plant cells do other jobs. We say that they are **specialised** to do particular jobs.

20 What do you think each of the cells on the right is specialised to do? Choose from:

- support
- making new cells
- storage
- photosynthesis (making food)

Give a reason for each answer.

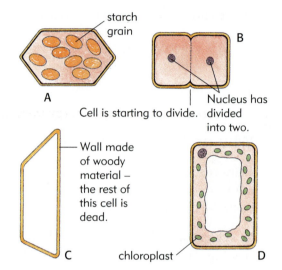

You should be able to relate the structures of plant cells to their functions in different tissues and organs.

How do plants get their food?

All living things need food to survive.

Green plants can make their own food. Animals cannot do this. They have to eat other animals or plants for food.

1 Why do animals have to eat plants or other animals?

Animals have to eat plants or other animals to stay alive. They cannot make their own food.

Plants make their own food from water and carbon dioxide.

2 If all the plants died, what would happen to animals like sheep?

3 What would happen to us? Explain why.

What kind of food do plants make?

Plants use light energy to make **glucose**. Glucose is a kind of sugar.

Plants can use glucose to make other foods. One of these other foods is another sugar called **sucrose**. This is the kind of sugar you use at home.

4 Where do plants get the energy from to make their own food?

5 What type of food do plants make first using light energy?

6 What plants do we grow for sugar
(a) in hot countries?
(b) in Britain?

Sugar cane grows in hot countries.

Sugar beet grows in cooler countries.

Where your sugar comes from.

What happens to the glucose plants make?

Plants make glucose. Then they use it in different ways. The diagram shows the ways plants use glucose.

7 Plants use glucose for respiration.
What does this word mean?

8 In what form do plants store the food they make?

9 Write down three other ways plants use glucose.

Some is reacted with oxygen. This releases energy for cells to live and grow. This release of energy is called respiration.

Some is made into sucrose.

Glucose

Some is made into starch. This is a food store.

Some is made into other substances.

potatoes

How do plants make starch?

Plants join together lots of the small glucose molecules they have made. This makes long **starch** molecules.

Starch is stored in the plant as a 'food store'. Starch is a good way of storing sugar because it is **insoluble** (it will not dissolve).

10 Explain how plants make starch.

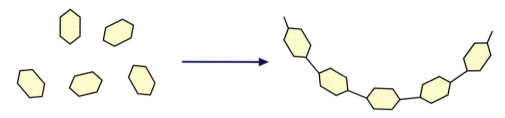

glucose molecules   part of a long starch molecule

Food factories – the leaves

Plants use light energy to make food.
This is called **photosynthesis**.

You cannot see this happening, but you can prove that
plants make food by testing leaves for the starch that they
make. The diagrams show you how you can do this.

11 List the steps for testing a leaf for starch.

12 Plants use light energy to make food.
What do we call this process?

Take a leaf which has been in the light.

ethanol   hot
water

Remove the green colour by putting
the leaf in hot ethanol.

iodine solution

Test with iodine solution.
Any starch in the leaf goes black.

Do all plant cells make food?

Carol tested a leaf for starch. These are her results.

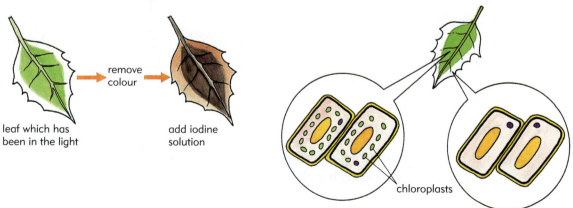

leaf which has
been in the light

remove
colour

add iodine
solution

chloroplasts

13 Which parts of the leaf had starch in them?

14 What is the difference between the cells in the two parts of the leaf?

Where do plants make their food?

The cells in the green parts of a plant contain **chloroplasts**. These are filled with a green substance called **chlorophyll**. Chlorophyll takes in light energy. The cells use this energy for **photosynthesis**.

15   Why are the leaves of the potato plant green?

16   Which parts of the potato plant could not photosynthesise? Explain why you think this is.

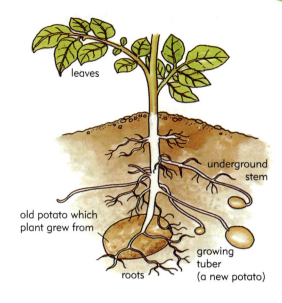

Why are leaves good at photosynthesis?

Most of a plant's chloroplasts are in its leaves.
Leaves are usually broad and flat.

17   How does the shape of a leaf help it to photosynthesise? (Hint: imagine you are sunbathing.)

18   There are more chloroplasts in the palisade cells than in the cells in the spongy layer. Why?

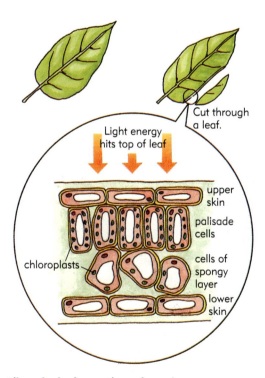

Slice of a leaf seen through a microscope.

What do plants make sugar from?

The diagram shows what plants make the sugar from.

1 What <u>two</u> substances do plants use to make sugar?

2 Where does each substance come from?

3 What else does the plant produce as it makes sugar?

4 Where does this other substance go?

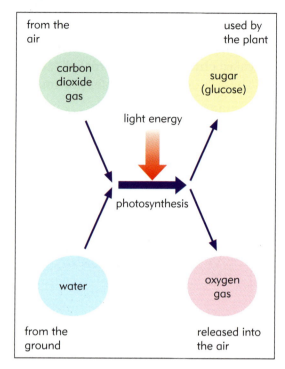

We can write down what happens during photosynthesis as a word equation.

carbon dioxide   +   water   +   light energy   ⟶   glucose   +   oxygen

Why do plants need light to make sugar?

You may have used a Bunsen burner to give you the energy needed to join chemicals together. Plants need energy to join carbon dioxide and water together.

Plants get the energy they need for photosynthesis from light.

5 Where do plants usually get this light energy from?

6 Copy the graph. Then copy these labels on to the right place on the graph:

| sunset | | sunrise |

7 No photosynthesis takes place at night. Explain why.

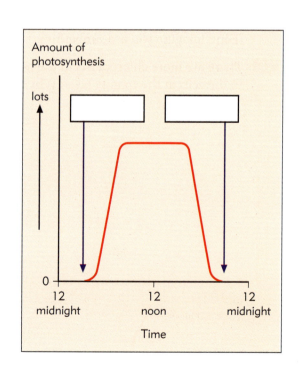

Investigating photosynthesis

A group of students learned that plants need light and carbon dioxide for photosynthesis. They decided to find out if this is true.

They set up four plants in different conditions. After 24 hours the students tested a leaf from each plant to see if it had starch in it.

8 (a) In which of the plants, A–D, did the students <u>not</u> find starch?
(b) Explain why each of these plants did not produce starch.

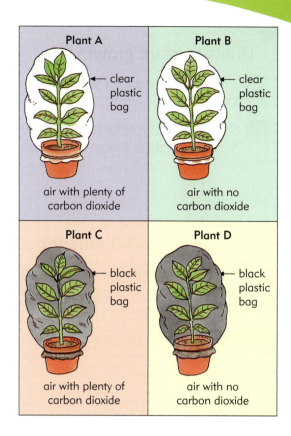

Plant A — clear plastic bag — air with plenty of carbon dioxide

Plant B — clear plastic bag — air with no carbon dioxide

Plant C — black plastic bag — air with plenty of carbon dioxide

Plant D — black plastic bag — air with no carbon dioxide

Higher

What do plants use glucose for?

Plants release energy from glucose when they respire. They use the energy released to build other sugar molecules into larger molecules such as starch, cellulose, lipids (fats or oils) and amino acids. They need nitrates and other minerals, as well as glucose, to make amino acids.

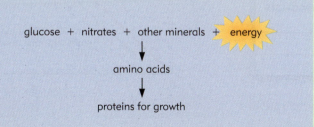

glucose + nitrates + other minerals + energy
↓
amino acids
↓
proteins for growth

cell wall

glucose molecules

The cellulose in cell walls is made from lots of glucose molecules.

9 Make a list of ways that plants use the energy released in respiration.

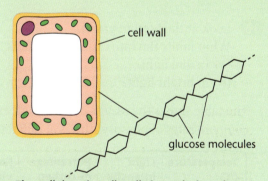

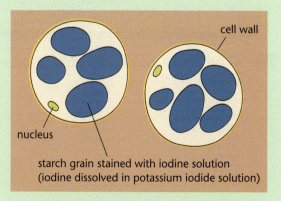

These cells from a potato tuber change glucose into starch for storage.

nucleus

cell wall

starch grain stained with iodine solution (iodine dissolved in potassium iodide solution)

Limits to plant growth

Even in Britain, wheat crops grow faster when farmers give them extra water.

10 To grow faster, a plant needs to make food faster.
Write down <u>three</u> things that a plant needs to do this.

What affects plant growth?

The graph shows how the amounts of light and carbon dioxide affect how fast a plant makes food. We call this the **rate** of photosynthesis.

11 Describe the change in the rate of photosynthesis:
(a) between A and B;
(b) between B and C.

12 Line AD shows photosynthesis in the same plant.
Write down:
(a) <u>one</u> way it differs from line ABC;
(b) <u>one</u> reason for the difference.

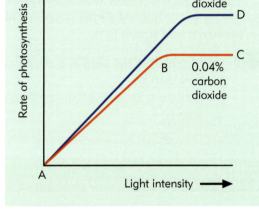

13 Look at the bottom graph.
What limits the rate of photosynthesis:
(a) in dim light?
(b) in bright light?

The table shows the results of a different experiment to compare rates of photosynthesis at different temperatures.

Temperature (°C)	Light intensity	Percentage of carbon dioxide	Rate of photosynthesis
20	3	0.04	15
25	3	0.04	18

14 As the temperature varies, what happens to the rate of photosynthesis?

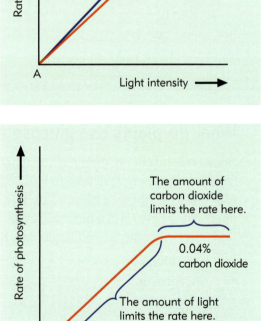

What else affects plant growth?

Tai grew the plants in the pictures from cuttings taken from one plant. He wanted to find out if the type of soil affected their growth.

1 Tai grew all the cuttings in exactly the same light and temperature. Why?

The soil that plant Q is growing in hasn't got enough **nitrates** in it. Plants need nitrates to make proteins and, like you, they need proteins for growing.

2 Write down <u>two</u> differences between plants P and Q.

3 Explain why plant P grew faster than plant Q.

Plants R and S are also short of **minerals**. These minerals are the ones they need to make chlorophyll.

4 Write down the names of the minerals which plants need to make chlorophyll.

5 Why do you think that the leaves of plants R and S are not as green as those of plant P?

6 When a plant cannot make much chlorophyll, it cannot grow well. Explain why.

Plant P in soil with plenty of minerals.

Plant Q in soil short of nitrates.

Plant R in soil short of magnesium.

Plant S in soil short of iron.

Higher

More about minerals

If a plant can't get enough minerals, its growth may be stunted. Lack of a particular mineral will have a particular effect. This is because different minerals do different jobs. The 'big three' minerals are:

- nitrates;
- phosphates;
- potassium.

Nitrates are used to make the amino acids and proteins needed for growth.

Phosphates are needed for energy transfers in photosynthesis and respiration. Plants which don't get enough phosphates have tiny leaves and their younger leaves are purple.

Potassium helps photosynthesis and respiration by making sure the enzymes can do their jobs well. Plants which lack potassium have yellow leaves with dead spots.

7 Look at the photographs.
  (a) Write down <u>one</u> effect of a lack of nitrates on older leaves.
  (b) Is the unhealthy vine leaf from a plant short of phosphates or short of potassium?

8 (a) The yield of wheat in a farmer's best field was less than usual. He thought that the soil <u>might</u> be short of nitrates, so he decided to add some next year. His son suggested that he should have the soil analysed first. Why was this good advice?
  (b) The farmer could add a natural fertiliser such as manure instead of an artificial nitrate fertiliser. Why is natural fertiliser more likely to solve any problem?

9 Plants need magnesium and iron to make chlorophyll. Explain exactly which parts of a plant are affected by a lack of these minerals.

10 Some crops take large amounts of one particular mineral out of the soil. If a farmer grows the same crop on a field year after year, the soil then lacks that mineral. The crop yield gets smaller.

Explain why shortage of just one mineral affects the yield of a crop.

Effect of shortage of nitrates on older leaves.

Effect of shortage of a mineral.

How minerals get into plants

Remember

Plants need to take in water, minerals and carbon dioxide.

Substances diffuse through cell membranes to get into cells, including the transport tissues xylem and phloem.

$$\text{high concentration} \xrightarrow{\text{diffusion}} \text{low concentration}$$

Diffusion and concentration gradients

Molecules in liquids and gases move in all directions.
So, they spread out.

Look at the diagrams. There are more molecules of dye at
A than B. So, more of them move towards B than A. We
say that they diffuse along a **concentration gradient**.

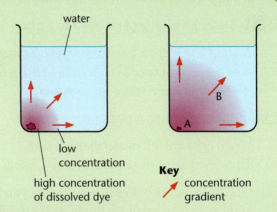

Substances such as potassium ions diffuse from one
cell to another. There are more potassium ions in cell A
than in cell B. So, more of them move from A to B than
from B to A.

11 Copy and complete the diagram of cells A, B and C.
Don't forget to draw some potassium ions in cell C.

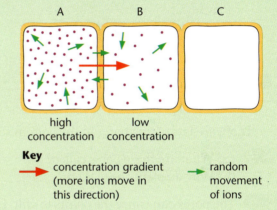

Suppose a cell needs to take in potassium ions from a very
dilute solution such as soil water. The ions cannot diffuse
in against the concentration gradient.
The diagram shows how the cell absorbs them.

12 (a) Explain, as fully as you can, how cells absorb ions
against a concentration gradient.
(b) Why do we call this process 'active transport'?

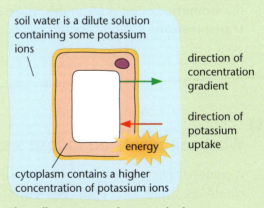

The cell uses energy from respiration to
absorb potassium ions by active uptake
(active transport).

Plants and water

If plants don't get enough water, they **wilt**.

1 Describe what a plant looks like when it wilts.

A plant wilts when its cells lose water faster than it can take it in. Plants need water to hold them up. We say they need water for **support**.

2 What else does a plant need water for?

When cells are full of sap they are firm. They lose their firmness as they lose water.

It is the same with 'freeze pops' before you freeze them. Full freeze pops are firm. But freeze pops which aren't full flop over.

This plant has plenty of water.

This plant is short of water. It has wilted.

How do plants lose water?

There are lots of tiny pores in the bottom surface of leaves.

3 (a) What are these pores called?
(b) Why do plants need them?

4 The pores can cause the plants to wilt. How?

Water evaporates from leaves, mainly through the **stomata**. Losing water in this way is called **transpiration**.

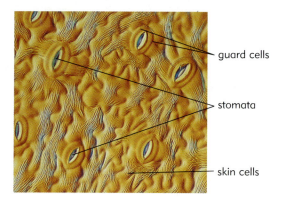

guard cells

stomata

skin cells

What can plants do to lose less water?

A plant wilts if it loses water faster than it takes it in. Some plants lose water faster than others. Most leaves have a waxy layer to slow down water loss.

5 Plants which grow in dry areas usually have the thickest waxy layers. Why do you think this is?

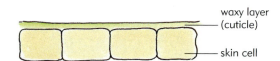

waxy layer (cuticle)

skin cell

Leaves have special cells to close their pores. We call them **guard cells**.

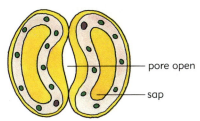
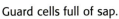

Guard cells full of sap.

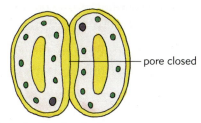

Guard cells after losing water.

6 (a) Describe the changes to the guard cells and leaf pores when the guard cells lose water.

(b) Explain how these changes benefit the plant.

When do plants lose most water?

Sue and Raj are trying to find out when plants lose most water. The diagram shows what they do.

The plant takes in water to replace the water it loses. The faster the plant takes in water, the faster the bubble moves.

7 Sue and Raj's results are shown in the table. They forgot to fill in two of the figures. What could the missing figures be? Give your reasons.

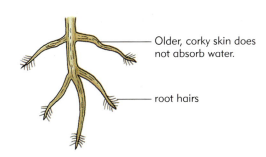

The leaves lose water.

The plant takes in water.

We close this tap to start the experiment.

scale

The bubble moves along the tube.   bubble

	Conditions	Average time (minutes) for bubble to move 100 mm
moist air around the plant	hot, windy	3
	hot, still	
	cool, windy	9
	cool, still	18
dry air around the plant	hot, windy	1
	hot, still	2
	cool, windy	3
	cool, still	

8 In which conditions did the plant take in (and lose) most water?

Which parts of a root take in water?

Plants get water from the soil. The water is taken in by the roots. We say the roots **absorb** the water. Most of the water goes in through the **root hair** cells. Root hairs increase the surface area of the root cells for absorption.

Older, corky skin does not absorb water.

root hairs

Each root hair grows out from just one cell. Root hairs soon get worn away as the root grows through the soil.

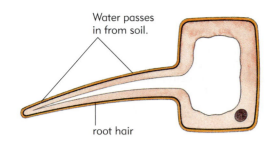

Water passes in from soil.

root hair

9   Where are the root hairs on a plant?
Describe this as clearly as you can.

10   How big do you think root hairs are?
Give a reason for your answer.

11   Older parts of roots do not have root hairs.
Why do you think this is?

How do cells take in water?

Cell membranes are **permeable** to water. This means that water can diffuse through them from where it is in high concentration to where it is in low concentration.

Soil water is a very dilute solution of mineral ions. So, water diffuses from the soil into root cells.

12   Look at the diagram.
     (a) Explain why the <u>water</u> concentration is higher in a dilute solution of mineral ions than in a concentrated solution of mineral ions.
     (b) What is a solute?
     (c) Write down <u>one</u> example of a solute.

Dilute solution	**Concentrated** solution
– fewer solute particles	– more solute particles
– more water molecules	– fewer water molecules

Key
- solute particles (ions or molecules)
- water
- water diffuses in this direction

Cell membranes control the passage of dissolved substances (solutes) such as mineral ions. So we say they are **partially permeable membranes**. The diffusion of water through a partially permeable membrane is called **osmosis**.

13   Explain, as fully as you can, how water passes by osmosis from the soil into a cell. Use the information and the diagram in this section to help you.

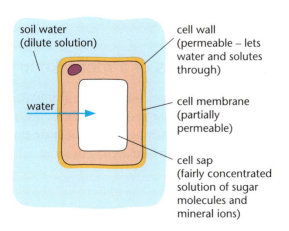

soil water (dilute solution)

water

cell wall (permeable – lets water and solutes through)

cell membrane (partially permeable)

cell sap (fairly concentrated solution of sugar molecules and mineral ions)

What happens in plant roots?

Plant roots take in water and minerals.

14 Put the following sentences in the right order to describe where water goes in a plant.
The first one has been done for you.

- Water passes through the cell membrane into a root hair cell by osmosis.

- Water goes into the xylem.

- Water passes into the cell sap.

- Water passes to the stem and leaves.

- Water passes from cell to cell by osmosis.

Soil water also covers the outside of the root. The diagram doesn't show this.

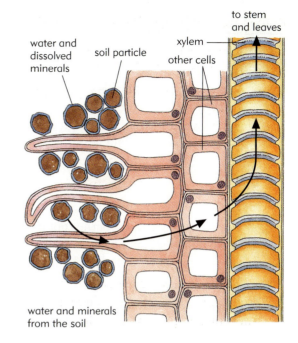

Higher

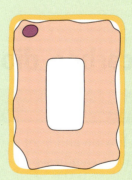

What happens when a plant doesn't take in enough water?

For your body to work properly, the water, sugar and ion contents are important. These things are also important for plants.

When a plant loses water it wilts. Its leaves and leaf stalks become soft and bendy. This is because its cells have lost more water than they have taken in. If plant cells take in enough water, they remain firm or rigid. We say that they have **turgor**.

This cell is short of water. Its contents aren't pressing against the wall. So the wall is not firm. When its cells are like this, a plant is wilting.

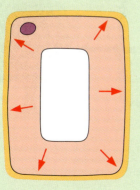

This cell has taken in enough water. Its contents are pressing against the cell wall. So the wall is firm and rigid. Rigid cells provide a plant with support.

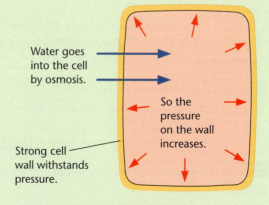

Water goes into the cell by osmosis.

So the pressure on the wall increases.

Strong cell wall withstands pressure.

15 Draw a flow chart to show how a cell maintains its turgor (stays firm)

How a cell's turgor is maintained.

16 Look at the picture.
(a) The guard cells take in minerals by active uptake from the cells next to them. Explain why water then passes into the guard cells.
(b) As water passes into the guard cells, the pore opens. Explain why.

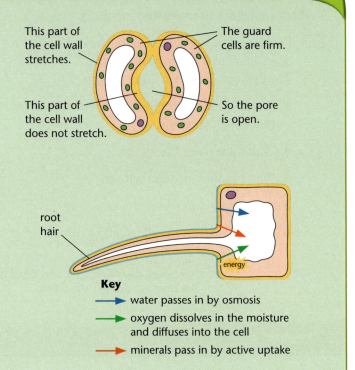

This part of the cell wall stretches.

The guard cells are firm.

This part of the cell wall does not stretch.

So the pore is open.

17 Explain <u>three</u> features of the root hair cell which make it a good exchanger.

root hair

energy

Key
→ water passes in by osmosis
→ oxygen dissolves in the moisture and diffuses into the cell
→ minerals pass in by active uptake

Plants and carbon dioxide

Carbon dioxide is a gas in the air. Like other gases, it spreads out everywhere it can. We call this **diffusion**.

1 Explain these two facts:
(a) carbon dioxide diffuses away from burning fuel;
(b) it diffuses towards trees during the day.

2 Why do green leaves take in carbon dioxide?

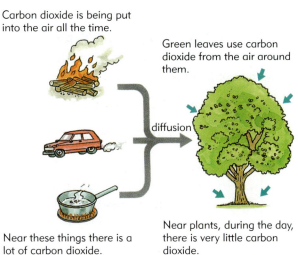

Carbon dioxide is being put into the air all the time.

Green leaves use carbon dioxide from the air around them.

diffusion

Near these things there is a lot of carbon dioxide. The **concentration** of carbon dioxide is quite high.

Near plants, during the day, there is very little carbon dioxide. The concentration of carbon dioxide is low.

How does carbon dioxide get inside leaves?

Carbon dioxide gas diffuses through tiny holes or pores in the skin of a leaf. We call these holes **stomata**.
The diagram shows where the carbon dioxide goes.

3 Copy the drawing. Then write down the following sentences in the right order:

- Carbon dioxide diffuses through the spaces between the cells.

- Carbon dioxide reaches the leaf by diffusion through the air.

- Carbon dioxide diffuses into the cells.

- Carbon dioxide diffuses into the leaf through tiny pores called stomata.

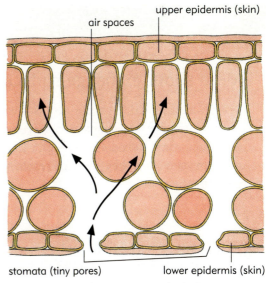

Carbon dioxide diffusing into a leaf.

How does carbon dioxide get into cells?

Carbon dioxide <u>gas</u> diffuses through the air spaces inside a leaf. But only <u>dissolved</u> carbon dioxide diffuses into cells.

4 Look at the diagram.
All the cells inside a leaf have a moist surface like this. Explain way.

So, inside a leaf the exchange surface:

- is very thin;
- is moist;
- has a large area.

These are all features of an efficient exchange surface.
The thin flat shape of a leaf also helps. Gases haven't far to diffuse and light can reach all the cells.

5 Why do leaf cells need light?

6 Look at the diagram of a squashed ball of Plasticine.
Think of it as a model of a leaf.
In which shape leaf will:
(a) carbon dioxide reach cells more quickly?
(b) light reach more cells?

7 Look at the diagram of part of a leaf.
Which <u>two</u> tissues are in the veins of a leaf?

Note that xylem carries water and minerals from the roots. Phloem carries sugar to growing points, storage organs and other parts.

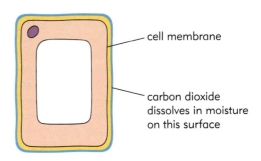

Substances can only pass through cell membranes if they dissolve first.

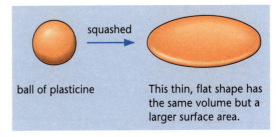

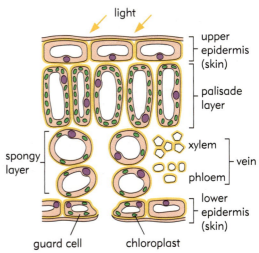

More about exchange surfaces

Leaves are adapted to exchange gases with the air. In the dark, they take in oxygen for respiration and give out carbon dioxide. In daylight, they make more oxygen than they need for respiration, so they give out the rest.

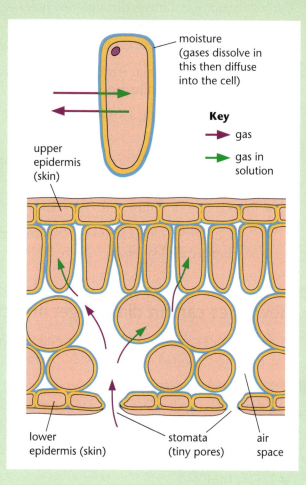

moisture (gases dissolve in this then diffuse into the cell)

upper epidermis (skin)

Key

→ gas

→ gas in solution

lower epidermis (skin)   stomata (tiny pores)   air space

8 (a) List the parts that a molecule of carbon dioxide passes through on its way from the air into the cytoplasm of a palisade layer cell.

(b) Where exactly are the exchange surfaces of leaves, and what makes them suitable for their job?

Good exchange surfaces:

- are thin;
- are moist;
- have a large surface area.

9 The drawings show simplified 'leaves'.

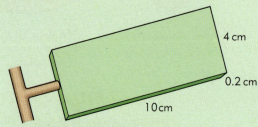

4 cm

0.2 cm

10 cm

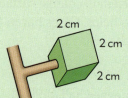

2 cm

2 cm

2 cm

(a) Draw a table to show the surface area and volume of the leaves.

(b) Which of the two is the better shape for gas exchange?
Explain your answer.

> You should be able to explain how gas and solute exchange surfaces in plants are adapted to their jobs.

How do plants know which way to grow?

Plants take in water through their roots and take in light energy through their leaves. So the roots of plants have to grow to where they can find water. The top parts of plants, the **shoots**, have to find light.

Shooting for the light

You may have noticed how a plant on a window sill grows towards the brightest light. This is because the shoot of the plant is sensitive to light.

Plants naturally grow straight up.

1 Plant shoots benefit from growing towards the light. Explain this as fully as you can.

Do plant shoots know which way is up?

The Earth pulls everything down towards it. We call this the **force of gravity**. Jack thinks that plant shoots grow away from the force of gravity.

He decides to do an experiment to test this idea. He sets up his experiment in a dark box.

2 Why did he do this experiment in the dark?

3 What did he find out from his experiment?

At the start.   A few days later.   The force of gravity pulls this way.

Jack's experiment with plants in the dark.

Do plant roots know which way is down?

Leena has four bean seedlings with tiny roots and shoots. She wants to find out if gravity also affects plant <u>roots</u>.

The diagrams show how she plants the seedlings. She then leaves the seedlings for a week to grow.

4 Where do you think Leena should put the seedlings to grow?
Give a reason for your answer.

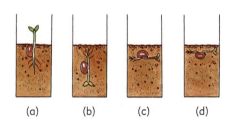

(a)   (b)   (c)   (d)

Leena's experiment with bean seedlings.

These diagrams show what happens in Leena's experiment.

5 What does this tell us:
(a) about the plant roots?
(b) about the plant shoots?

6 Draw what you would expect to see in jar (d).

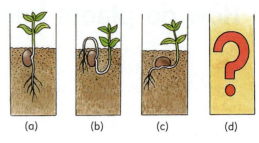

(a)   (b)   (c)   (d)

One week later.

Can we confuse a seedling?

The diagrams show how you can make roots and shoots grow sideways.

7 (a) What can you do to make roots and shoots grow sideways?
(b) Why does this work?

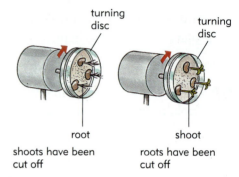

turning disc   turning disc

root   shoot

shoots have been cut off   roots have been cut off

The discs turn slowly, so that gravity pulls on all sides of the seedlings.

Why do shoots grow up and roots grow down?

The diagram shows how plants normally grow.

8 Why is it important that plants grow with their leaves above the ground?

9 Why should their roots grow below the ground?

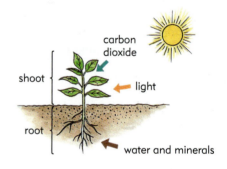

carbon dioxide

shoot

light

root

water and minerals

A plant growing normally.

Can plants find water?

Plant roots can usually find water by growing down. But sometimes there is more water to the side of a plant.

10 Sarah and Raj set up the experiment shown in the diagram.
What do you think they were trying to find out? How would they know whether they were right?

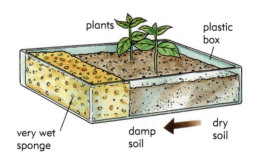

plants   plastic box

very wet sponge   damp soil   dry soil

What part of a plant detects light?

Charles Darwin thought that plant shoots can detect light. Over 100 years ago, he did an experiment to find out.

The diagrams show his results.

11 Which part of a shoot detects light?
Give reasons for your answer.

What makes the shoot bend towards the light?

Special chemicals affect the speed at which plants grow. We call these chemicals **hormones**.

12 When lit from one side only, a shoot grows faster on the side away from the light.
(a) What makes the shoot grow faster on that side?
(b) What is the effect of this faster growth on the direction of growth?

Hormones to ripen fruit

Scientists can now make the hormones that plants naturally make. We can use these hormones to make plants do what we want them to do.

For example, we can use hormones to ripen fruit.

Sometimes farmers pick fruit such as tomatoes before it is ripe. This is because it is hard and doesn't get damaged. Later it is sprayed with hormone to make it ripen.

13 Why do farmers pick their fruits before they are ripe?

14 Why are the fruits treated with hormones?

Hormones as killers

2,4-D is a hormone that kills weeds (unwanted plants). It makes plants grow so fast that they die.

2,4-D does not kill plants with narrow leaves.

The picture shows some of the weeds growing on a lawn of grass.

15 Why is 2,4-D a good weedkiller to use on the lawn?

16 Which plants will 2,4-D get rid of?
Give a reason for your answer.

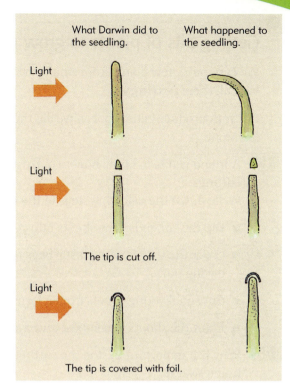

What Darwin did to the seedling.   What happened to the seedling.

Light

Light

The tip is cut off.

Light

The tip is covered with foil.

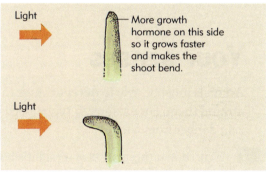

Light — More growth hormone on this side so it grows faster and makes the shoot bend.

Light

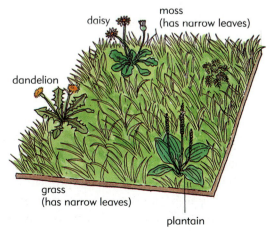

daisy   moss (has narrow leaves)

dandelion

grass (has narrow leaves)   plantain

Plants on a lawn.

Getting bits of plants to grow roots

We can make new plants from small parts of older plants. We call these **cuttings**.

Cuttings produce better roots if we dip them in rooting hormone.

17 A friend wants to know how to make new plants from cuttings.
Write down the following steps in the right order.

- Dip the cut end of the shoots into rooting hormone.

- Cover the shoots with plastic bags to prevent them drying out.

- Cut small young shoots from a larger plant.

- Plant the shoots in compost and water them.

18 Why is it important to stop the cuttings losing too much water?

Dip in rooting hormone.

Place in compost, water and cover with plastic bag.

Remove plastic bag when new roots have grown.

Your senses

Animals must be able to **detect** things that are going on around them. For example, a rabbit needs to know when it is in danger from a fox.

1 Write down <u>three</u> parts of a rabbit's body that it can use to detect a fox.

Anything the rabbit detects with its senses is called a **stimulus**. For example, the scent of a fox is a stimulus. (When there is more than one stimulus we call them **stimuli**.)

2 Write down <u>two</u> other stimuli a rabbit can use to detect a fox.

How do we detect stimuli?

We detect stimuli with our eyes, ears, tongue, nose and skin. We call these our **sense organs**. Sense organs contain special cells called **receptors**. The receptors detect the stimuli. For example, receptors in your eyes detect light.

3 Make a table to show some stimuli and the sense organs that detect them.

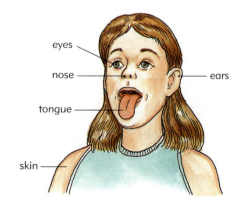

eyes
nose
ears
tongue
skin

Your sense organs.

From sense organ to brain

The receptors in your sense organs are connected to your **nervous system**.

4 Look at the diagram.
Write down the parts of the nervous system.

Nerves carry information from receptors to the brain and spinal cord. The brain and spinal cord make sense of the information.

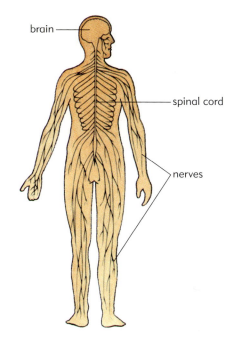

Your nervous system.

Taste and smell

There are thousands of receptors on your tongue. These are your taste receptors and they detect chemicals dissolved in water. The taste receptors are in groups called **taste buds**. There are four different kinds of taste buds. The diagram shows where they are.

5 List the <u>four</u> different tastes that your tongue can detect.

Receptors inside your nose detect chemicals in the air (smell). You need to taste and smell to get the full flavour of food.

6 Food seems tasteless when you have a cold. Why do you think this is?

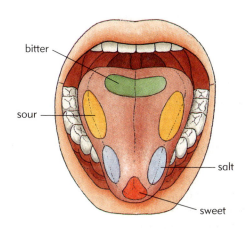

Skin deep

Your skin also contains receptors. The diagram shows three different kinds of receptors in your skin.

7 What stimuli is your skin sensitive to?

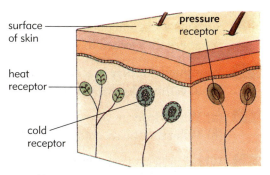

Your skin.

Ears aren't only for hearing with

Your ears contain two kinds of receptors.

8 (a) Write down the <u>two</u> places in your ear where there are receptors.
(b) In each case say what the receptors are for.

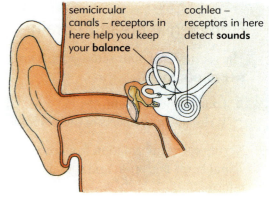

Your ear.

Eyes – your window to the world

Your eyes are sense organs. They contain light receptors. Information from these receptors passes along nerve fibres to your brain.

Your eye from the outside

Your eyes are in hollows, or **sockets**, in your skull. The bone protects your eyes from bumps and bangs.

9 Your eyes are protected in other ways. Describe <u>three</u> of these ways.

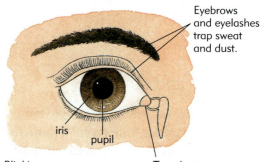

A look into your eye

The diagram shows all the different parts inside your eye. Use the diagram to answer the questions below.

10 (a) What is the white of your eye called?
(b) Why do you think it has to be tough?

11 The cornea and lens are **transparent**.
(a) What does this word mean?
(b) Why do they need to be transparent?

12 Suspensory ligaments hold the lens in place. What are they joined to?

13 Your pupil is a hole in a coloured ring. What is this coloured ring called?

14 (a) Which layer of your eye contains receptors?
(b) What are the receptors sensitive to?

15 What carries information from light receptors to your brain?

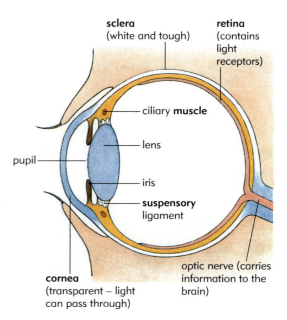

Section through your eye.

Getting the right amount of light

Light gets into your eye through your pupil. You may have noticed that your pupil isn't always the same size. The muscles in the iris cause these changes in the size of the pupil.

In bright light your pupil gets smaller.
This stops too much light getting into your eye.

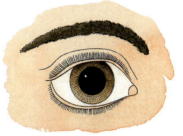

In dim light your pupil gets larger.
This lets enough light into your eye.

16 (a) Describe the change in size of your pupil as you go from a dimly lit into a brightly lit room.
(b) Why is this change useful?

How can we see things?

The diagram shows what happens when you see a tree.

17 Look at the diagram.
Then put these sentences in the right order.

- The cornea and lens bend the rays of light.

- Information is sent along the optic nerve to the brain.

- Receptor cells detect the image.

- Light from the tree reaches your eye.

- An image of the tree forms on the retina.

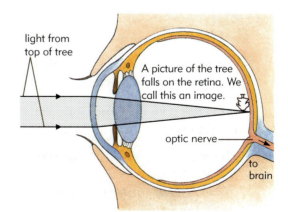

light from top of tree

A picture of the tree falls on the retina. We call this an image.

optic nerve

to brain

You need to be able to label the following on a diagram of the eye:

optic nerve, lens, cornea, ciliary muscles, iris, suspensory ligaments, sclera, retina, pupil.

> **Remember**
>
> The cornea and lens focus images on the retina.
>
> The retina is made of light sensitive cells.
>
> These cells convert light energy into nerve impulses.

Producing a clear image

You have to focus a lens to get a clear image. With a glass lens you do this by changing the **distance** between the lens and the screen.

Your eyes need to focus but you cannot change the distance between the lens and the retina. However, the lens can change **shape**. Lenses of different shapes focus at different distances. So, you can use the same lens to focus on near and distant objects.

18 Look at the diagrams.
Make a table to show how the ciliary muscle, suspensory ligament and lens differ when your eye focusses on near and distant objects.

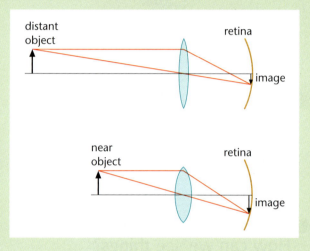

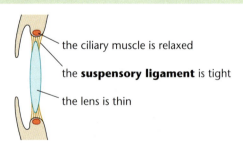

The eye is focussed on a distant object.

- the ciliary muscle is relaxed
- the **suspensory ligament** is tight
- the lens is thin

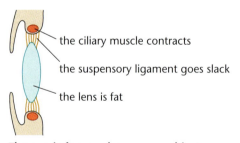

The eye is focussed on a near object.

- the ciliary muscle contracts
- the suspensory ligament goes slack
- the lens is fat

19 Some people say that cats can see in the dark.
(a) Explain why this is not true.
(b) What do they really mean?

20 Good advice to readers is to look into the distance from time to time to relax their eyes.

Explain why looking at a book can strain your eyes but looking into the distance relaxes them.

Coordination

Suppose you see a £10 note lying on the floor. You bend down and pick it up. You can do this because of your nervous system.

Your sense organs send information to your brain all the time. Your brain sorts out this information. It also controls what you decide to do. We say that your brain **coordinates** your actions.

How does your brain know about the money?

1 Look at the diagram. What type of stimulus travels from the money to your eyes?

The receptors in your eyes send information to your brain. These are called **nerve impulses**.
They travel very quickly along a nerve to your brain.

What does your brain do next?

Your brain lets you **react** or **respond**. You decide what to do. Then your brain sends impulses to the muscles in your body to carry out your decision.

2 What is your response to seeing the £10 note?

3 Which parts of your body produce this response?

Your brain coordinates the actions of all the muscles you use to pick up the money. Your brain also makes sure that you do not fall as you bend over.

How information is carried to and from your brain

Nerve fibres need to be very long because they carry information to and from your brain. These fibres are parts of cells called **neurones**.

Sensory neurones carry impulses from your receptors to your brain. **Motor** neurones carry impulses from your brain to your muscles.

4 Describe the difference between the job of a sensory and a motor neurone.

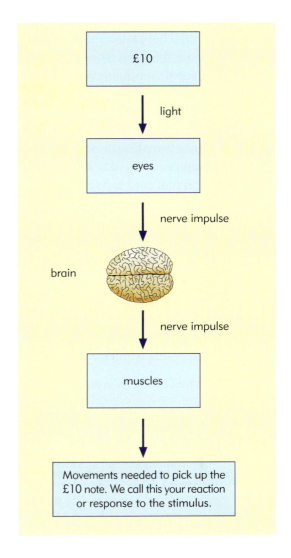

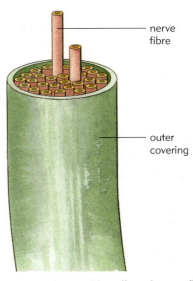

Nerves are made up of bundles of nerve fibres.

Ouch! That's hot

If you touch a hot plate, you move your hand away quickly without even thinking. The diagram shows what happens. We call a rapid, automatic response like this a **reflex** action.

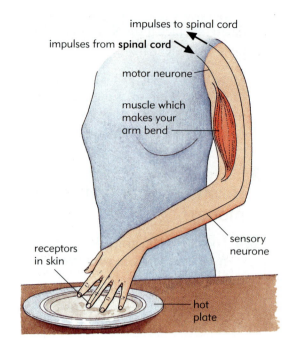

5 Put the following sentences in the right order to explain what happens.

- A muscle makes the arm bend away from the hot plate.
- Receptors in the skin detect that the plate is hot.
- The spinal cord sorts out the information.
- Impulses travel along sensory neurones to the spinal cord.
- Impulses from the spinal cord travel along motor neurones to the arm muscle.

6 For each stimulus listed below, write down the automatic response:
(a) dust in eye;
(b) bright light shone at eye;
(c) food enters windpipe.

Athletes must set off as fast as they can when they hear the starting pistol.

7 Describe, step by step, what happens inside an athlete's body when the starting pistol is fired.

Higher

Pathways in your nervous system

Some reactions, such as picking up a £10 note, are **voluntary** – you decide whether or not to do them. Other reactions don't involve thinking – if you touch a hot pan you don't want to waste time thinking about it before you move your hand. This quick, automatic reaction is called a **reflex** action.

In both voluntary and reflex actions, the pathway of the impulse is similar.

8 (a) What is a synapse?
(b) How does a nerve impulse cross it?

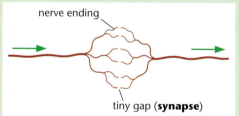

There is a tiny gap between one neurone and the next. A chemical released at the end of one neurone causes an impulse to start in the next one.

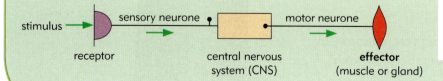

Higher

As well as sensory and motor neurones, a reflex action usually involves a **relay** neurone in the **central nervous system**, or CNS (the brain or spinal cord).

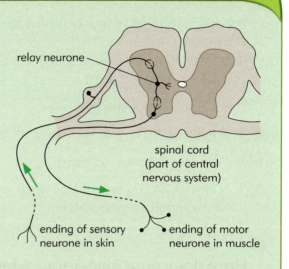

relay neurone

spinal cord
(part of central
nervous system)

ending of sensory
neurone in skin

ending of motor
neurone in muscle

9   Look at the diagram. Then, write down the following sentences in the correct order.
   ● The effector (a muscle or a gland) responds.
   ● Impulses from a receptor pass along a sensory neurone.
   ● A muscle responds by contracting, a gland by releasing (secreting) chemical substances.
   ● In the central nervous system, impulses pass from a sensory to a relay neurone, then to a motor neurone.

10 When Liam saw his dinner, his mouth started to 'water' (his brain caused his salivary glands to secrete saliva).
   (a) Why is this reflex action useful?
   (b) For this reflex action, what is:
      ● the stimulus?   ● the effector?   ● the receptor?
      ● the response?   ● the coordinator?

Keeping things the same inside your body

To work properly your body must be at just the right temperature. Your blood must also contain just the right amounts of water, sugar and other substances.

Everything inside your body has to be kept at a constant level. You don't have to think about this.
Your body controls these things automatically.

1   Explain how each part labelled on the body diagram helps to keep conditions inside the body constant.

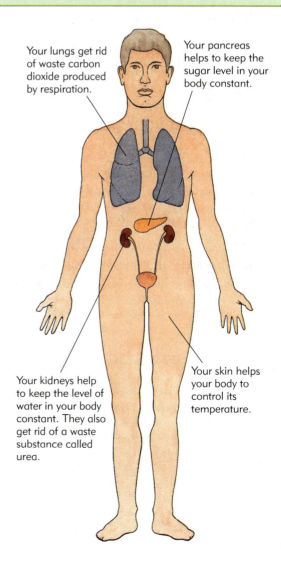

Your lungs get rid of waste carbon dioxide produced by respiration.

Your pancreas helps to keep the sugar level in your body constant.

Your kidneys help to keep the level of water in your body constant. They also get rid of a waste substance called urea.

Your skin helps your body to control its temperature.

How your skin can help keep you cool

Try licking the back of your hand and blowing on it.
The wet part of your hand feels colder than the dry part.
Heat from your body makes the water evaporate.
This helps to cool you down.

The diagram shows how it does this.

2   What happens in your skin when you are hot?

3   Why does this cool you down?

4   Do you think you sweat much when you are cold?
    Give a reason for your answer.

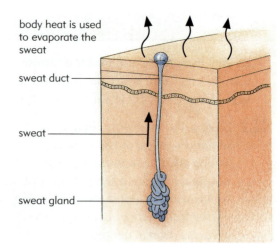

When you are hot, your skin produces sweat.

How does your skin know when to sweat?

Your body constantly checks what temperature it is at.
The diagram shows how it does this.

5   Which part of your body checks your temperature?

6   What should the temperature of your body be?

7   What happens if your body temperature is higher
    than this?

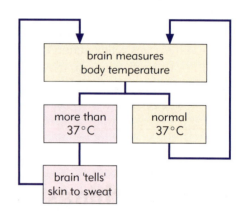

Why must your body be at 37 °C?

Chemical reactions happen all the time in your body.
Enzymes are substances that make these chemical
reactions happen. Your enzymes work best at 37 °C.

8   Copy the graph. Mark where you think the
    temperature 37 °C should be.

9   Why don't enzymes work as well:
    (a) above 37 °C?
    (b) below 37 °C?

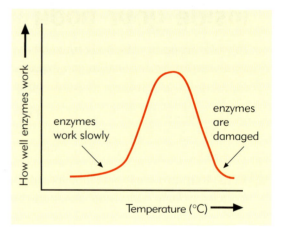

Helping your body stay at 37 °C

In very hot or very cold weather you can make it easier for
your body to stay at 37 °C.

10   Explain how you can do this.

11   You need to drink more when the weather is hot.
     Why is this?

Keeping your blood glucose concentration constant

It is important that the amount of sugar (glucose) in your blood is kept constant.

Read the newspaper article. It tells you what happened to Barbara when the amount of sugar in her blood was too low.

12 What effect did low blood sugar have on Barbara?

13 What did Ben do to solve this problem?

Does your blood glucose concentration change?

The amount of sugar in your blood goes up after a meal. It falls when you exercise. Usually your body detects these changes and returns the amount of sugar to normal.

14 Look at the graph. What do you think is the normal blood glucose concentration?

15 At what time of day did this person have a meal? How do you know?

16 How long did it take for the blood glucose concentration to return to normal?

17 At what time of day did this person take exercise? How do you know?

How do you keep your blood glucose concentration constant?

If your blood glucose concentration changes, your pancreas releases special chemicals called **hormones**. These hormones make your blood glucose concentration go back to normal.

18 Describe what happens when the blood glucose concentration:
(a) rises higher than normal;
(b) falls below normal.
Use the diagram to help you.

Ben, 4, saves life of coma mother

Barbara, 25, told how four-year-old Ben kept her alive.

He knew exactly what to do when he found her unconscious.

He poured a bottle of his strawberry drink with extra sugar down her mouth then some glucose jelly, kept in the fridge for emergencies and finally some sugar-rich cough mixture.

Making the first telephone call of his life he dialled 999.

When the police arrived, Ben told them his mother was diabetic.

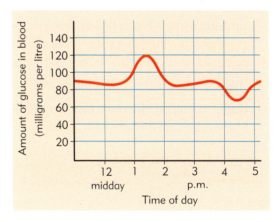

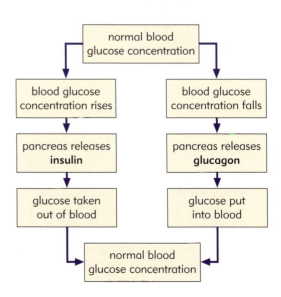

How do the hormones work?

Hormones are chemical messengers. Each hormone is produced by one organ in your body but travels through your blood and affects a different organ.

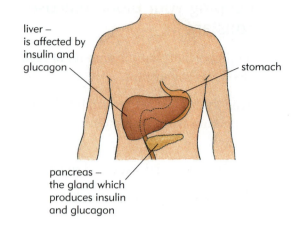

liver –
is affected by insulin and glucagon

stomach

pancreas –
the gland which produces insulin and glucagon

19 Look at the diagram. Which organ do the hormones insulin and glucagon affect?

20 How do the hormones reach this organ?

We say that the liver is the **target organ** for insulin and glucagon.

What happens if your pancreas is not working properly?

Barbara's pancreas does not work properly. It does not produce enough insulin.

21 Without insulin, what happens to Barbara's blood glucose concentration?

Barbara's disease is called **diabetes**. She must take regular injections of insulin. This lowers the amount of sugar in her blood. She must also be careful about how much carbohydrate (sugar and starch) she eats.

Barbara's blood glucose concentration fell too low. She could have died.

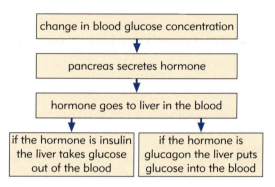

change in blood glucose concentration

↓

pancreas secretes hormone

↓

hormone goes to liver in the blood

↓

| if the hormone is insulin the liver takes glucose out of the blood | if the hormone is glucagon the liver puts glucose into the blood |

22 How can you help someone in a diabetic coma? (Remember what Ben did.)

Higher

More about blood glucose concentration

Having too little or too much glucose in your blood is dangerous. Either of these can cause coma and death. Your pancreas monitors your blood glucose concentration. If this concentration is too high (or too low) the pancreas releases insulin (or glucagon) into your blood. We say that the pancreas **secretes** these hormones.

Both hormones affect your liver cells. Without them, your liver cells can't control the concentration of glucose in your blood. **Insulin** enables liver cells to take glucose out of your blood and store it as insoluble glycogen. **Glucagon** causes the reverse to happen. It enables liver cells to change glycogen into soluble glucose which is released into your blood.

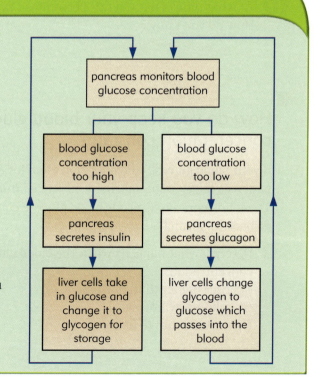

pancreas monitors blood glucose concentration

blood glucose concentration too high	blood glucose concentration too low
pancreas secretes insulin	pancreas secretes glucagon
liver cells take in glucose and change it to glycogen for storage	liver cells change glycogen to glucose which passes into the blood

Higher

23 Your pancreas and your liver control your blood glucose concentration. Explain, as fully as you can, the part played in lowering the concentration of glucose in your blood by:
(a) your pancreas;
(b) your liver.

Controlling temperature

Your inner body temperature or **core temperature** should be about 37 °C. It is safe for your skin and some other parts to get hotter or colder than this. Part of your brain acts like a thermostat. We call it the **thermoregulatory centre**. It checks the temperature of the blood passing through your brain. It also receives impulses from temperature receptors in your skin.

If your core temperature is too high or too low, your thermoregulatory centre sends impulses to your skin and muscles. These affect muscles, sweat glands and the blood vessels to your skin. Evaporation of sweat cools your body. Shivering muscles are contracting muscles. To contract, they use the energy from respiration. Some of this energy is released as heat.

24 (a) Where is your thermoregulatory centre?
(b) Explain how it helps to control your body temperature.

25 List the changes to your body when your core temperature is:
(a) too high;
(b) too low.

26 What is shivering and why does it warm you?

27 Look at the diagram that shows why you lose less heat when the blood vessels in your skin constrict. Draw a similar diagram to show what happens when these blood vessels dilate.

28 Why does your skin look flushed when you are hot?

We call the maintenance of a constant internal environment **homeostasis**.

You will see how your kidneys act as organs of homeostasis as well as excretory organs on pages 72 and 73.

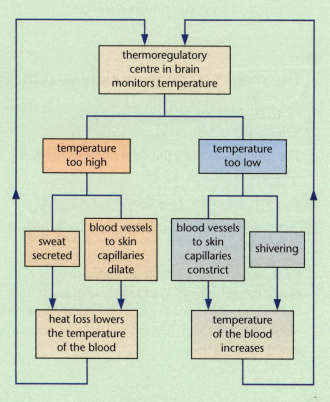

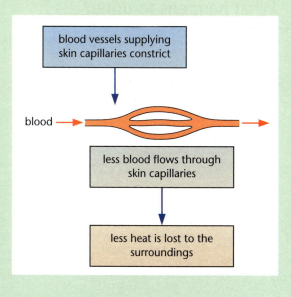

The kidneys

Cleaning blood and balancing water

The blood in your body is constantly passing through your kidneys. Your kidneys remove the poisonous substances which your cells produce. If your blood is not cleaned in this way you will soon die.

1 Write down <u>two</u> ways we can save the life of a person whose kidneys have stopped working.

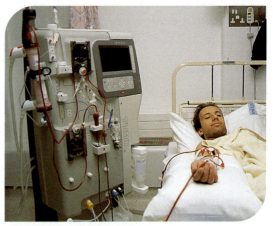

This man's kidneys have stopped working. His blood is cleaned by a kidney machine. He may be lucky enough to get a kidney transplant one day. Then he won't need the kidney machine any more.

The main poison your kidneys remove

Our bodies break down the protein in our food into amino acids.

The diagram shows what then happens to these amino acids.

2 What does your body use amino acids for?

3 What happens to any amino acids your body doesn't use?

Urea is made in the liver from amino acids. Urea is a poisonous substance. Your kidneys remove it from your blood. They also remove excess salts (ions) from your blood. The urea and salts are dissolved in water to make a liquid called **urine**.

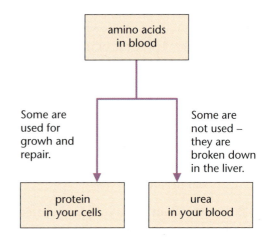

amino acids in blood

Some are used for growh and repair.

Some are not used – they are broken down in the liver.

protein in your cells

urea in your blood

What happens to urine?

The diagram shows what happens to the urine your kidneys produce.

4 (a) Name the tubes that carry urine to your bladder.
(b) Why do you need a bladder?

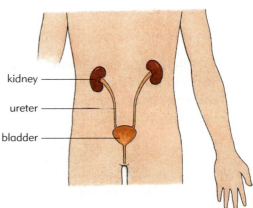

kidney

ureter

bladder

Your bladder stores urine.

A balancing act

Your body is about two-thirds water.

The amount of water you lose must balance the amount you take in each day. The diagram shows the different ways you gain and lose water.

5 If you weigh 45 kg, how much of this is water?

6 (a) Your body loses water in <u>four</u> main ways. Write them in a list, including the amounts of water lost in each way.
(b) Make a similar list of the <u>three</u> main ways that your body gains water.

7 Marathon runners breathe hard and sweat a lot for a few hours. Explain why they must drink plenty of water.

The kidneys and water control

Your kidneys control the amount of water in your body.

8 The amount of water in your body affects the volume and colour of your urine. Explain the differences shown in the diagrams.

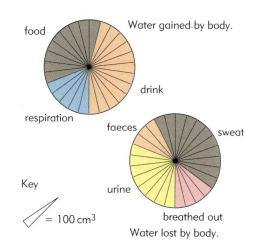

Daily water balance.

How much urine you make in one day …

if your body has a lot of water.

if your body has a little water.

Your kidneys filter your blood

Your kidneys get rid of urea and the excess water and salts from your blood. Over one litre of blood passes through your kidneys every minute. Urine trickles out of them continuously and travels down tubes called **ureters** into your bladder.

Each kidney is made of millions of tiny **tubules** with walls only one cell thick. Water, ions, glucose and urea are filtered from the blood into the tubules at high pressure. Your body needs to keep all the glucose and some of the water and ions. So, these are re-absorbed into your blood.

9 Write a list of the substances that your kidney tubules:
(a) filter from your blood;
(b) re-absorb into your blood.

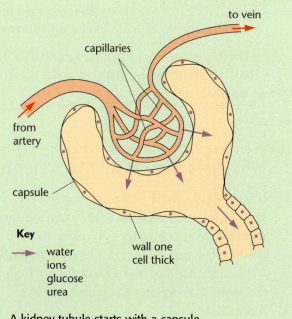

A kidney tubule starts with a capsule that is like a cup.

Higher

Then they re-absorb the things you need

The concentration of ions and glucose in your kidney tubules is as high as or higher than their concentration in your blood. This means that your kidneys re-absorb them against a concentration gradient. They need energy to do this, so re-absorption is by active uptake.

As glucose and ions pass into your blood, the fluid in the tubules becomes more dilute. So, water then passes into your blood by osmosis.

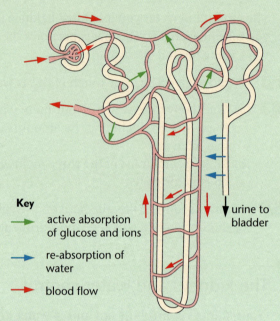

Key

→ active absorption of glucose and ions

→ re-absorption of water

→ blood flow

↓ urine to bladder

One kidney tubule and its blood supply.

10 Explain how glucose, water and dissolved ions are re-absorbed into your blood.

11 Look at the diagram. Three tubes, A, B and C are connected to the kidney.

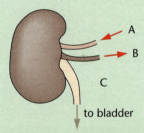

A
B
C
to bladder

(a) Which one carries unfiltered blood?
(b) Which one carries filtered blood?
(c) Which one carries urine?
(d) Which one should contain the least urea?
(e) Which one should not contain glucose?

12 Organs suitable for exchanges have thin, moist surfaces, a large surface area and a good capillary supply. Describe how the kidneys satisfy all these conditions.

Why does the amount of water re-absorbed vary?

If the concentration of ions in your blood is too high, your kidneys get rid of the extra ions and re-absorb as much water as possible. So, you produce less urine, but it is more concentrated. If the water content of your blood is too high, less water is re-absorbed and your urine is more dilute.

13 What would be the effect of the following things on your urine volume and concentration:
(a) drinking four mugs of unsweetened tea?
(b) eating lots of salted crisps?
(c) playing a football match on a hot day?

What controls the re-absorption of water?

The **pituitary gland** in your brain monitors the amount of water in your blood. It releases a hormone called **ADH**. Its effect is to make your kidney tubules re-absorb water into your blood. The more water you have in your blood, the less ADH is secreted.

14 Look at the diagram. Draw a similar diagram to show what happens when you have too much water in your blood.

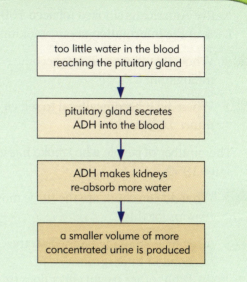

too little water in the blood reaching the pituitary gland

↓

pituitary gland secretes ADH into the blood

↓

ADH makes kidneys re-absorb more water

↓

a smaller volume of more concentrated urine is produced

People and drugs

A **drug** is a substance that can change the way your body works. Some drugs, such as painkillers and antibiotics, are useful. We use antibiotics to kill bacteria in our bodies. But other drugs, such as solvents, alcohol and chemicals in tobacco, may harm the body.

Drugs aren't a new problem. People have been using drugs for thousands of years.

Alcohol – an old drug

We know that people made beer in Babylon 8000 years ago. The Egyptian picture writing from nearly 4000 years ago warns people not to drink too much.

1 What dangers of drinking alcohol are mentioned in this writing?

2 (a) Make a copy of this picture.

(b) What do you think it means?

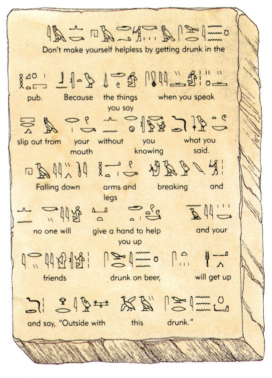

Don't make yourself helpless by getting drunk in the
pub. Because the things you say slip out from your mouth without you knowing what you said. Falling down arms and legs breaking and no one will give a hand to help you up and your friends drunk on beer, will get up and say, "Outside with this drunk."

Egyptian picture writing (hieroglyphics).

Smoking milestones

Native Americans smoked tobacco 1000 years ago. Look at the information about smoking in England.

3 When did tobacco houses become common in England?

4 When did it become illegal to sell cigarettes to those under 16 years old?

The number of people who smoke has gone down a lot since 1950.

5 What do you think are the <u>three</u> most important reasons for this?

6 (a) Why is smoking banned on the London Underground?
(b) What other useful effect does this ban have?

Another ancient habit that is growing

People have sniffed substances for thousands of years. Over 3000 years ago Greeks sniffed oils, herbs and spices as part of their religion. Some Greeks still burn olive leaves in their homes as part of their religion.

In the 1700s some people sniffed a gas called **nitrous oxide**. Nitrous oxide puts people to sleep. Smaller amounts make people feel relaxed and happy. They tend to laugh a lot.

7 (a) Why do you think people sniffed nitrous oxide?
(b) Why is nitrous oxide sometimes called 'laughing gas'?

During the last 100 years some people have started sniffing **solvents**, such as those used in glues and paints.

Using solvents in this way can harm people and even kill them. Sniffing solvents is called **solvent abuse**.
The graph shows how the number of deaths from solvent abuse changed from 1971 to 1990.

8 What was the number of deaths from solvent abuse:
(a) in 1972?
(b) in 1987?

9 How did the number of deaths from solvent abuse change from the early 1970s to the late 1980s?

People can become **dependent** on or **addicted** to a drug. They get **withdrawal** symptoms when they cannot get the drug.

1492 European sailors see cigars being smoked in Cuba

1625 Many tobacco houses open in England

1908 Illegal to sell cigarettes to children under 16 years of age

1947 43 per cent rise in cigarette tax

1951 Medical report shows smoking causes lung cancer

1965 Ban on cigarette advertising on TV

1971 Message on cigarette packs: 'Warning by HM Government: Smoking can damage your health'

1974 Tobacco tax increased by 20 per cent

1985 London Transport bans smoking on the Underground after a fire which killed many people

2002 Government passes Act to ban tobacco advertising and to strengthen messages on cigarette packs

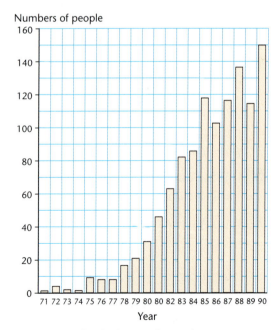

Number of deaths from solvent abuse.

The dangers of sniffing solvents

Solvents are found in many products. They are in aerosol sprays, glues, some correcting fluids and petrol.

Solvents and the body

When you breathe a solvent in, the fumes go into your lungs. The solvent passes into your blood and travels to other parts of your body. Some of the effects are shown in the diagram.

10 What are the main organs which solvents damage?

11 How does a solvent get to the brain, the lungs and the liver?

12 What other effects do solvents have?

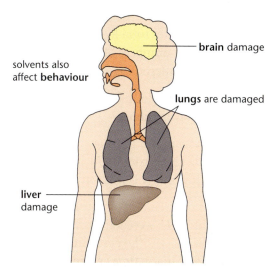

solvents also affect **behaviour**

brain damage

lungs are damaged

liver damage

Organs that are damaged by solvents.

How can you tell if someone is sniffing solvents?

Some of the warning signs are shown on the right.

13 You need to look for a few of these signs and not just one. Why do you think this is?

14 Which <u>three</u> signs of solvent abuse could you spot just by looking at someone?

Warning signs of solvent abuse

- the person smells of solvent
- red, glassy or watery eyes
- wide pupils
- slurred speech
- loss of appetite
- lack of concentration
- does poorer work at school
- bad coordination

Darren died of solvent abuse

Many people die after the first time they sniff a solvent.

Darren was 16 years old when his mother found him dead. He died of solvent abuse.

15 Why do you think this mother wants to tell others how she feels?

16 How do you think the families of those who use solvents feel?

"I will never stop grieving over the death of my son. I will never stop crying when I remember he will never have another birthday, never have another girlfriend, never ever be able to dash out and meet his friends"

Darren's mother

Who are the main solvent abusers?

The graph gives total figures for a 20-year period.
It shows the numbers of people who died after sniffing
solvents at different ages.

17 Which age group had the largest number of deaths
from solvent abuse?

18 (a) How many 18-year-olds died of solvent abuse in the
20 years from 1971 to 1991?
(b) How many is this, on average, each year?

19 (a) 12 per cent of all those who died of solvent abuse
in 1991 were females.
What was the percentage of males?
(b) How many times more males than females died of
solvent abuse?

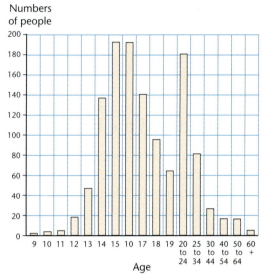

Deaths in the UK from solvent abuse (1971–91).

Are all parts of the UK affected?

Some parts of the UK have more of a problem
than others.

20 Which part of the UK has the greatest problem?

21 Make a table to show the numbers of deaths from
solvent abuse in the regions of the UK shown on
the map.
Put the region with the highest number of deaths at
the top of your table and the region with the lowest
number of deaths at the bottom.

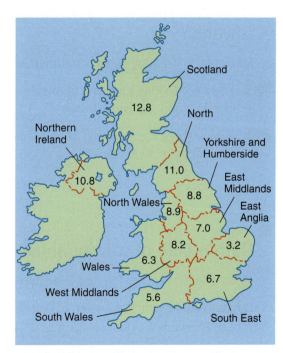

Number of 10- to 24-year-olds dying of solvent
abuse every year for every million people.

What's your poison – alcohol?

Beers, wines and spirits like whisky contain **alcohol**.
Alcohol is a drug. Its effects on the body have been known
for a long time.

'One drink and you act like
a monkey; two drinks, and
you strut like a peacock;
three drinks, and you roar
like a lion; and four drinks –
you behave like a pig.'
Henry Vollam Morton (1936)

1   Look at Morton's description of how alcohol
affects behaviour. Explain what he meant.

Effects of alcohol on the body

Small amounts of alcohol slow your reactions.
Larger amounts can make you unconscious or even go
into a coma.

Your liver changes the alcohol into harmless substances.
However, too much alcohol can damage organs in
your body.

2   What does alcohol do to your body that makes it
dangerous to drink and drive?

3   What can happen if you drink too much alcohol?

4   Which organs in your body can be damaged by alcohol?

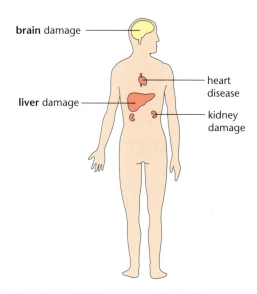

brain damage

heart disease

liver damage

kidney damage

If you keep drinking too much alcohol.

How much alcohol is there in different drinks?

The drinks shown in the diagram all contain the same
amount of alcohol.

This is called one **unit**.

5   Tony drinks two pints of ordinary beer.
Alex drinks one pint of strong beer and
one single whisky.
Nassia drinks two glasses of wine.

How many units of alcohol does each person drink?

1 glass of
table wine

1 single
whisky

$\frac{1}{2}$ pint
ordinary
beer or
cider

$\frac{1}{3}$ pint
strong
brew

1 glass
of sherry

How much drink gives one unit of alcohol.

What is 'reasonable drinking'?

Alcohol is a poison. It slows your reactions and it affects your behaviour.

You can become dependent on alcohol or addicted to it. This means that you can't do without it. But small amounts of alcohol can help some people to relax.

The diagram shows how many units of alcohol many health workers think it is safe to drink each week.

6 (a) What is the maximum number of units a man should drink each week?

(b) How much strong beer is this each day?

7 A woman drinks three glasses of wine and a single whisky each day.

(a) How many units of alcohol is this in a week?

(b) What advice would you give the woman?

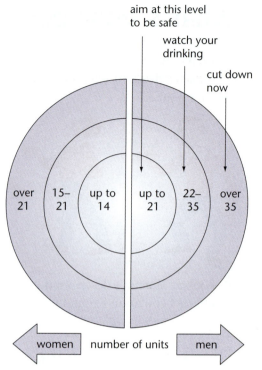

Safe drinking – units per week.

Legal but harmful – tobacco

What is in cigarette smoke?

Cigarette smoke contains many harmful chemicals. Some of these are shown in the table.

1 Which chemical is an addictive drug?

2 Which substance contains chemicals that cause lung cancer?

3 When carbon monoxide is in your blood your heart has to work harder. Why do you think this is?

It is the carbon monoxide in car exhaust fumes that can kill people. If gas fires aren't working properly they can also release carbon monoxide.

Nicotine is a natural poison found in many plants. The plants produce it to stop insects eating them. Some insecticides contain nicotine.

4 What effect does nicotine have on your body?

5 Why is nicotine useful to gardeners?

Chemical	Addictive drug	Poison	Effect on body
tar (1000 different chemicals)		✓	causes cancer
nicotine	✓	✓	increases blood pressure, raises pulse rate
carbon monoxide		✓	blood can carry less oxygen

Remember

You can become addicted to some drugs. This means you feel that you can't do without them.

The dangers of smoking

6 Use the information on the diagram to make a table about some dangers of smoking. Use the headings:

- Organ affected
- Name of disease
- What the disease does.

If you smoke you have …

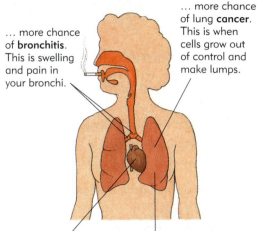

… more chance of **bronchitis**. This is swelling and pain in your bronchi.

… more chance of lung **cancer**. This is when cells grow out of control and make lumps.

… more chance of a **heart attack**. This is when your heart stops working.

… more chance of **emphysema**. This is when alveoli break down because of coughing. You have less lung surface to take in oxygen.

Does it matter whether you smoke?

The table shows how smoking affects your chance of getting lung cancer.

Number of cigarettes smoked per day	Increased chance of cancer compared with non-smokers
5	4 ×
10	8 ×
15	12 ×
20	16 ×

7 Write down, as carefully as you can, what the table tells you.

8 Look at the photograph. What message is the photograph trying to get across to people?

Would you like a cigarette?

No!

9 The fetus (developing child) inside a pregnant woman who smokes is likely to get less oxygen than it should. Why is this?
(Hint: Look at the table on page 78.)

10 Write down <u>one</u> effect of lack of oxygen on the baby.

Facts and figures

- Out of every 1000 teenagers who smoke, about 250 will eventually die of smoking related diseases.

- On average, smokers die 10 to 15 years earlier than non-smokers.

- Over 90 per cent of people who die from lung cancer are smokers.

- The babies of pregnant women who smoke weigh, on average, 200 g less than the babies whose mothers do not smoke. We say that the babies have a **low birth mass**.

- Pregnant women who smoke are more likely to have a miscarriage or still birth.

Carbon monoxide and your body

When fuels such as petrol and natural gas burn completely, they produce **carbon dioxide** and water. If there isn't enough oxygen to burn them completely, some **carbon monoxide** is produced too. Remember that there is carbon monoxide in cigarette smoke.

Carbon monoxide is a poisonous gas. We can neither see it nor smell it. That is why many people now have detectors for it in their homes.

11 Look at the pictures. Describe the symptoms of carbon monoxide poisoning.

How does carbon monoxide poison you?

The haemoglobin in your red cells absorbs carbon monoxide more easily than it absorbs oxygen. So it soon picks up any carbon monoxide that you breathe in.

Carbon monoxide combines with haemoglobin to form cherry-red carboxy-haemoglobin. This reaction is <u>irreversible</u>. So this haemoglobin can no longer carry oxygen. The brain is the first organ to suffer from lack of oxygen.

12 Write a word equation for the reaction between haemoglobin and carbon monoxide.

13 Explain how this reaction is different from the reaction between haemoglobin and oxygen.

14 Explain the symptoms of carbon monoxide poisoning.

15 A chemical was added to some blood to stop it clotting. Then it was divided between three beakers. A different gas was bubbled into each beaker. The gases were carbon monoxide, carbon dioxide and oxygen. The pictures show the results.
   (a) Which gas was bubbled into which beaker?
   (b) Oxygen was then bubbled into all three beakers. The colour changed in only one of them. Which one was it and why?

"My head aches and I feel confused."

"I'm finding it hard to breathe."

"He's in a coma."

The flue from Carl's gas heater is blocked. The pictures show what happened to him.

haemoglobin + oxygen ⇌ oxyhaemoglobin
(dull red)   (bright red)

In your tissues, oxyhaemoglobin gives up its oxygen. The reaction is reversible.

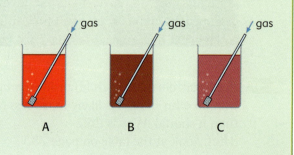

gas   gas   gas

A   B   C

Smoking and lung cancer

Sir Walter Raleigh brought tobacco to Britain in the sixteenth century. It was smoked mainly in pipes. Cigarettes became popular during and after the 1914–1918 war. Few women smoked before 1920.

Deaths from lung cancer increase

In the years after World War I, the number of men dying from lung cancer increased. At first many people suggested that the figures weren't really true. It was just that doctors were getting better at recognising lung cancer. This could have explained some of the increase. But people began to look for other reasons.

16 Look at the graph.
What was the death rate from lung cancer in:
(a) 1920?
(b) 1940?

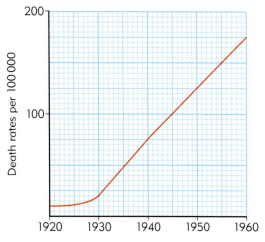

Death rates of men aged 45–64 from lung cancer in England and Wales.

Why were deaths from lung cancer increasing?

About 50 years ago in Britain, Professor Richard Doll and Dr Bradford Hill thought that there might be a link between cigarette smoking and lung cancer.
Other scientists were thinking about and working on similar ideas in the United States. This idea is not strange to us now.

17 Why did Doll and Hill think that smoking and lung cancer were linked?

Doll and Hill decided to test their idea by doing a survey. They asked a group of men with lung cancer about their smoking habits. They selected a group of men of similar ages and backgrounds who didn't have lung cancer. They asked the men the same questions. These men were the **control** group.

18 What was the idea or **hypothesis** that Doll and Hill were testing?

19 Why did they need a control group?

20 Look at the table.
Do the results support the hypothesis that there is a link between smoking and lung cancer?

	Lung cancer patients	Control group
smokers	99.7%	95.8%
non-smokers	0.3%	4.2%

Looking for more evidence

In 1951 Doll and Hall decided to do some longer-term studies. They recorded the smoking habits, the health and the death rates of a group of doctors over 5 years.

21 Look at the table.
   (a) Which group of doctors were at most risk from lung cancer?
   (b) How many times greater was the risk of lung cancer for heavy smokers than for non-smokers?

	Deaths of doctors from lung cancer (per 100 000)
non-smokers	7
light smokers	47
moderate smokers	86
heavy smokers	166

At the same time as Doll and Hall were researching the health of doctors, scientists at New York University began testing the chemicals in cigarette smoke to find out if any of them caused cancers in animals. They found that many did.

In 1962 a report summarising the research was published in Britain. It concluded that there was a relationship between lung cancer and smoking. Not everyone agreed. Many people looked for other explanations.

22 Who produced the report and what was it called?

Following the report, there was only a 4% fall in the number of cigarettes smoked. One group that did take notice of the report was the doctors. Large numbers of doctors gave up smoking. The death rates of doctors from lung cancer fell.

23 What action did large numbers of doctors take that provided further evidence of a link between smoking and lung cancer? Explain your answer.

24 Some groups of people did not want to accept that there was a link between smoking and lung cancer. Think of two reasons for this.

By the end of the twentieth century, even the tobacco companies had to agree that tobacco smoking increases the chances of developing lung cancer.

SMOKING
AND
HEALTH

A report of The Royal College of Physicians on smoking in relation to cancer of the lung and other diseases

"My factory makes millions of cigarettes every year."

"Not all smokers get lung cancer."

"Non-smokers can get lung cancer too."

You do not need to remember any information from this topic. You must be able to interpret information you are given in the same sort of way.

3 Environment

Staying alive

Remember from Key Stage 3

It is hard to survive in some places on Earth.
In some places it is sometimes too hot or too cold.
Or there might not be enough light. Sometimes there
might not be enough water or oxygen or carbon dioxide.

We call all of these things <u>physical</u> factors. These factors
vary according to the time of day and time of year.

1 Write down a list of physical factors which affect plants
and animals.

2 Between which temperatures do most plants and
animals live?

3 Why do you think not many organisms live below 0 °C?

4 Not many organisms live above 40 °C. Why?

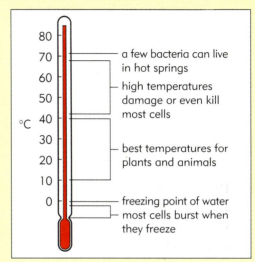

a few bacteria can live in hot springs

high temperatures damage or even kill most cells

best temperatures for plants and animals

freezing point of water

most cells burst when they freeze

Day and night

During the day, the Earth gets heat and light from
the Sun.

5 Explain why the temperature at the Earth's surface
rises during the day and falls at night.

These daily changes in temperature and light affect
animals and plants.

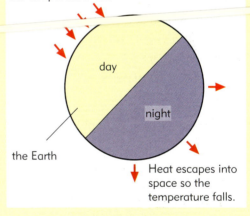

Heat from the sun makes the temperature rise.

day

night

the Earth

Heat escapes into space so the temperature falls.

Seasons of the year

In tropical rainforests there are many different kinds of
plants and animals which live, grow and breed there all
year round. In countries like Britain, some plants and
animals can be seen all the year round. But others can
be seen only at some times of the year. This is because
they live, grow and reproduce only when conditions
are right for them.

6 The diagram shows how the temperature in Britain
changes with the seasons. Write down <u>two</u> other
physical factors that change with the seasons.

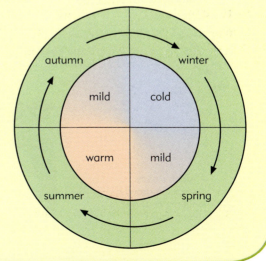

autumn   winter

mild   cold

warm   mild

summer   spring

How plants survive the winter

Frost can damage plants, and plants cannot get enough water when it is frozen in the soil. Different plants have different ways of surviving in cold weather.

Some plants, like the poppy, complete their life cycle in one growing season. They are called <u>annuals</u>.

Some plants, including trees, live for many growing seasons. They are called <u>perennials</u>. Different trees and other woody plants have different ways of surviving frost and water shortage.

7 Which part of an annual plant survives the winter?

8 Why do broad-leaved trees lose their leaves in winter?

9 Evergreen trees do not need to lose their leaves in winter. Write down <u>two</u> reasons why.

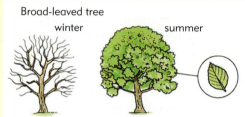

Winter	Spring	Summer	Autumn
seeds	seeds germinate	flowers	seeds are spread, plants die

Annual plants

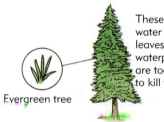

Broad-leaved tree

winter   summer

These trees lose their leaves in autumn. This stops the tree losing water from the leaves.

These trees lose less water because their leaves are small and waterproof. The leaves are too tough for frost to kill them.

Evergreen tree

winter and summer

How animals survive the winter

Animals also have problems in winter. It is cold and food is in short supply.

10 Write down <u>three</u> different ways animals can survive the winter in Britain.

11 Explain how each of the animals shown in the pictures survives the winter in Britain.

Some animals *hibernate*. They do not feed or move. They live on the fat stored in their bodies.

Ladybirds *hibernate* in a sheltered place.

Hedgehogs *hibernate* in piles of leaves.

A rabbit's body *changes*. It grows a longer coat of hair to keep it warm.

Swallows *migrate* to South Africa where it is summer.

Frogs *hibernate* under rocks and stones.

Surviving in different places

Humans live in most parts of the world. In many places they can survive only by making clothes, houses and many other things.

1. How is the Inuit in the picture able to live in the extreme cold?

People in other parts of the world survive in different ways.

2. Write down some of the ways that people use to survive in the desert.

Different kinds of animals and plants live in different places. They live, grow and reproduce in places where conditions are suitable for them. We say that they are **adapted** to the conditions in which they live.

Inuit.

How animals are adapted to cold places

The Arctic fox is adapted to live in the cold, but the Fennec fox is adapted to live in the heat of the Sahara.

3. Look at the pictures. Write down <u>two</u> differences between the Arctic and Fennec foxes.

Fennec foxes have long bodies and ears. This means that they have a large surface area. They have hardly any fat and very short fur. All these things mean that lots of heat can escape from their bodies. They stay cool!

Arctic foxes have a shorter body, small ears, lots of fat and thick fur. The small surface area, and thick fur and fat cut down on heat loss. They stay warm!

Fat and fur are good for stopping heat escaping. We say that they are good **insulators**.

4. How do you think these features help Arctic foxes to live in cold climates?

Some animals have colours which match their surroundings. We say that they are **camouflaged**.

5. Arctic foxes have white coats in winter and dark coats in summer.
Explain how this helps them to survive.

Arctic fox.

Fennec fox.

How animals and plants are adapted to desert life

In the desert there is very little water.
Gerbils survive by living in burrows in the desert.
The burrows stay cool during the day. Droplets of
water (condensation) collect on the walls. Each day
a gerbil produces only one or two drops of very
concentrated urine.

The leaves form spines so they lose less water.

The roots of the cactus spread a long way so they can take in a lot of water when it rains.

6 Write down <u>two</u> ways that a gerbil can survive the shortage of water in the desert.

7 The cactus is a desert plant. Write down <u>two</u> ways the cactus is adapted to live in dry places.

8 The camel is well adapted to life in the desert. Look at the picture. Explain how each adaptation shown helps the camel to survive.

A camel's hump is a fat store. It can break down fat to release water.

A camel can drink large amonts of water.

Coarse wool on top of its body protects the camel from the sun

Its mouth is tough so it can eat thorny plants like cacti.

Short hair underneath the camel lets heat escape.

Its big flat feet stop it sinking into the sand.

You need to be able to explain how adaptations of animals to desert and Arctic conditions are related to:

● body size and surface area;
● amount of insulating fur and fat;
● camouflage.

Surviving in water and on land

Animals have features which help them to survive in the conditions in which they live. They are adapted to their **habitats**. Some animals are adapted to live in water. Other animals are adapted to live on land.

Life in water

Like all other animals, fish need oxygen to survive. Fish live in water, so they must get the oxygen they need from the water.

9 How do fish get oxygen from water?

10 Write down <u>three</u> ways that fish are adapted to move easily through the water.

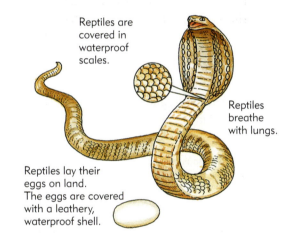

Streamlined shape to move easily through water.

Powerful tail muscles for swimming.

Fins steer the fish through the water.

Fish produce eggs with no shells. They are laid in water.

Gills with large surface area. They take oxygen from water into the blood.

Life on land

Some snakes and other reptiles live on land. All animals which live on land must stop too much water escaping from their bodies. Reptiles and birds lay their eggs on land. Nearly all birds are adapted for flying.

11 How do reptiles stop water escaping from their bodies?

12 How do reptiles take in oxygen from the air?

13 Why do you think that reptiles and birds lay eggs with waterproof shells?

Reptiles are covered in waterproof scales.

Reptiles breathe with lungs.

Reptiles lay their eggs on land. The eggs are covered with a leathery, waterproof shell.

14 Write down <u>three</u> ways in which birds are adapted for flying.

15 What else do feathers help birds to do?

Wing with feathers for flying.

Feathers for insulation to keep warm.

streamlined body

light, hollow bones

Birds lay eggs with shells so they do not dry up on land.

Many mammals are adapted to life on land. The picture shows some of the features of a land mammal.

16 Make a table to show the features of a land mammal and how they help it survive.

It has hair/fur for insulation to keep warm.

The mammal's babies develop inside the mother's body. This is a safe, warm, moist place.

It has lungs for breathing air.

It has legs for movement.

Life on land and in water

Some animals, like the frog, start their lives in water but then spend a lot of time on land. They go back to water to mate and lay eggs.

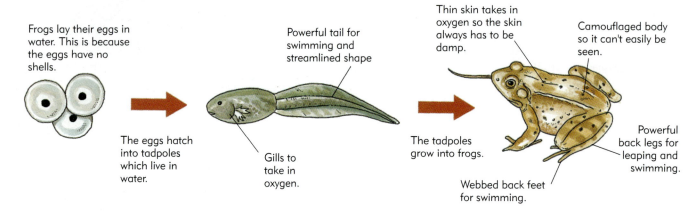

Frogs lay their eggs in water. This is because the eggs have no shells.

The eggs hatch into tadpoles which live in water.

Powerful tail for swimming and streamlined shape

Gills to take in oxygen.

The tadpoles grow into frogs.

Thin skin takes in oxygen so the skin always has to be damp.

Camouflaged body so it can't easily be seen.

Powerful back legs for leaping and swimming.

Webbed back feet for swimming.

17 What would happen to a frog's eggs if it laid them on land?

18 Write down two ways that a tadpole is adapted to living in water.

19 We usually find frogs in damp places. Why is this?

20 Write lists to show how a frog is adapted to life:
(a) on land;
(b) in water.

You should be able to look at a picture of an animal and say how it is adapted to its surroundings.

Different places, different plants

Plants and animals live in places where conditions are suitable. The place where a plant or animal lives is called its **habitat**. A plant or an animal is adapted to survive in its habitat. For example, oak trees survive only in well-drained soils. Alder trees can survive in wet soil. They cannot compete with other trees in dry soil.

A walk in the country

21 Imagine you are walking up the hill from the river in the drawing below. Describe the three types of land you walk through.

22 Explain, as fully as you can, why:
(a) oak trees grow on the hill but not by the river;
(b) alder trees grow by the river but not on the hill.

steep hill
(thin, dry soil)

flat, marshy
ground
(very wet soil)

river

pasture
(grassy field)
(deep, moist soil)

Why don't trees spread to the pasture?

Sheep are eating the grass in the field. They are grazing. Seeds from the trees fall on the pasture and some of them start to grow. Sheep eat the tops of the tiny trees. Look at the drawings of the tree seedling and the grass plant.

23 Sheep usually kill young trees. Explain why.

24 Grass can survive grazing by sheep. Explain why.

25 What would happen to the pasture if there were no sheep on it for several years? Explain your answer.

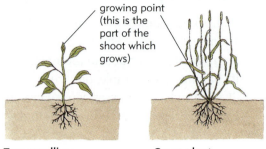

growing point (this is the part of the shoot which grows)

Tree seedling   Grass plant

Into the woods

It gets darker as you go into the wood. The plants are different from those in the field. They are adapted to different conditions.

26 Write down two differences between conditions in the wood and in the field.

14 °C

25 °C

To the mountains

As you go over the top of the hill you can see mountains in front of you. The trees above you are all **conifers**.

As you walk up the mountain it gets colder and more windy.

Water evaporates more quickly in windy places. Trees living in windy places can easily lose too much water. To survive they have to lose as little water as possible.

27 Write down two ways conifers are adapted to stop them losing water from their leaves.

28 Higher up the mountain there are no trees at all. Why do you think this is?

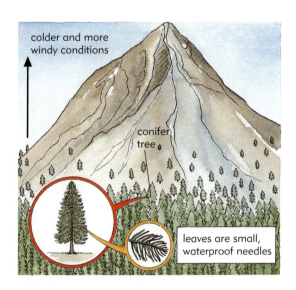

colder and more windy conditions

conifer tree

leaves are small, waterproof needles

You need to be able to explain why particular plants or animals live where they do.

Competition and predation

What are weeds?

Weeds are plants which are growing where they are not wanted. People plant poppies in their flower gardens, but poppies are weeds in a farmer's crop.

1 Write down the names of <u>three</u> weeds in the picture.

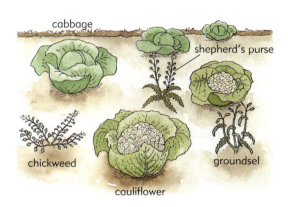

cabbage

shepherd's purse

chickweed

groundsel

cauliflower

What is wrong with weeds?

Gardeners don't like weeds because they take the things flowers, fruit and vegetables need to grow. They **compete** with the gardeners' plants. So, gardeners try to get rid of weeds so that their plants can grow better.

They:

- pull them up;
- cut them down;
- use weedkillers.

2 Write down <u>four</u> things the crops and weeds compete for. (Use the remember box to help you.)

3 Write down <u>three</u> ways of getting rid of weeds.

> **Remember**
>
> Plants need the right conditions to grow well. They need water and nutrients from the soil, carbon dioxide from the air, and light.

Weeding is a constant battle

A garden is overgrown.
The weeds are all crowded together.

A family clears the weeds and plants some vegetables.
The last picture shows the same garden a few months later.

How weeds can start to grow again.

  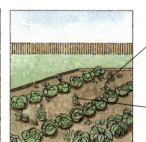

This weed grew from a seed blown into the garden.

This weed grew from a bit of root that was left in the soil.

4 Describe <u>two</u> differences between the weeds in the first and last pictures.

5 Why are there weeds in the garden a few months later?

Competition between weeds

The groundsel plants on the right were taken from the garden shown on the last page. Both of these plants took three months to grow.

6 Describe <u>one</u> way in which they are similar.

7 Describe <u>one</u> difference between them.

The first plant has grown tall because the garden was overcrowded. The plant was competing with many other plants.

8 What are the plants competing for when they grow tall?

9 Which parts of the plant would you expect to grow bigger in the competition for water and nutrients?

Plants compete for light as well as for water and nutrients from the soil. Plants can get what they need more easily if they have plenty of space.

Groundsel from a very weedy part of the garden.

Groundsel from a less weedy part of the garden.

What happens if we never weed a garden?

The picture shows the same garden after it has not been weeded for many years.

10 What plants have taken the place of many of the smaller weeds?

11 Why do you think the smaller weeds died out?

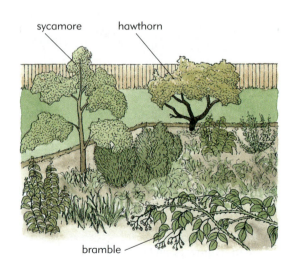

sycamore   hawthorn

bramble

> You should be able to suggest what plants are competing for if you are given information about a particular habitat.

Competition between animals

Animals often have to compete with each other for the things they need to stay alive.

If you keep gerbils as pets, you have to supply them with food, water, nesting material and shelter. If you keep too many gerbils in one cage they will be overcrowded and uncomfortable. This is cruel.

12 What would the gerbils be competing for in an overcrowded cage?

Animals most often compete with each other for food, water and space.

Competition between fish

Animals which don't compete very well for food and other things they need may die.

The graph shows how the number of fish in a tank affects the number of young fish which die. If you are a fish farmer, the more fish you rear the more profit you make.

13 Why is it a bad idea to keep too many fish in one tank?

14 Why do you think more young fish die in an overcrowded tank?

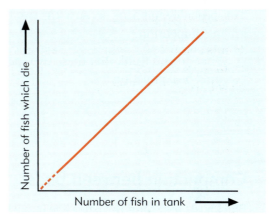

Competition between flies

Animals which don't compete very well for food and other things they need may not be able to breed.

Scientists often keep fruit flies for breeding experiments. The graph shows what happens if they keep too many flies in the same space.

15 What happens to the number of eggs that are laid as the number of fruit flies goes up?

16 Why do you think this happens?

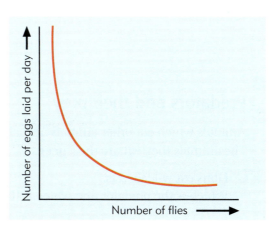

Competition between squirrels

So far we have talked about animals of the same kind, or **species**, competing with each other. But as well as competing with members of their own species, animals compete with those of other species.

If two different species compete for exactly the same things they cannot live together. One of the two species will win the competition and so survive.

The graph shows what happens when grey squirrels arrive in an area where red squirrels live.

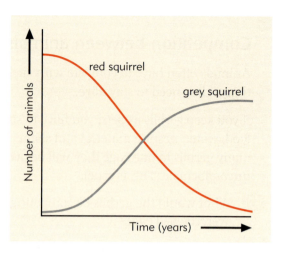

17 Describe what happens to the numbers of the two types of squirrels.

This probably happens because the two types of squirrel are in competition.

18 What do you think that these two species of squirrel could be competing for?

Competition between birds

Blackbirds and song thrushes live in parks and gardens. Although they eat similar food they can still live together in the same place.

19 What <u>two</u> differences in their diet mean that blackbirds and song thrushes can live in the same place?

You need to be able to suggest what animals are competing for when you are given information about a particular habitat.

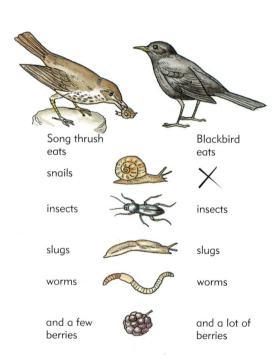

Song thrush eats		Blackbird eats
snails		✗
insects		insects
slugs		slugs
worms		worms
and a few berries		and a lot of berries

Predators and their prey

Animals which eat other animals are called **predators**. The animals they eat are their **prey**.

20 Lions eat wildebeests.
Which is the predator and which is the prey?

Lioness and wildebeest.

Blackbird and earthworm.

21 Write down the names of <u>two</u> other predators and their prey.

Usually predators feed on more than one kind of animal. For example, a lion doesn't just eat wildebeest.

22 Write down the name of another animal it can eat.

An animal can be preyed upon by several predators. For example, mice are not just eaten by owls.

23 Write down the names of <u>two</u> other animals which eat mice.

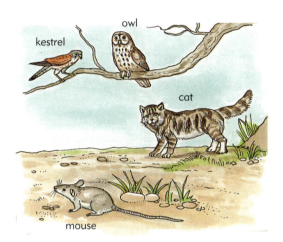

Stoats and rabbits

Stoats are predators. They eat rabbits and other animals. We call all the rabbits which live in one place a **population**. The size of the population is limited by the amount of food available.

24 Look at the graph. What happened to the size of the rabbit population between 1930 and 1950?

If the population of rabbits increases, there is more food for the stoats.

25 (a) What do you think happened to the number of stoats as the population of rabbits went up?
(b) Give a reason for your answer.

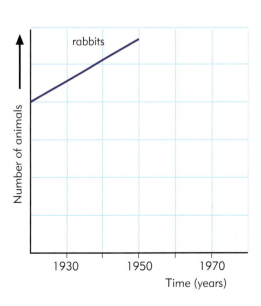

A disaster for rabbits (and for stoats)

In the 1950s rabbits were infected with a disease. Most of them died. The stoats did not catch this disease, but 95 per cent of them still died.

26 Explain why there was a fall in the population of:
(a) rabbits;
(b) stoats.

27 The graph shows what happened to the rabbit population after 1953. Copy the graph and continue your line for the stoats as far as 1960.

In the 1960s the number of stoats began to go up again. This is because they started to eat different food such as a lot more small birds, mice and even earthworms instead of rabbits.

28 Draw a line on your graph to show this increase. Label the line with a reason for the increase.

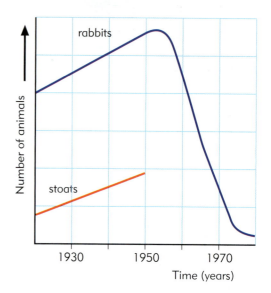

Predators of the Great Barrier Reef

The Great Barrier Reef is off the east coast of Australia. It is made of coral. Coral is an animal which feeds on microscopic sea-life called **plankton**.

The crown-of-thorns starfish feeds on coral. So the starfish is a predator, and the coral is its prey. Starfish themselves are prey for other predators.

29 Write down the names of <u>three</u> predators of the starfish. Use the diagram to help you.

Many people visit the reef and collect shells. They also catch trigger fish, Napoleon fish and puffer fish. So numbers of all these have gone down.

30 The number of starfish is going up. Why is this?

31 What effect does the increase in the number of starfish have on the coral population?

32 What happens to the prey population when the predator population increases?

The Great Barrier Reef.

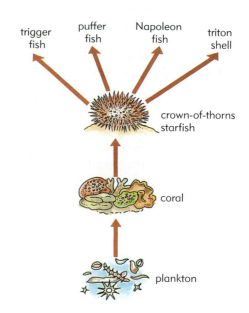

Look at the flow diagram on the right. In some parts of the reef, the starfish have eaten most of the coral animals. Only the skeletons are left. Scientists are worried because they think the coral will take 20 years to grow back.

Some scientists think that people should not be allowed to collect shells or to fish on the reef. A ban on fishing and collecting could help to save the coral.

33 There is an increase in the number of trigger fish. Draw a flow diagram, similar to the one on the right, that shows how this affects the starfish population.

The populations of different species which live in the same place are called a **community**.

Population of predators
(starfish) increases

↓

So they need more food

↓

So the population of prey
(coral) goes down

The balance of nature

In nature, the numbers of predators and prey stay in overall balance. On the Great Barrier Reef, people have upset this balance.

34 What did they do to upset it?

The balance in nature doesn't always mean steady numbers. Sometimes the numbers of predators and prey go up and down every few years. This is what happens in Canada with the lynx and the snowshoe hare. The lynxes prey on the hares.

The graph shows how the numbers of lynxes and hares changed over 30 years.

35 Describe what happened to the population of hares during the first 10 years.

During this time, the population of lynxes also changed.

36 Describe **one** similarity between the population change for the lynx and that for the hare.

37 Describe the differences between the two population changes.

38 The population changes for the lynx arc always behind those for the hare by about two years. Why do you think this is?

Lynx (predator) and snowshoe hare (prey).

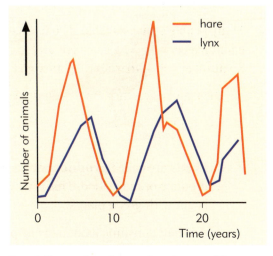

Population cycles of snowshoe hare and lynx.

Food chains, webs and pyramids

Remember from Key Stage 3

Food chains

Food chains are a way of showing what animals eat. They always begin with green plants.

animal
↑
animal
↑
plant

Each arrow in a food chain means 'is eaten by'. The food chain on the right tells you that the grass is eaten by the rabbit and the rabbit is eaten by the fox.

1 What does this food chain tell you?

barley ⟶ mouse ⟶ owl

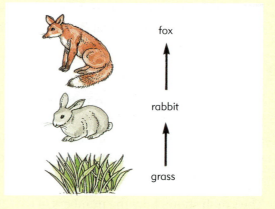

fox

rabbit

grass

Drawing your own food chains

If you know what different animals eat, you can draw your own food chains.

2 Look at the sets of pictures below. Draw a food chain for each set.

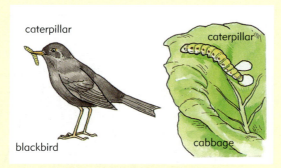

caterpillar

caterpillar

blackbird

cabbage

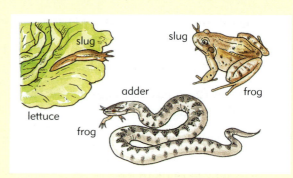

slug

slug

adder

frog

lettuce

frog

Plants are called **producers** because they produce or make food. This is why food chains begin with green plants.

Animals consume food made by plants or they eat animals which have eaten plants. So they are called **consumers**.

Remember from Key Stage 3

The diagram shows what the words mean in a food chain.

3 Go back to the food chains you drew for question 2. Write 'producer', '1st consumer' or '2nd consumer' next to each member of the chains.

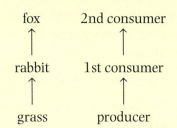

fox   2nd consumer
↑   ↑
rabbit   1st consumer
↑   ↑
grass   producer

Food webs

Most animals don't eat just one thing. So they belong to more than one food chain.

When different food chains contain the same animals we can join them together. We then get a **food web**.

4 Draw a food web to show:
(a) caterpillars, earthworms and rabbits eat leaves;
(b) hedgehogs and blackbirds eat caterpillars and earthworms;
(c) foxes eat earthworms and rabbits;
(d) falcons eat rabbits and blackbirds.

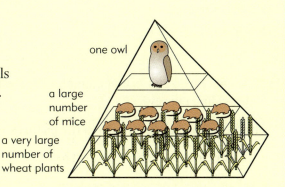

one owl

a large number of mice

a very large number of wheat plants

Pyramids in food chains

5 Explain what the pyramid of numbers tells you about the food chain

$$\text{wheat} \longrightarrow \text{mice} \longrightarrow \text{owl}.$$

The numbers of plants and animals in this food chain make a pyramid. We call this a **pyramid of numbers**. The diagram shows a simple way of drawing the pyramid:

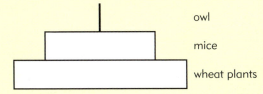

owl

mice

wheat plants

6 The diagram shows this food chain in a field:

$$\text{daisy} \longrightarrow \text{caterpillar} \longrightarrow \text{blackbird}$$

(a) Count the number of each of daisies, caterpillars and blackbirds. Set out your results in a table.

(b) Draw a pyramid of numbers for this food chain. Use a scale of 1 mm to represent one plant or animal.

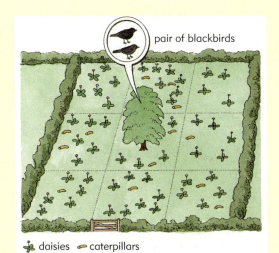

pair of blackbirds

🌿 daisies   🐛 caterpillars

Food and energy chains

To stay alive, plants and animals need a constant supply of energy. The diagrams show how they get the energy they need.

7 How do plants get the energy they need?

8 What do plants do with the energy once they have got it?

Plants only use a small amount of the energy which reaches them from the Sun. Some of the energy that they trap is then passed along the food chain.

9 How do animals get the energy they need?

10 What else do plants and animals use food for besides providing energy?

This is the food chain shown in the pictures:

lettuce ⟶ snail ⟶ thrush

Food, or stored energy, passes along the chain:

lettuce ⟶ snail ⟶ thrush

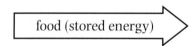

food (stored energy)

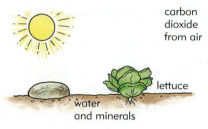

carbon dioxide from air

lettuce

water and minerals

Plants capture **light** energy from the sun. They use this to make **food**. Food is a store of chemical energy.

Animals eat plants. They use some of the energy from food to move. They use some of the food to grow. Energy is then stored in the animals' bodies.

thrush

Animals may be eaten by other animals. Part of the energy stored in the food is used to move or to keep warm. Some food is used to grow so the energy is stored in the animals' bodies.

Less and less energy

Animals use a lot of the energy from their food to move about. Some animals also use energy from food to keep warm. All of this energy ends up in the surroundings as heat.

Animals use some of their food to grow. This means that the energy in the food is stored in the animals' bodies.

The diagram shows what happens to the energy stored in the bodies of plants and animals as you move up a food chain. The higher you go, the less energy there is.

11 Draw, and label, a similar diagram for the food chain:

cereals ⟶ chickens ⟶ humans

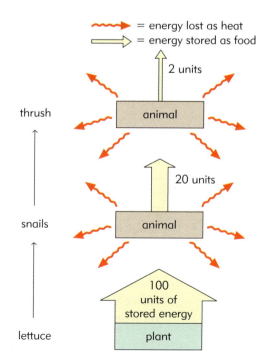

⟿ = energy lost as heat

⟹ = energy stored as food

thrush

animal — 2 units

snails

animal — 20 units

lettuce

plant — 100 units of stored energy

Energy loss and food supplies

In some countries many people don't get enough to eat. If we reduce the number of stages in a food chain, we can feed more people. We call it 'eating lower in the food chain'. It means eating plants instead of meat.

12 Explain why more people could be fed if we all ate lower in the food chain.

The amount of land needed to feed one person on:

● chicken

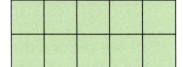

● grain

Why is there less energy at each stage of a food chain?

If there is less material at each stage of a food chain, there is also less energy. So, a pyramid of biomass is also a **pyramid of energy**. Look at the diagram. It shows what happens to the food that a chicken eats. Some of the food is digested and absorbed into its body as sugars, fats and amino acids. All these contain energy. The undigested waste also contains energy.

The cells of the chicken release energy from food when they respire. Chickens, like other birds (and also like mammals), have constant body temperatures. In Britain, this body temperature is usually higher than the temperature of their surroundings. This means that chickens are constantly transferring thermal energy to their surroundings. So, much of the energy released from their food ends up as thermal energy in the surroundings. The chicken stores the rest of the food in its cells or uses it for growing. Only these parts of its food can be passed on to the next animal in the food chain.

13 (a) Draw a pyramid of energy for the food chain:

grain → chicken → human

(b) Describe what happens to the biomass and the amount of energy along this food chain.

14 (a) What does the Sankey diagram show?
(b) for every 100 joules from grain, write down the number of joules:
   (i) of wasted energy;
   (ii) of used energy;
   (iii) remaining in the chicken's flesh.

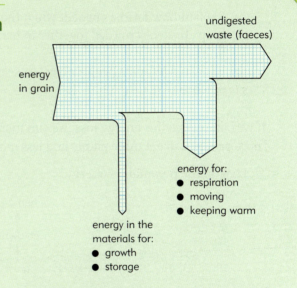

This Sankey diagram shows what happens to the energy in a chicken's food.

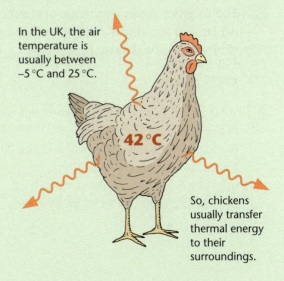

In the UK, the air temperature is usually between −5 °C and 25 °C.

42 °C

So, chickens usually transfer thermal energy to their surroundings.

Higher

Can we reduce this energy loss?

Most farm animals are mammals or birds. These animals use a lot of the energy in their food for moving and keeping warm. They use less energy if we keep them in warm conditions and we stop them moving about very much.

Then they use a bigger proportion of their food for growing.

The chickens haven't much room to move about. They keep each other warm.

15 (a) Explain how farmers benefit when they keep chickens crowded together in barns or in cages.
(b) Explain why some people object to keeping chickens like this.

If we can reduce this energy loss, we can increase food supplies. Another way of feeding more people is to eat more plants and less meat. We call this 'eating lower in the food chain'.

16 Look at the pyramids of energy.

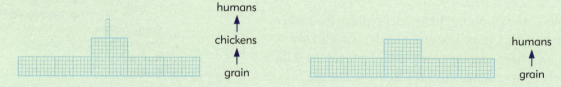

(a) How many times more food do humans get if they eat grain instead of feeding it to chickens?
(b) So, how many more times more people can be fed?

17 The diagram shows a pyramid of numbers.
(a) Draw a pyramid of energy for the same food chain.
(b) Explain why the two pyramids are different shapes.

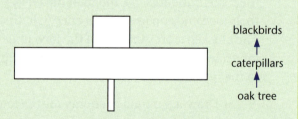

blackbirds
↑
caterpillars
↑
oak tree

18 A farmer insulated the floor, roof and walls of her chicken huts.
(a) Explain the advantages of keeping farm animals in warm conditions.
(b) Why is insulating huts better than heating them with electricity?

19 Fish do not keep their body temperature constant. Birds and mammals do.

The conversion rate (amount of growth) for each 100 g of food is:
    fish   25 g;
    chicken   10 g.

(a) What is the difference in the conversion rates of food for fish and for chickens?
(b) Explain the reason for this difference.

Another kind of pyramid

The materials which animals and plants are made of store energy. We call these materials **biomass**.

At each stage in a food chain there is less and less energy. So there is also less biomass. We can show this by drawing a **pyramid of biomass**.

20 One thrush has a bigger mass than one snail. But the pyramid of biomass shows a bigger mass for the snails. Why is this?

The diagram shows the pyramid of numbers and the pyramid of biomass for another food chain.

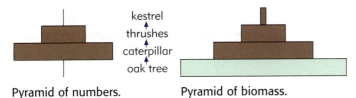

Pyramid of numbers.   Pyramid of biomass.

21 What difference is there between the pyramids? Explain the reason for the difference.

You need to be able to draw and to interpret pyramids of biomass.

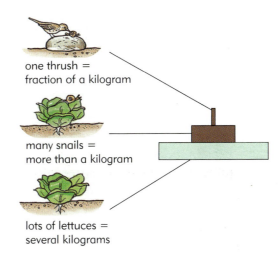

one thrush = fraction of a kilogram

many snails = more than a kilogram

lots of lettuces = several kilograms

Pyramid of biomass.

Getting more food to the consumer

A lot of fruit and vegetables are grown a long way from our homes. They have to be harvested, packed and transported to the shops. We don't want them to be bruised, damaged, mouldy or over-ripe when we buy them.

Scientists and technologists have researched the best conditions for harvesting, storing and transporting fruit. They have also developed new varieties that stay in good condition for longer after they have been picked. They did this by **selective breeding** and **genetic engineering**. Some people object to this.

22 Look at the spider diagram.
(a) Write down one change to the tomato itself that makes it last longer.
(b) Explain why market gardeners usually harvest tomatoes before they are ripe.

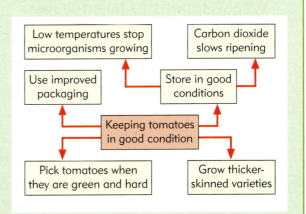

Low temperatures stop microorganisms growing

Carbon dioxide slows ripening

Use improved packaging

Store in good conditions

Keeping tomatoes in good condition

Pick tomatoes when they are green and hard

Grow thicker-skinned varieties

Tomatoes are picked and stored when they show the first signs of a pink colour. Extra carbon dioxide in the air stops them ripening. When it is time to ripen them, the plant hormone **ethene** is used. The tomatoes become softer and sweeter as they ripen. But they are not as sweet as those left to ripen on the plant.

The new breeds of tomatoes with thick skins stay fresh for a long time, but they were not bred for flavour. Also some people don't like the thick skins.

23 Some people pay extra for tomatoes that were ripened on the plant. Explain why.

Write down <u>two</u> disadvantages of the new breeds of tomatoes.

24 Explain the advantages and disadvantages of picking tomatoes when they are green:

25 (a) for the growers and distributors;
(b) for consumers.

> You need to be able to point out the advantages and disadvantages of managing food production and distribution, and to recognise the need to compromise.

Recycling materials

Recycling minerals

It isn't only people who produce waste. Waste is produced all the time in nature.

Look at the picture of the wood. The woodland floor is covered in dead leaves. This is called **leaf litter**.

1 What sort of things do we usually call litter?

2 As well as dead leaves, other things are found in leaf litter. List <u>four</u> of these things.

What happens to fallen leaves?

Leaves fall from trees and then decay or rot. If they didn't, leaf litter would get deeper and deeper every year. Animal droppings and the dead bodies of plants and animals in the leaf litter also decay. All these things break down into simple chemicals as they decay.

3 Write down the names of <u>three</u> substances which are made when dead plants and animals decay.

4 Leaves disappear as they rot. But the chemicals they break down into can't just disappear. Where do they go to?

animal droppings   acorns   woodlice   twigs   leaves

carbon dioxide goes into the air

dead leaves

water and **minerals** go into the soil

Dead leaves decay into simple substances.

What causes decay?

Tiny living things called <u>microorganisms</u> make plant and animal waste decay. Microorganisms digest the waste and take it into their cells. It is their food. Like you, they need food for energy and growth.

Some of the microorganisms which digest waste are bacteria. Others are fungi.

5 The microorganisms which break down waste are often called **decomposers**.
Why do you think this is?

Does all waste decay?

Microorganisms break down waste from living things. We call this **biodegradable** waste.

Unfortunately, microorganisms cannot break down some of the litter we drop. This litter is **non-biodegradable**.

6 Make a list of things in litter that are:
(a) biodegradable (will rot or decay);
(b) non-biodegradable (will not rot or decay).

6 months later

Why is decay important?

All living things depend on plants for food. So new plants must grow all the time. When plants grow they take minerals from the soil. But the soil doesn't run out of minerals, because the same minerals are used over and over again. We say that the minerals are **recycled**.

7 Look at the recycling diagram. Describe, as fully as you can, how the minerals locked up in dead plants and animals are released for use by living plants.

In a stable community such as natural woodland, there is a balance between processes that:

- remove materials from the environment;
- replace materials into the environment.

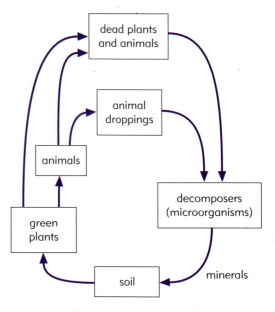

How minerals are recycled.

Microorganisms in soil

Just one teaspoonful of soil may contain a billion bacteria and 100 metres (m) of fine threads of fungi.

Look at the diagrams.

8 How many rod-shaped bacteria will fit end to end in 1 centimetre (cm)?

9 How can you get so many bacteria in a teaspoonful of soil?

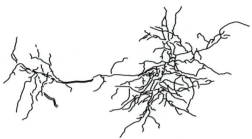

These threads of fungi are magnified 125 times.

This bacterium is magnified 2000 times.

These bacteria and fungi feed on waste such as dead plants and animal droppings. We call this **organic** waste. The diagrams show what the microorganisms do to this waste.

10 What do we call the decayed plant material in soil?

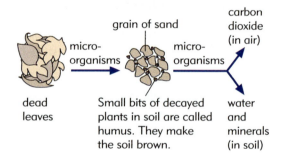

dead leaves

micro-organisms

grain of sand

micro-organisms

Small bits of decayed plants in soil are called humus. They make the soil brown.

carbon dioxide (in air)

water and minerals (in soil)

Recycling waste in the garden

Many gardeners use compost heaps to recycle plant waste from kitchens and gardens.

11 List <u>ten</u> things you could put in a compost heap. (Remember they must be biodegradable.)

12 (a) Describe how compost forms from waste.
(b) What happens to compost after the gardener puts it into the soil?

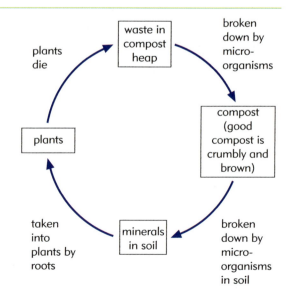

plants die

waste in compost heap

broken down by micro-organisms

plants

compost (good compost is crumbly and brown)

taken into plants by roots

minerals in soil

broken down by micro-organisms in soil

How to make a good compost heap

Microorganisms digest waste in a compost heap faster in warm conditions. They produce heat as they do this.

13 Look at the pictures. Which heap, A or B, is better for keeping this heat in? Explain why.

14 Mice sometimes make their nests in the middle of compost heaps. Why do you think this is?

Most microorganisms digest waste faster when they have plenty of oxygen. They also need moisture. But no air gets in if a heap gets too wet.

15 Look at the design of heap B. How does air get in?

16 What could the gardener do to stop the heap getting too wet in winter?

17 Compost is dark and crumbly when the microorganisms have done their work.
You cannot see what it was made from. Why not?

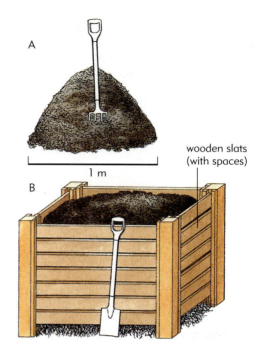

A

1 m

wooden slats (with spaces)

B

Why do gardeners need to add fertiliser to soil?

Gardeners often pick flowers, fruit and vegetables. So the minerals from these things do not naturally go back into the soil. Plants cannot grow without minerals, so gardeners have to replace them.

18 Write down <u>two</u> ways that gardeners can put minerals back into their soil.

19 Manure is animal droppings, often mixed with straw. What happens to manure in soil?

20 Chemical fertiliser can be used straight away by plants. Explain why.

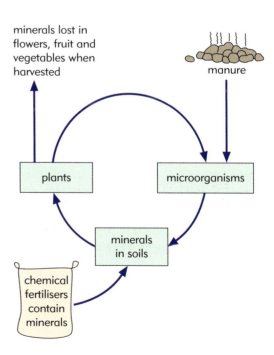

minerals lost in flowers, fruit and vegetables when harvested

manure

plants

microorganisms

minerals in soils

chemical fertilisers contain minerals

Recycling watery waste

Towns and cities produce huge amounts of watery waste. We call this **sewage**.

21 Write down <u>three</u> ways that we produce sewage in our homes.

22 Where does the sewage go to when it leaves our homes?

To stop sewage polluting our rivers and seas we usually treat it in sewage works.

There, organic waste in sewage is broken down by microorganisms.

23 (a) What percentage of sewage in Britain is treated in sewage works?
(b) What happens to the rest of the sewage?

24 Which part of the sewage from our homes contains most organic waste (contains the most carbon compounds)?

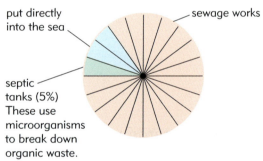

Where Britain's sewage goes.

Separating the sewage into parts

First, different parts of the sewage are separated from each other. The diagram shows how this is done.

25 What happens in these parts of a sewage works:
(a) the screen?
(b) the grit tank?
(c) the settlement tank?

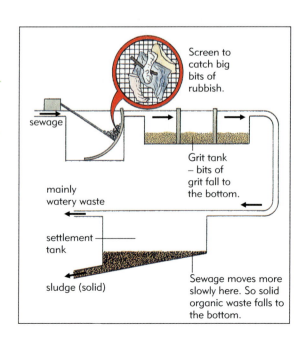

Cleaning the mainly watery waste

The watery waste in sewage still contains small bits of organic material. We use microorganisms to break this down.

The diagrams show two ways of doing this. Both ways give the microorganisms plenty of oxygen. This means that they can digest the waste more quickly.

26 Two methods of cleaning the watery waste are shown in the diagrams. For each method:
(a) Where are the microorganisms that digest the waste?
(b) How do the microorganisms get the oxygen that they need?

After treatment some sludge settles out. The water is clean enough to put into the river.

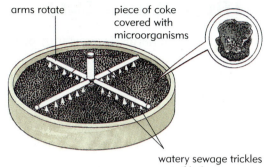

arms rotate   piece of coke covered with microorganisms

watery sewage trickles out of holes

Trickle filter method.

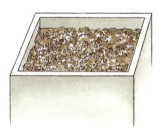

The watery waste is stirred up by the air bubbling through it.

This process uses microorganisms already in the sewage.

Activated sludge method.

What happens to the sludge?

Sludge can be dried and burnt, or dumped in the sea. In many sewage works different microorganisms break sludge down in a **digester**. These microorganisms do not need oxygen. They make the sludge safe to use as fertiliser. They also make methane gas which can be used as fuel. Some sewage works use it for heating, others bottle it and use it to run their vans.

27 (a) Write down two things that can be made from sewage sludge.
(b) How much of the sludge is used in this way?
(c) What happens to the rest of the sludge?

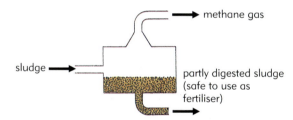

methane gas

sludge

partly digested sludge (safe to use as fertiliser)

A digester.

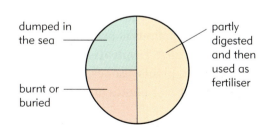

dumped in the sea

partly digested and then used as fertiliser

burnt or buried

What happens to the sludge.

The carbon cycle

Carbon, like minerals, is constantly being recycled. All living things are made from substances called **carbohydrates**, **fats** and **proteins**. These substances all contain carbon. We call them **carbon compounds**.

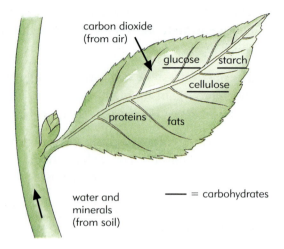

Foods made in a leaf.

Where do carbon compounds come from?

Plants take carbon dioxide from the air. They join carbon dioxide with water to make glucose. Plants use light energy to do this so we call it **photosynthesis**.

Next, plants use the glucose to make other carbohydrates as well as fats and proteins. Plants use these carbon compounds to grow.

1　Write down <u>three</u> carbohydrates plants make.

2　(a)　What substances do plants use to make glucose?
　　(b)　Which <u>one</u> of these substances contains carbon?

The diagram shows plants taking in carbon dioxide from the air. This is just one part of the <u>carbon cycle</u>.

3　Name the process by which plants use carbon dioxide.

What happens to carbon.

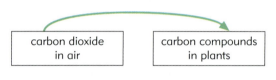

Why isn't all the carbon dioxide used up?

There is only a very small amount of carbon dioxide in the air. Plants keep using carbon dioxide to grow. It doesn't all get used up because animals and microorganisms feed on plants. They use some of the carbon compounds in plants to supply energy. This process is called **respiration**. It puts carbon dioxide back into the air. Plants also respire. So when there isn't enough light for photosynthesis, they put carbon dioxide into the air.

Decomposers also use animal wastes and dead animals for food.

4　Write down <u>three</u> groups of organisms that put carbon dioxide back into the air.

5　Name the process in living cells in which carbon dioxide is released.

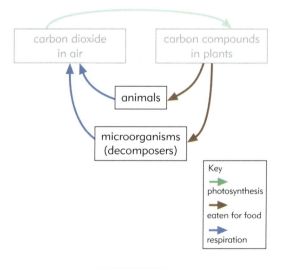

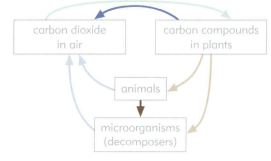

Nature's balanced carbon cycle

In nature the amount of carbon dioxide in the air stays the same. The same carbon is used over and over again. This is why we call it the carbon **cycle**.

6 Look at the diagram.
Why does the amount of carbon dioxide in nature stay the same?

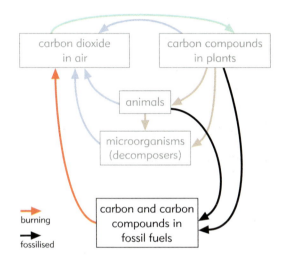

The two sides **balance**. So the amount of carbon dioxide in the air stays the same.

Carbon dioxide in nature.

How humans upset the balance

Coal, gas and oil are the fossil remains of plants and animals from millions of years ago. These fossil fuels contain a huge amount of carbon. When we burn these fuels, we add a lot of extra carbon dioxide to the atmosphere.

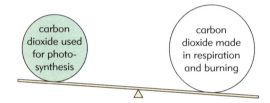

Human effect on carbon dioxide.

7 Humans have affected the balance of carbon dioxide in nature.
(a) Describe the change.
(b) Explain how this change has happened.

Higher

The nitrogen cycle

Remember

Nitrogen is a gas in the air.

Ammonium compounds, nitrates and proteins all contain nitrogen.

Plants use nitrates to make proteins. The proteins in plants pass along food chains.

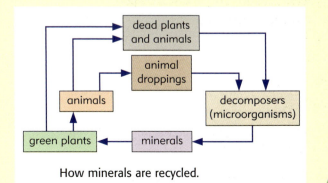

How minerals are recycled.

Higher

Why don't plants run out of nitrates?

An **ecosystem** is made up of all the living and non-living things in an area. Farms and woods are both ecosystems. When farmers harvest crops, they take nitrogen away in the plants' proteins. If they don't put more nitrates into the soil, the next crop doesn't grow so well. So, farmers add fertilisers to the soil. In a natural ecosystem, such as a wood, nitrogen compounds are used over and over again. We say they are **recycled**. Animal waste, and dead animals and plants, are the fertilisers.

Animals such as earthworms, dung beetles, woodlice and maggots eat this waste or **detritus**. So, we call them detritus feeders. The waste gives them materials and energy to grow.

Putrefying bacteria and fungi also get their energy and materials from this waste. They release ammonium compounds into the soil. These contain nitrogen, but plants cannot use them. Other bacteria change the ammonium compounds to nitrates.

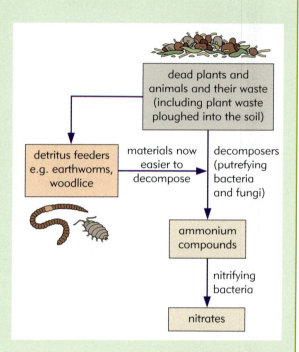

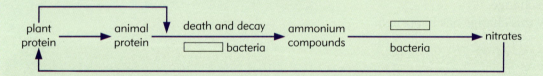

1 The two blank boxes in the diagram above represent two groups of bacteria. Which bacteria change:
(a) animal protein into ammonium compounds?
(b) ammonium compounds into nitrates?

2 Detritus feeders and decomposers help to recycle dead plants and animals and their waste in soil. Describe how <u>each</u> of these two groups helps.

3 Much of the nitrogen in soil is in the form of complex organic molecules which have come from the dead bits of plants. Explain, as fully as you can, how these will become useful to next year's crop.

What happens to the energy in the waste?

When microorganisms break down waste, they release energy as well as nitrogen compounds. So, by the time all the nutrients in waste have been recycled, all the energy in them will have been transferred. It is not recycled. It ends up as thermal energy in the surroundings. The diagram shows some of the energy transfers in an ecosystem.

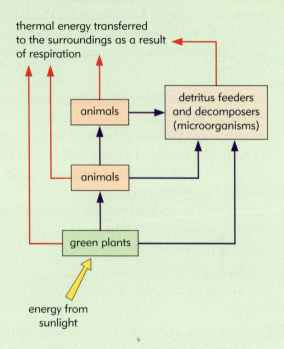

thermal energy transferred to the surroundings as a result of respiration

animals

detritus feeders and decomposers (microorganisms)

animals

green plants

energy from sunlight

4 (a) Where does the energy entering an ecosystem come from?
 (b) Which living things capture this energy?

5 Name two kinds of living things to which the energy can be transferred.

6 Name the process in which living things release energy.

7 The diagram shows what happens to the nitrates used as fertiliser on a crop of wheat. Make a table to show the percentages of nitrates going to each place.

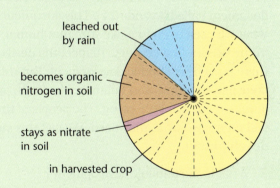

leached out by rain

becomes organic nitrogen in soil

stays as nitrate in soil

in harvested crop

8 The diagram shows how nitrogen-fixing bacteria add nitrates to soil.

Some of these bacteria live in soil.
Others live in the root nodules of leguminous plants such as peas or beans.

Denitrifying bacteria change nitrates in the soil back to nitrogen.

(a) Explain why organic farmers like to grow a leguminous crop every few years as part of their crop rotation.
(b) What is the difference between nitrogen-fixing and denitrifying bacteria?
(c) Write down two ways of changing nitrogen from the air to nitrates.

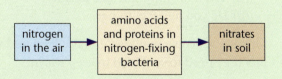

nitrogen in the air

amino acids and proteins in nitrogen-fixing bacteria

nitrates in soil

Sustainable development

In the rest of this section you will look more closely at some of the ways in which humans harm the environment. As the population increases and as we all try to improve our standard of living, the problem gets worse.

In 1983 an international commission on environment and development was set up. Gro Harlem Brundtland, former Prime Minister of Norway, was in charge. The Commission produced a report called 'Our Common Future' in 1987. The report talked about the needs of human beings, the environment and the availability of resources. The findings pointed out that there was a difference between what humans want and what they actually need to live comfortably.

1　Look at the pictures. Which one should be labelled 'I need' and which 'I want'? Explain your answer.

In 1992, the Earth Summit in Rio de Janeiro agreed on a document called 'Agenda 21'. This document sets out what we need to do to make sure that humans can survive in the twenty-first century. To meet the needs of the poor, there must be economic development. But this must be done in a way that won't damage the Earth and so that development can keep going. The people of the future will then have what they need. We call this **sustainable development**.

2　What is 'sustainable development'?

3　Brundtland said 'It is both futile and indeed an insult to the poor to tell them that they must live in poverty to protect the environment.'
(a) What do you think she meant?
(b) Do you agree with her?

Millions of people don't get enough food and water or the fuel they need to cook and to keep warm.

"I need some Nike trainers."

"I haven't eaten today. I need some food."

Getting the energy we need

We burn fuels for cooking, heating and to make electricity.

4 We can burn wood or fossil fuels or biogas, which is made from organic waste. Which <u>one</u> of these is <u>not</u> sustainable? Explain your answer.

5 (a) Why is it more sustainable to make electricity from wind or solar energy than by burning fuel?
   (b) Write down <u>one</u> use that is better than burning for each of fossil fuels, wood and organic waste.

Getting the food we need

Collecting and growing food has energy costs too. To survive, people must gain more energy from their food than they use getting it. Hunter-gatherers in Africa get 5 to 10 times more energy from their food than they use getting it. So they can easily collect enough food for a family.

A European farmer grows enough food for many families. The energy in the food is many times more than the energy the farmer uses. But we must not forget that the farmer will have used up lots of energy from fossil fuels.

Organic farmers don't use weedkillers, so weeding provides extra jobs on organic farms.

6 Look at the picture. Write down <u>three</u> ways in which modern farming uses the energy from fossil fuels.

7 Write down <u>one</u> way in which the way the hunter-gatherer gets food is more sustainable than that of the farmer.

8 Write down <u>two</u> advantages for local people of organic farming.

It takes lots of energy to make and use equipment and chemicals. Some chemicals harm the environment too.

You need to be able to <u>form judgements</u> about environmental issues, including the importance of sustainable development.

Think about sustainable development as you study the environmental issues in the rest of this section.

Increasing population

More people, more problems

Nowadays, people have a much bigger effect on the environment than they used to. One reason for this is that there are a lot more people. We say that the human population is increasing.

How the human population has grown

The graph shows the change in world population since 1850.

1 (a) What was the world population in 1850?
  (b) Predict the population for 2025.

2 What does the shape of the graph tell you about how fast the human population is rising?

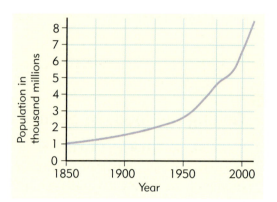

More people means fewer trees

More people need more land for growing food. The last great forests in the world are in danger. They are being cut down to provide land for farming and wood to make things with.

3 (a) What are these forests called?
  (b) How long will they last if we keep chopping them down at the present rate?

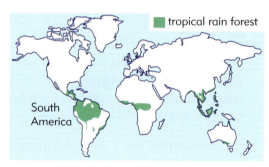

The biggest area of tropical rain forest is in South America. About 1 per cent is being cut down each year.

More farming means more pollution

Using more land for farming means that we pollute water more.

4 There is much less pollution of water when land is covered with trees. Explain why as fully as you can.

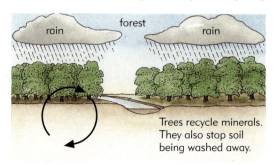

Trees recycle minerals. They also stop soil being washed away.

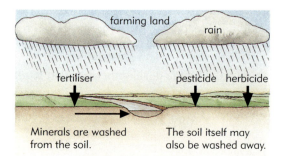

Minerals are washed from the soil.

The soil itself may also be washed away.

We all use more things

There are a lot more people in the world than there used to be. Many of these people have a lot more things like fridges, cars and TV sets. We say they have a high **standard of living**.

This also affects the world around us.

5 (a) Where do we get the raw materials to make all these things?
(b) What will eventually happen to the supply of these raw materials?

6 The things we use eventually wear out. What further problem does this cause?

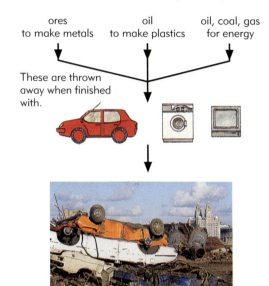

Raw materials from the ground. (These will eventually run out.)

ores to make metals | oil to make plastics | oil, coal, gas for energy

These are thrown away when finished with.

Old products are thrown away. They can pollute the land.

More energy means more air pollution

We don't just need energy to make new things. We also need energy to use them. For example, people travel in cars, trains and planes and all these use fuel. The more fuel we burn, the more waste gases we produce and the more we pollute the air.

7 In a table, list the waste gases we produce next to the environmental problems they cause.

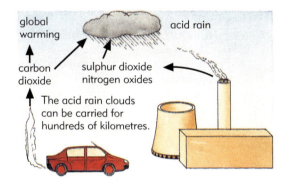

global warming
acid rain
carbon dioxide
sulphur dioxide nitrogen oxides
The acid rain clouds can be carried for hundreds of kilometres.

Polluting the whole Earth

Humans have always affected the local area where they live. But there are now many more people, and they are having a bigger effect.

8 (a) How does burning fuels in Britain affect nearby countries?
(b) How does it affect the whole Earth?

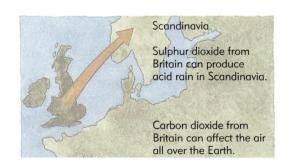

Scandinavia

Sulphur dioxide from Britain can produce acid rain in Scandinavia.

Carbon dioxide from Britain can affect the air all over the Earth.

How humans have changed the landscape

People don't just affect water and air.
They also change the land that they live on.

After the last Ice Age, 60 per cent of the British Isles was woodland. The maps show how this had changed by the year 1086.

1 Describe the change in the percentage tree cover in Britain between 7000 BC and AD 1086.

2 Was more woodland lost from the north or the south of Britain?

What happened to the trees?

- From 7000 BC   The land was cleared of trees for animals to graze and then to grow crops.

- From 500 BC   Wood was first used to make charcoal to get iron from iron ore.

The number of people increased. More land was cleared for farming. More wood was needed for building.

- From AD 1086   As the population continued to increase, more land was needed for farming and building. More trees were cut down.

3 Write down <u>three</u> reasons for cutting down trees.

Twentieth-century forests

Millions of acres of trees were cut down during the First World War. Timber was needed for pit-props in coal mines, paper and building. The war ended in 1918. By this time trees covered only 4 per cent of the British Isles.

The Forestry Commission was set up after the First World War. Large areas of new forests were planted.

About 10 per cent of Britain is now covered by trees.

4 What was done, after 1918, to increase the numbers of trees in Britain?

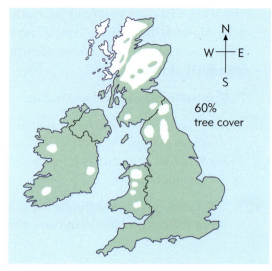

7000 BC – the British Isles after the last Ice Age.

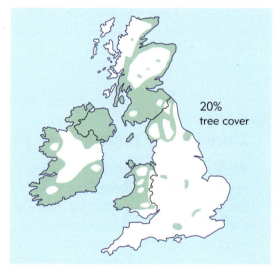

AD 1086 – Domesday survey.

Effects of losing the trees

Many woodland plants and animals can survive only in shady and damp conditions. When woods are cut down, there are fewer places for them to live.

5 Some woodland plants survive in hedges. Why do you think this is?

6 During the last 50 years many hedgerows have been pulled up to make bigger fields. Why do you think many people are worried about this?

What else do people use land for?

People use land for other things besides farming. We use land for buildings and roads. We also use land for quarries for stone, and for landfill sites where we dump rubbish.

When land is used in these ways, many plants and animals can no longer live there. We say we have destroyed their habitats.

The map shows land use in a small area.

7 What is most of the land in this area used for?

8 What is shown in square B4?

9 Land is also used for quarrying and for dumping waste. In which squares can you find evidence for these activities?

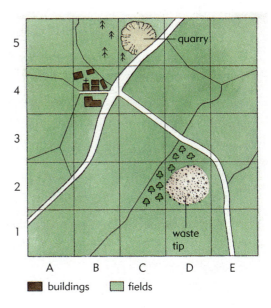

You need to be able to describe the effects of humans on landscapes when you are given information about them.

How humans affect water

We treat most of our sewage so that it doesn't pollute rivers and seas. The diagram shows what can happen when untreated sewage gets into a river.

1 What happens to fish in a river polluted with large amounts of sewage?

2 Why isn't there enough oxygen in the water?

We also pollute water with the chemicals we use on farms and in factories.

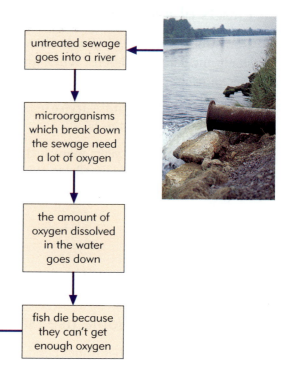

untreated sewage goes into a river

↓

microorganisms which break down the sewage need a lot of oxygen

↓

the amount of oxygen dissolved in the water goes down

↓

fish die because they can't get enough oxygen

Chemicals we use on farms

Farmers spray chemicals on to their crops. Pesticides and herbicides are two examples of these chemicals.

Farmers use **pesticides** to kill the insects and other animals that feed on their crops.

They use **herbicides** to kill weeds.

Pesticides and herbicides are <u>toxic</u>. This means that they are poisonous.

3 Why do farmers want to get rid of weeds?

4 Organic farmers try not to use chemicals on their crops. Why is this?

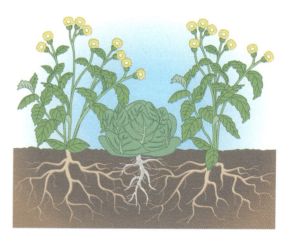

Weeds compete with the cabbage for light, water, minerals and space.

How pesticides and herbicides can pollute water

The diagrams show what can happen to the herbicides and pesticides that farmers spray on their crops.

5 How do pesticides and herbicides get into streams and rivers?

6 What other chemicals can also get into streams and rivers?

7 Where do the chemicals eventually end up?

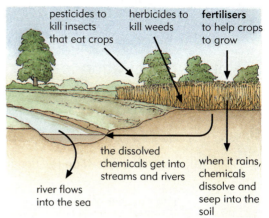

pesticides to kill insects that eat crops

herbicides to kill weeds

fertilisers to help crops to grow

the dissolved chemicals get into streams and rivers

when it rains, chemicals dissolve and seep into the soil

river flows into the sea

Farmers spray chemicals on to their crops.

Higher

How do fertilisers get into rivers?

Some of the fertiliser that farmers add to soil gets washed out by rain. Natural fertilisers are broken down by microorganisms so the minerals are released into the soil a bit at a time. Artificial fertilisers dissolve in water straight away. So, they can be washed down through the soil more easily. We call this **leaching**.

These nitrates are fertilisers for the plants in the rivers too. These plants grow faster than usual. We call this type of excessive plant growth **eutrophication**.

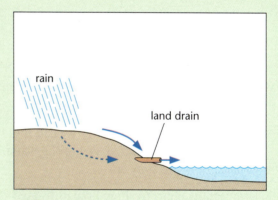

rain

land drain

Fertilisers often leach into rivers and lakes.

8 Crops don't take in all the fertiliser that they are given. How does the excess fertiliser get into rivers and lakes?

9 The minerals in natural fertilisers are less likely to be leached than those in artificial fertilisers. Explain why.

10 (a) The water plants in the two photographs grew very rapidly. Why was this?
    (b) What do we call rapid plant growth in nutrient-rich water?

Fertilisers drained into this river. It is almost choked with plants.

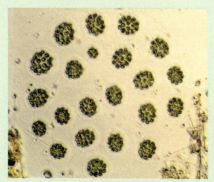

Under a microscope, the water was full of green algae.

Higher

How eutrophication leads to pollution

The flow chart shows what happens when the plants die.

> Microorganisms feed on the dead plants.
> ↓
> The number of microorganisms increases because there is more food.
> ↓
> Microorganisms use oxygen to respire.
> ↓
> There is less oxygen in the water.
> ↓
> Fish and other animals can't get enough oxygen, so they die.

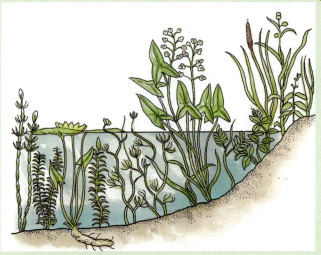

These plants are competing for light. So, many of them die sooner than they normally would and are decomposed by bacteria.

Microorganisms also feed on the organic material that is present in untreated sewage.

11 Eutrophication leads to an increase in the amount of dead plant material in the water. Explain how this happens.

12 Draw a flow chart to show how untreated sewage in a river can cause the death of fish.

13 Some students wanted to know if the amount of nitrates in the river water varied throughout the year. They knew that some of their drinking water came from the river and that too much nitrate is bad for your health.

Look at the chart of the students' results.
(a) Suggest reasons why the amount of nitrates in the river was higher in the winter than in the summer.
(b) Explain, as fully as you can, why the water had the most algae and other plants in it in September.

14 (a) Plot the data in the table on a bar chart.

(b) Which fish will survive best in water polluted with sewage? Explain your answer.

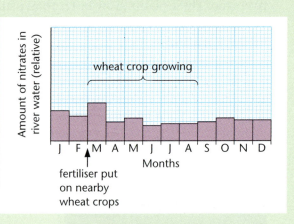

Fish	Amount of oxygen needed for survival (mg/dm³)
Tench	0.7
Roach	0.8
Perch	1.2
Trout	3.7

Industrial waste

Factories make waste that can cause pollution.

In the 1960s a chemicals factory in Minamata Bay, Japan, let out poisonous waste containing mercury into the sea. By 1969 many people were ill and 68 people had died.

15 Look at the diagram.
How did the mercury poison get into people's bodies?

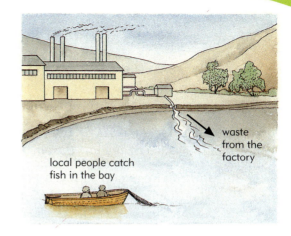

waste from the factory

local people catch fish in the bay

Polluting the air

Many lakes in Europe now have no fish because the water has become too acidic.

1 Why have the fish in these lakes died?

Acid rain makes lakes acidic. The rain is more acidic than it should be because people have polluted the air.

How humans affect the air

We are constantly putting waste gases into the air around us.

2 Write down <u>three</u> places where we produce waste gases.

3 How are these waste gases produced?

These waste gases pollute the air.

They can cause acid rain and many other problems.

waste gases

waste gases

waste gases

Power stations use coal, oil or gas.

Cars and lorries use petrol or diesel.

We often use coal, oil or gas to heat our homes.

Burning fuels produces smoke and waste gases. These go into the air.

What causes acid rain?

Fuels often contain sulphur. When fuels are burnt, this sulphur produces a gas called **sulphur dioxide**.

When we burn fuels, nitrogen and oxygen from the air also react to produce **nitrogen oxides**. The diagram shows how these gases make acid rain.

4 (a) Write down <u>two</u> gases that cause acid rain.
(b) Describe <u>two</u> effects of acid rain.

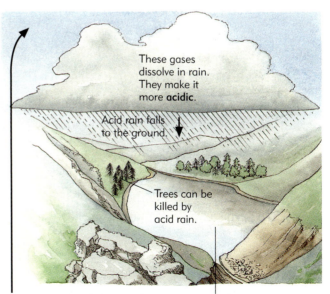

These gases dissolve in rain. They make it more **acidic**.

Acid rain falls to the ground.

Trees can be killed by acid rain.

Nitrogen oxides and sulphur dioxide go into the air.

Animals and plants cannot live in very acidic lakes and rivers.

What can we do about acid rain?

To make rain less acid we must put less sulphur dioxide and nitrogen oxides into the air.
The diagram shows some ways of doing this.

5 Write down <u>two</u> ways of reducing the amount of sulphur dioxide we put into the air.

6 (a) How can we remove nitrogen oxides from car exhaust fumes?
(b) What harmless gases are produced from this?

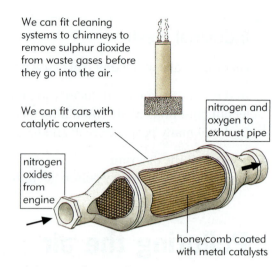

We can fit cleaning systems to chimneys to remove sulphur dioxide from waste gases before they go into the air.

We can fit cars with catalytic converters.

nitrogen and oxygen to exhaust pipe

nitrogen oxides from engine

honeycomb coated with metal catalysts

Helping to reduce acid rain.

Asthma

Tom has asthma. When he has an asthma attack he cannot breathe in enough air. Tom uses his inhaler to get medicine inside his lungs when he has an asthma attack. A serious attack could kill him.

Some scientists believe that pollution in the air could be one thing that can cause these attacks.

Children with asthma who live near a main road seem more likely to have serious attacks.

7 Why do you think this could be?

There is more carbon dioxide in the air

Coal, gas and oil all contain a lot of carbon. When they burn a gas called **carbon dioxide** is produced.

8 Look at the graph. What has happened to the level of carbon dioxide in the air since 1750?

Some scientists think that the rising level of carbon dioxide in the air may be making the Earth a warmer place.

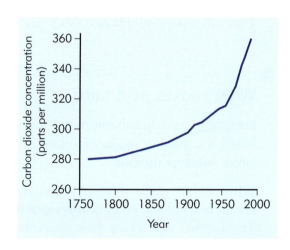

Are we changing the climate?

The amount of carbon dioxide in the atmosphere depends on:

● the amount of fuel (including wood) burned;
● the amount of decay of waste by microorganisms;
● the amount of photosynthesis.

The carbon dioxide taken up by tree leaves for photosynthesis 'locks up' carbon in wood for many years.

Microorganisms growing without oxygen put methane into the atmosphere. These microorganisms live in rice fields, marshes and the guts of animals such as cattle.

The size of the human population affects all of these things.

1 The pictures tell the story of a family in Java. Explain how each of the things in the pictures affects the amount of carbon dioxide or methane in the atmosphere.

2 Look at the pie charts.
Write down the <u>two</u> main sources of:
(a) carbon dioxide;
(b) methane.

We cut down some trees and sold them for timber.

We burnt the rest of the trees to get rid of them.

We grow rice on some of the land.

We also keep some cattle.

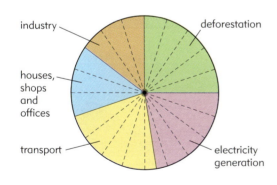

Carbon dioxide

industry
houses, shops and offices
transport
deforestation
electricity generation

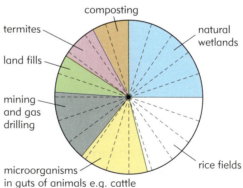

Methane

termites
land fills
mining and gas drilling
composting
natural wetlands
rice fields
microorganisms in guts of animals e.g. cattle

The pie charts show the percentage of the increase in greenhouse gases in the atmosphere caused by various activities.

3 Many people blame deforestation in tropical countries for the increasc in the amount of carbon dioxide in the atmosphere. Are they right to do so? Use evidence from this page to support your ideas.

Is the amount of these gases a problem?

No-one is sure about the answer to this question. Some scientists think that the increasing amounts of carbon dioxide and methane are causing an increase in the average temperature of the Earth. They call this **global warming**.

The Earth's atmosphere controls the amount of heat energy that reaches and leaves the Earth. If more energy reaches the Earth than escapes, the Earth will warm up. We call this the **greenhouse effect**. Carbon dioxide and methane are two of the gases that are particularly good at keeping the heat in. So we call them **greenhouse gases**.

4 (a) What is the greenhouse effect?
(b) Write down the names of <u>two</u> gases that cause the greenhouse effect.
(c) What is global warming?

Is the Earth getting warmer?

5 Write down <u>one</u> piece of evidence <u>from the charts</u> which supports the idea that:
(a) global warming is happening;
(b) the rise in temperature is part of a natural cycle of temperature changes.

Some effects of global warming

The temperature will only have to increase by a few degrees Celsius to cause big changes on Earth. It may melt a lot of the ice so that the sea level rises. It may affect the amount of rain and where that rain falls.
It may affect the winds and the lengths of the seasons. Climate change may also affect the types of plants that grow in different parts of the Earth.

6 (a) Write down <u>three</u> effects of global warming.
(b) Write a few sentences about how these changes may affect humans.

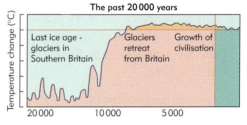

The past 20 000 years

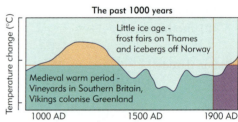

The past 1000 years

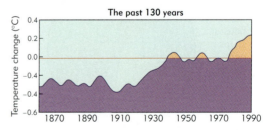

The past 130 years

- The red lines show the average temperature in the second half of the twentieth century.
- Notice that the scales on the three graphs are different.

You need to be able to consider evidence about environmental issues such as global warming.

More about climate change

Without its atmosphere, the average temperature of the Earth would be about 38 °C cooler than it is.

7 Look at the diagrams.
   (a) What kind of rays does the Earth radiate?
   (b) Write down <u>two</u> things that can happen to this radiation.
   (c) How does this affect the average temperature of the Earth?

But, the amount of the greenhouse gases in the atmosphere is increasing. Some scientists think that this will make the Earth warmer than it otherwise would be.

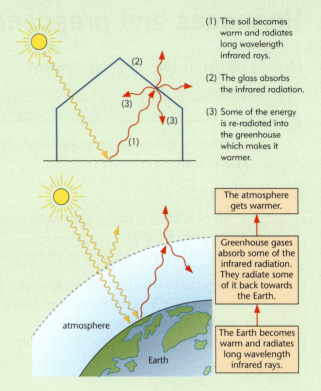

(1) The soil becomes warm and radiates long wavelength infrared rays.

(2) The glass absorbs the infrared radiation.

(3) Some of the energy is re-radiated into the greenhouse which makes it warmer.

The atmosphere gets warmer.

Greenhouse gases absorb some of the infrared radiation. They radiate some of it back towards the Earth.

The Earth becomes warm and radiates long wavelength infrared rays.

atmosphere

Earth

The table shows how different greenhouse gases can affect global warming. They are based on comparing the effect of 1 kg of a gas with 1 kg of carbon dioxide.

Gas	Carbon dioxide	Methane	Other gases
concentration before industry	280 ppm	0.8 ppm	0.28 ppm
concentration now	353 ppm	1.72 ppm	0.31 ppm
years it stays in the atmosphere	50–200	10	65–150
global warming potential 20 years	1	63	5800
100 years	1	21	5400
500 years	1	9	3000
percentage contribution to global warming now	55%	15%	30%

8 (a) Which gas causes the most warming now?
   (b) If no more greenhouse gases are added to the atmosphere, which gas would lose its effect first?
   (c) Explain why 'other gases' have a higher warming potential than carbon dioxide, but cause much less warming.
   (d) Which gases will cause problems for the longest time?

4 Inheritance and selection

Hormones and pregnancy

How a woman becomes pregnant

Remember from Key Stage 3

A woman can become pregnant if one of her eggs is fertilised by a sperm.

About every 28 days:

- an egg ripens in one of a woman's ovaries and it is then released;
- a thick lining is prepared inside her womb.

The egg may be fertilised by a sperm as it travels down a tube called the egg duct, or oviduct. If it is fertilised it will settle into the thick lining of the womb and start to grow into a baby.

1 Where are human eggs:
   (a) made?
   (b) fertilised?

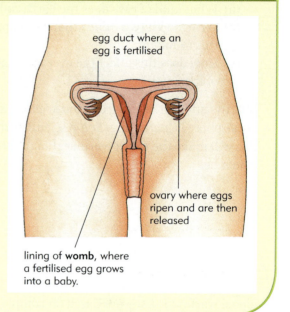

egg duct where an egg is fertilised

ovary where eggs ripen and are then released

lining of **womb**, where a fertilised egg grows into a baby.

How does everything happen at the right time?

The lining of the womb thickens at the same time as an egg is ripening. It is then ready for the egg if the egg is fertilised.

Chemical messengers make sure that everything happens at just the right time. We call these chemical messengers **hormones**.

2 (a) The pituitary gland makes <u>two</u> hormones that affect pregnancy. What do these hormones do?
   (b) What are the effects of the <u>two</u> hormones that the ovaries make?

3 Where in the woman's body is her pituitary gland?

4 How do hormones get from where they are made to where they make things happen?

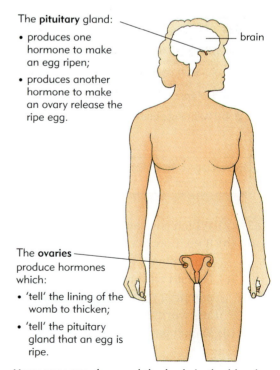

The **pituitary** gland:
- produces one hormone to make an egg ripen;
- produces another hormone to make an ovary release the ripe egg.

brain

The **ovaries** produce hormones which:
- 'tell' the lining of the womb to thicken;
- 'tell' the pituitary gland that an egg is ripe.

Hormones travel around the body in the blood.

What happens next?

What happens next depends on whether the egg has been fertilised.

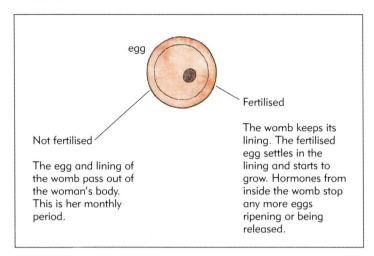

egg

Not fertilised

Fertilised

Not fertilised

The egg and lining of the womb pass out of the woman's body. This is her monthly period.

Fertilised

The womb keeps its lining. The fertilised egg settles in the lining and starts to grow. Hormones from inside the womb stop any more eggs ripening or being released.

If the egg is not fertilised, a hormone from the pituitary gland tells another egg to start ripening.
This starts the process all over again.

The cycle takes about a month.
It is called a **menstrual cycle**.

5 What happens to the egg and the lining of the womb if:
(a) the egg is fertilised?
(b) the egg is not fertilised?

6 (a) Where are the hormones that control the monthly cycle of a woman made?
(b) If an egg is fertilised, a different hormone stops the monthly cycle. Where is that hormone made?

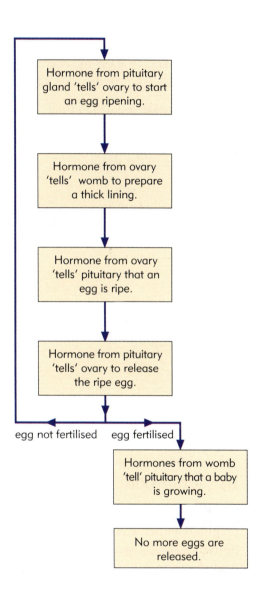

Hormone from pituitary gland 'tells' ovary to start an egg ripening.

Hormone from ovary 'tells' womb to prepare a thick lining.

Hormone from ovary 'tells' pituitary that an egg is ripe.

Hormone from pituitary 'tells' ovary to release the ripe egg.

egg not fertilised   egg fertilised

Hormones from womb 'tell' pituitary that a baby is growing.

No more eggs are released.

Different hormones do different jobs

Hormones from the pituitary gland and ovaries control a woman's menstrual cycle. They make sure that the womb is ready to receive a fertilised egg.

The pituitary gland produces:
- **FSH** – to make an egg mature in an ovary.
  – to make the ovaries secrete oestrogens.
- **LH** – to make an ovary release an egg in the middle of the menstrual cycle.

7 Look at the picture.
Make a table to show the effects on the ovaries and the womb of the hormones FSH, LH and oestrogen.

8 Ovaries make a hormone called oestrogen.
After an ovary releases an egg they make another hormone called progesterone. This hormone prevents breakdown of the lining of the womb.

Look at the graph.
(a) Why is it important to prevent breakdown of the womb lining after ovulation?
(b) Which line, A or B, is the line for progesterone?

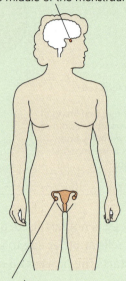

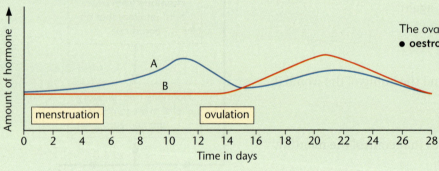

The ovaries produce:
- **oestrogen** – to inhibit FSH production.
  – to stimulate LH production.
  – to stimulate preparation of the womb lining.

Using hormones to control pregnancy

Some women want to have a baby but can't become pregnant. Other women don't want to become pregnant. Both these problems can sometimes be solved using hormones.

Hormones from the pituitary gland cause eggs to ripen and be released. More of these hormones can help some women become pregnant.

9 Look at the diagram.
Which hormones could be used:
(a) to help a woman become pregnant?
(b) to prevent a woman becoming pregnant?

These hormones are being used to control **fertility**.

Hormones from the womb stop eggs being released if a baby is developing. Taking these hormones can prevent a woman from becoming pregnant.

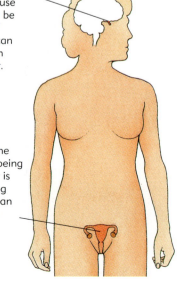

Using hormones to help a woman become pregnant

Janet and Carl want a baby.
They have been trying for a year.

Doctors at Janet's hospital have found out that her ovaries do not release eggs. There isn't much chance of her becoming pregnant. She is **infertile**.

Janet can be treated by having a hormone regularly injected into her blood. The hormone makes her ovaries release eggs.

We call this **fertility treatment**. It has helped many women to have the babies they want.

Fertility drugs don't always work, and sometimes they work too well! Several eggs can be released at once so a woman has several babies at the same time.

10 Why is Janet not becoming pregnant?

11 What will her fertility treatment do to help?

12 Write down <u>two</u> problems there can be with fertility drugs.

Higher

Fertility drugs

Janet can't become pregnant. The doctor thinks that this is because she doesn't make enough FSH. So, eggs don't develop in her ovaries. Her doctor can give her injections of FSH to make eggs mature. If her egg tubes and her partner's sperm are normal, she can then become pregnant.

If Janet's egg tubes are blocked, she will not become pregnant. Doctors can still help. They can take mature eggs from her ovaries and fertilise them in a dish. Then, they can transplant the embryos into her womb.

13 (a) Why is FSH called a 'fertility drug'?
(b) Explain how a 'fertility drug' works.

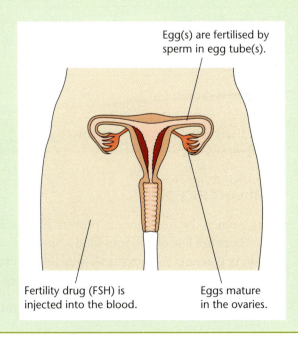

Egg(s) are fertilised by sperm in egg tube(s).

Fertility drug (FSH) is injected into the blood.

Eggs mature in the ovaries.

What's the chance of having twins?

The pie charts show how fertility drugs affect the number of twins, triplets, etc. that women have.
These are called **multiple births**.

14   Describe, as fully as you can, what the pie charts tell you.

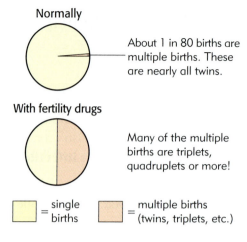

Normally

About 1 in 80 births are multiple births. These are nearly all twins.

With fertility drugs

Many of the multiple births are triplets, quadruplets or more!

☐ = single births   ☐ = multiple births (twins, triplets, etc.)

Using hormones to prevent pregnancy

Some couples want to have sexual intercourse but do not want the woman to become pregnant. The woman can take contraceptive pills. We call them **oral contraceptives**. Most of these contain hormones which stop the ovaries from releasing eggs.

However, the pills can have side effects, such as sickness and headaches. In a very few women the pills can cause serious heart problems or even death. Also, the woman must remember to take the pills regularly. If she doesn't, they may not work.

15   Write down <u>two</u> reasons why a couple might not want the woman to become pregnant.

16   Some women who don't want to become pregnant don't want to use pills. Suggest <u>two</u> reasons why.

> You should be able to describe and explain some of the problems and benefits of using hormones to control fertility.

Couples may want to plan when they have their children.

Some couples already have children and do not want any more.

Higher

Oral contraceptives

Some women use hormones to prevent pregnancy. They use contraceptive pills. There are lots of different kinds, but most contain **oestrogen**. This hormone prevents FSH production by the pituitary. As a result, eggs do not mature in the ovaries so the woman cannot become pregnant.

17   How do oral contraceptives prevent pregnancy?

Variation and inheritance

Who do you look like?

Children look like their parents in many ways. We say they have many of the same **features** or **characteristics**. For example, the child in the picture has two eyes and five fingers on each hand just like her parents and most other people.

But people's eyes can be different colours. They can have green, grey, brown or blue eyes. The child has blue eyes just like her parents.

Young animals and plants also look like their parents.

1 Write down:
   (a) <u>two</u> characteristics that the kitten shares with its parents and all other cats;
   (b) <u>two</u> features that the kitten shares with its parents but does not share with all other cats.

The characteristics which are passed on from parents to children are called **inherited** characteristics.

How are characteristics inherited?

This question was answered in the middle of the nineteenth century by Gregor Mendel (1822–1884). Mendel was a monk in a monastery in what is now the Czech Republic. He worked out the inheritance pattern of several characteristics of peas. Peas are easy to grow and to cross-pollinate. Also, many features of peas have a fairly simple inheritance pattern. So, it was lucky that he chose peas.

In one experiment, Mendel cross-pollinated flowers of lots of tall pea plants with those of lots of dwarf plants. He grew more pea plants from the seeds. He called these the **F1 generation**. Then he cross-pollinated the F1 plants with each other to produce the **F2 generation**.

2 Look at the diagram.
   (a) Describe a plant from the F1 generation.
   (b) How did Mendel produce the F2 generation?
   (c) Describe the plants from the F2 generation.

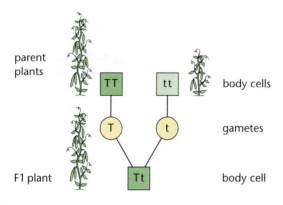

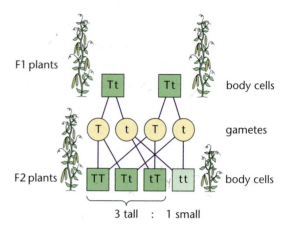

Explaining Mendel's results

Mendel explained his results in terms of what he called
<u>inheritance factors</u>. He realised that these factors must
come in pairs. The F1 plants shown in the diagram told
him that the tallness factor was more powerful than the
dwarfness factor. A pea plant only needs one tallness
factor to make it tall. So we say that the tallness factor is
dominant. A pea plant needs two dwarfness factors to
make it dwarf. So we say that the dwarfness factor is
recessive. Mendel called tallness, T, and dwarfness, t.

3 Describe what the plants with these factors look like:
   (a) T T;
   (b) T t;
   (c) t t.

Mendel wrote a scientific paper describing his work and
what he had found out. Like other scientific papers,
it was published in a scientific journal. The journal was
the Archives of the Brno Natural History Society. It was
published in 1865, but it was little known outside Brno.
It was 1901 before scientists recognised the importance of
Mendel's work.

4 Explain why the importance of Mendel's work was not
recognised until after his death.

> You need to be able to explain why Mendel suggested the idea of 'inheritance factors' and why the
> importance of his discoveries was not recognised until after his death.

More inheritance patterns

Mendel discovered some of the patterns of inheritance in peas. Other scientists did experiments using other plants and animals. They found similar patterns. They also found out about human patterns of inheritance. They did this by looking at what happened in families.

We now call Mendel's 'inheritance factors' **genes**. Different forms of a gene can produce different characteristics.

5 How do scientists investigate inheritance patterns in humans?

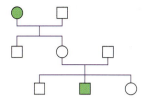

Family trees show how characteristics are passed on in humans.

How are genes passed on?

To produce a young animal, a sex cell from the father must join up with a sex cell from the mother. These sex cells are called **gametes**.

Plants can also produce young from sex cells.

The male sex cell of a plant is in a pollen grain. The female sex cell of a plant is in an ovule.

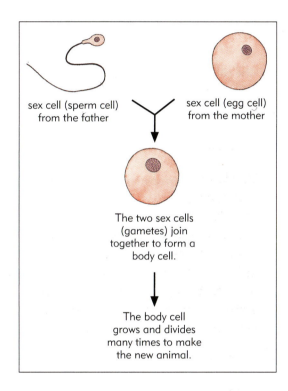

sex cell (sperm cell) from the father

sex cell (egg cell) from the mother

The two sex cells (gametes) join together to form a body cell.

The body cell grows and divides many times to make the new animal.

The female part of the flower contains the ovules.

The male part of the flower contains pollen.

6 (a) Name the male and female sex cells (gametes) in animals.

(b) Where are the sex cells in plants?

Sex cells carry information

Sex cells from parents pass on information. This information controls how the body cells of the young animal or plant develop.

7 Look carefully at the picture of a mother, father and their son.
Describe the variation in nose shape and in hair, eye and skin colour.

8 Which characteristics have been controlled:
(a) mainly by information from the egg?
(b) mainly by information from the sperm?
(c) by information from both egg and sperm?

Mother   Father

Their son Gary

Information is carried by genes

Information in cells is carried in units called genes.
Different genes control different characteristics.

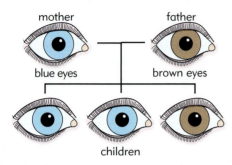

The children have either blue or brown eyes.
Just one gene controls their eye colour.

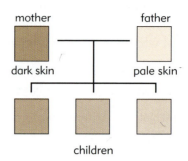

The children have skin with a range of colour.
Many genes control their skin colour.

9 Some characteristics can be controlled by just
one gene. Write down <u>one</u> example.

10 Most characteristics are controlled by several
genes. Write down <u>one</u> example.

Genes are made of DNA

All our genes, as well as the genes of all other animals and
plants, are made of a chemical called DNA. You can find
more about DNA on pages 170–172.

Why are we all different?

There are many different kinds of plants and animals.
We call these <u>species</u>. Plants or animals of one species
are the same as each other in many ways. They are also
different in some ways.
The differences are called **variation**.

Steve and Paul are not related to each other. This means
they have inherited many genes that are different from
each other. So many of their characteristics are different.

11 Write down <u>three</u> differences that you can see between
Steve and Paul.

12 Why do they have these different characteristics?

When variation is caused by different genes we say it has a
genetic cause.

> **Remember**
>
> Young plants and animals share
> characteristics with their
> parents. This is because of the
> genes that have been passed
> on to them.

Steve and Paul are unrelated.
They show lots of variation.

How can identical twins be different?

Sam and Jenny are **identical** twins. They have inherited exactly the same genes from their parents. Even so, Sam and Jenny are not exactly the same.

Sam has larger leg muscles than Jenny. Sam's body is different from Jenny's because different things have happened to it. Sam runs cross-country races, but Jenny doesn't. We say that the differences between them have **environmental** causes.

13 Look at the picture of Sam and Jenny.
Write down <u>three</u> differences between Sam and Jenny.

Sam and Jenny are identical twins. They show very little variation. Jenny cut her chin in a fall when she was small.

Genes and environment

Some differences have a mixture of genetic and environmental causes. For example, your genes control how tall you <u>can</u> grow. But if you don't get enough food as a child, you may not reach this height.

14 Ashfaq and David are not related. They are the same age. Ashfaq is 10 kilograms (kg) heavier. Give <u>two</u> possible reasons why you think they show this variation.

15 On average, people in Britain are a few centimetres taller than they were 50 years ago.
Why do you think this is?

16 The pictures show some variations between people. For each of (a), (b) and (c), write down <u>three</u> examples of variation caused by:
(a) genes;
(b) the environment;
(c) a mixture of genes and the environment.

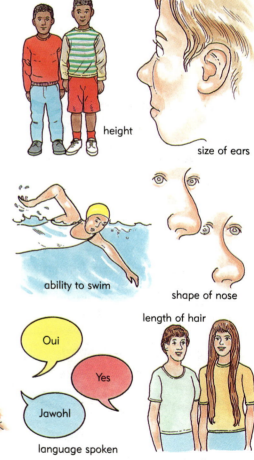

height

size of ears

ability to swim

shape of nose

length of hair

Oui

Yes

Jawohl

language spoken

strength

colour of skin

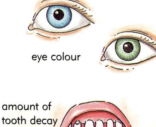

eye colour

amount of tooth decay

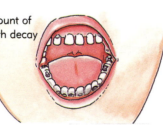

natural hair colour

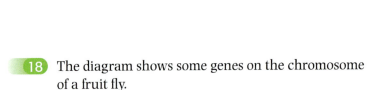

A human cheek cell.

Where are our genes?

Our cells contain information which tells them how to develop. This information is in units called **genes**.

Our genes are on parts of our cells called **chromosomes**.

17 Where exactly do we find genes?

18 The diagram shows some genes on the chromosome of a fruit fly.
Write down the feature controlled by:
(a) gene A;
(b) gene B;
(c) gene C.

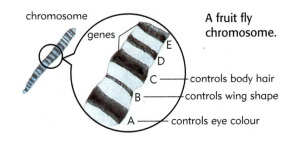

A fruit fly chromosome.

19 Look at the different fruit flies.
What features could genes D and E control?

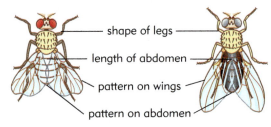

Fruit flies look different depending on their genes.

How many chromosomes?

The diagrams show the chromosomes in fruit fly cells and human cells.

The chromosomes of a fruit fly.

20 Chromosomes always come in pairs.
How many pairs of chromosomes are there in each nucleus of:
(a) a fruit fly?
(b) a human?

This is a pair of chromosomes.

The pairs of chromosomes of a human female.

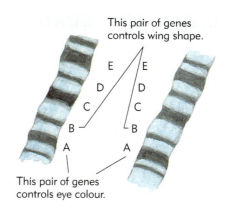

This pair of genes controls wing shape.

This pair of genes controls eye colour.

Pairs of genes

The diagram shows some of the genes in a pair of fruit fly chromosomes. Because chromosomes come in pairs, genes also come in pairs.

21 Say what you can about the control of eye colour in fruit flies.

Different forms of the same gene

You can get different forms of the same gene. These are called **alleles**. Different alleles produce different features. Look at the diagrams.

22 What colour eyes does a fruit fly have when it has:
(a) two A alleles?
(b) two a alleles?

A and a are **alleles** of the same gene

You can have one allele of a gene on one chromosome and a different allele on the other chromosome. Usually one of the alleles is stronger. This allele controls what happens. We say that it is <u>dominant</u>.

23 Is A or a the dominant allele of the gene for eye colour in the fruit fly? How do you know?

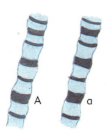

Genes and sexual reproduction

Sex cells

Look at the diagram on the right and then answer these questions.

1 (a) Where are sperm cells made?
  (b) Where are egg cells made?

2 In humans, how many chromosomes are there in each sperm cell and in each egg cell?

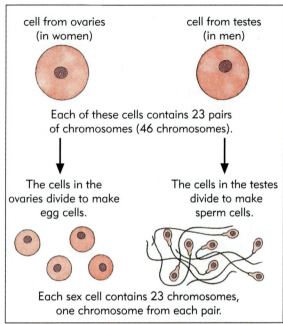

cell from ovaries (in women)

cell from testes (in men)

Each of these cells contains 23 pairs of chromosomes (46 chromosomes).

The cells in the ovaries divide to make egg cells.

The cells in the testes divide to make sperm cells.

Each sex cell contains 23 chromosomes, one chromosome from each pair.

Making sex cells.

Passing on life

The diagram shows what happens when a sperm cell and an egg cell join together.

3 How many chromosomes are there in:
  (a) each sex cell (the gametes)?
  (b) the cell produced during fertilisation?
  (c) the body cells of the baby?

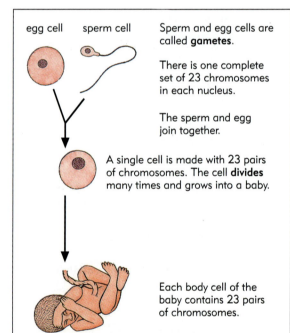

egg cell   sperm cell

Sperm and egg cells are called **gametes**.

There is one complete set of 23 chromosomes in each nucleus.

The sperm and egg join together.

A single cell is made with 23 pairs of chromosomes. The cell **divides** many times and grows into a baby.

Each body cell of the baby contains 23 pairs of chromosomes.

Mixing up the genes

All of the baby's cells have chromosomes from both parents. So they also have genes from both parents.

The parents may have different alleles of the same genes. So in their children these alleles can be mixed up in many different ways. This means that their children will be different from each other and from their parents.

The same thing happens with fruit flies.

Remember

Alleles are different forms of the same gene. So for a gene which controls eye colour, one allele could give blue eyes, another allele could give brown eyes.

4 Write down <u>three</u> features of the mother and father fruit flies which are different.

5 How many different types of young flies could they produce?

This shows that sexual reproduction produces a lot of **variation**. From two parents you can get a lot of different types of children with different features.

This applies to plants as well as animals:

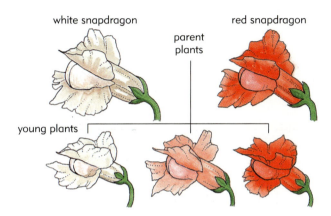

white snapdragon   red snapdragon

parent plants

young plants

6 Describe the results of breeding a white snapdragon with a red snapdragon.

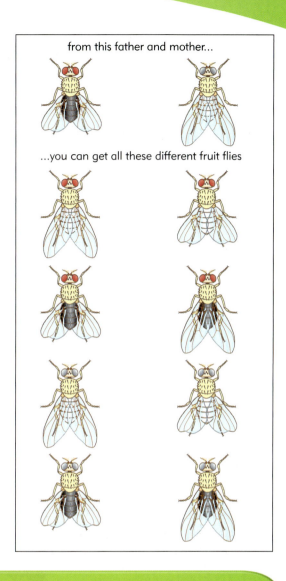

from this father and mother...

...you can get all these different fruit flies

Higher

Why do babies of the same parents vary?

The different ways that chromosomes combine can be shown as diagrams.

The diagrams show how four different eggs can be made from a cell with only two pairs of chromosomes. Then, how sixteen different offspring can be produced if these eggs join at random with four different sperm.

7 The diagram shows the chromosomes in a sperm-producing cell. Draw diagrams of the chromosomes in the sperm it could produce.

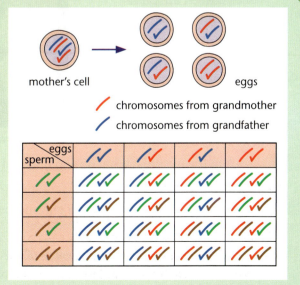

mother's cell   eggs

╱ chromosomes from grandmother

╱ chromosomes from grandfather

Higher

The information carried at a particular position on a chromosome is called a **gene**. Chromosomes come in pairs, so genes are also in pairs. One of each pair of genes in a baby comes from each parent.

The two members of a pair of genes can have different forms. They are called alleles. So, the gene for eye colour could be in the form of an allele for blue eyes or an allele for brown eyes. A baby may get different alleles from each parent. Different combinations of alleles produce different characteristics.

8 Copy the diagram.

B is the allele for brown eyes and b is for blue eyes. Fill in the missing alleles on your diagram.

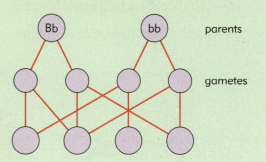

What makes you male or female?

9 Describe <u>one</u> difference between the two sets of chromosomes.

The last two chromosomes in each set carry the genes which make you male or female. We call them **sex chromosomes**. There are two different kinds of sex chromosome. They are called **X** and **Y** chromosomes.

Human body cells contain 23 pairs of chromosomes. They are in the nucleus of each cell.
Your 23 pairs of chromosomes look like this:

or like this:

10 Look at the photograph. What difference can you see between an X and a Y chromosome?

Each person has either one X and one Y chromosome or two X chromosomes.

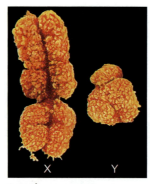

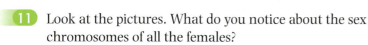
Sex chromosomes.

11 Look at the pictures. What do you notice about the sex chromosomes of all the females?

12 What can you say about the sex chromosomes of all the males?

13 Look back at the two sets of 23 pairs of chromosomes above. Which set belongs to a female and which belongs to a male?

Chromosomes in sperm and eggs

Sex cells (sperm and eggs) have <u>one</u> chromosome from each pair of chromosomes.

14 Write down the sex chromosome in a woman's
(a) body cell nuclei;
(b) sex cell nuclei.

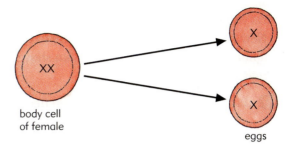

body cell of female

eggs

15 Write down the sex chromosome in a man's
(a) body cell nuclei;
(b) sex cell nuclei.

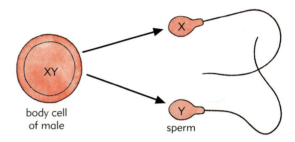

body cell of male

sperm

A boy or a girl?

Whether a baby is a girl or a boy depends on which sperm fertilises an egg.

Remember that all egg cells have one X chromosome.

16 What will be the sex of the child if the egg cell is fertilised by a sperm with:
(a) an X chromosome?
(b) a Y chromosome?

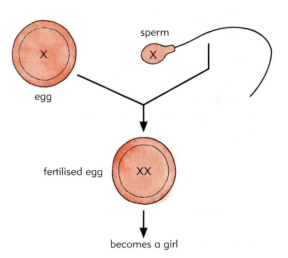

sperm

egg

fertilised egg

becomes a girl

During the first few weeks inside the womb, boys and girls are exactly the same. Then part of the Y chromosome starts working in a boy. It makes the growing sex organs develop into testes instead of ovaries.

How many girls, how many boys?

17 We would expect equal numbers of baby girls and baby boys to be born. Explain why.

18 In fact, 105 boys are born for every 100 girls. Why do you think this is?

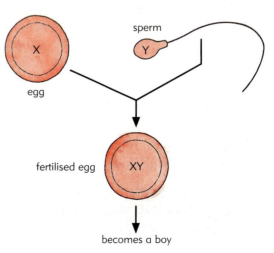

sperm

egg

fertilised egg

becomes a boy

Making more cells

As you grow you need to make more cells. New cells are needed for growth and to replace damaged or dead cells. The diagram shows how new cells are made.

19 Write down <u>two</u> reasons why you need to keep making new cells.

20 How many pairs of chromosomes are there in the nucleus of each body cell?

21 Each chromosome makes a copy of itself before a cell divides. Explain why.

Making new body cells.

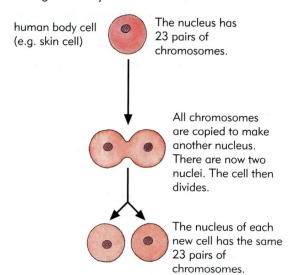

human body cell (e.g. skin cell) — The nucleus has 23 pairs of chromosomes.

All chromosomes are copied to make another nucleus. There are now two nuclei. The cell then divides.

The nucleus of each new cell has the same 23 pairs of chromosomes.

Higher

Producing genetically identical cells

In body cells the chromosomes are in pairs. So before a body cell divides, each chromosome makes a copy of itself to produce two strands of genetic information that are attached to each other. The two copies of each chromosome then separate and move apart. So, two new nuclei are produced. Each nucleus has 23 pairs of chromosomes exactly like those in the nucleus of the original cell. We call this type of division **mitosis**.

23 pairs of chromosomes.

Each chromosome copies itself.

Copies of chromosomes separate to form two sets of 23 pairs.

The cell divides.

MITOSIS for: ● making new body cells
● asexual reproduction

Producing sex cells

A different kind of division takes place in organs of sexual reproduction such as testes and ovaries. These organs make sex cells or gametes. Gametes have only <u>half</u> the usual number of chromosomes. So, in humans, gametes have 23 chromosomes, one from each pair. The chromosomes make copies of themselves in this kind of division too. But the cell then divides <u>twice</u> to produce four gametes. This is called **meiosis**.

22 (a) Write down the steps in the process of mitosis.
(b) Describe how meiosis differs from mitosis.

23 Offspring resulting from asexual reproduction are genetically identical. We call them **clones**. Are the cells of clones produced by meiosis or mitosis? Explain your answer.

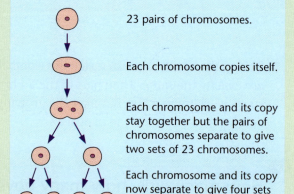

23 pairs of chromosomes.

Each chromosome copies itself.

Each chromosome and its copy stay together but the pairs of chromosomes separate to give two sets of 23 chromosomes.

Each chromosome and its copy now separate to give four sets of 23 single chromosomes.

MEIOSIS produces new cells with a single set of chromosomes (gametes).

Reproducing without sex

An **amoeba** is a tiny animal which has just one body cell. This cell divides just like your body cells do. Then it splits into two to make two amoebas.

The amoeba does not use sex cells to reproduce. So we call this **asexual** reproduction. (<u>A</u>sexual means <u>non</u>-sexual). Only <u>one</u> parent is needed.

24 (a) Describe how an amoeba reproduces.
(b) Explain why the two amoebas produced are identical.

25 What do we call this type of reproduction?

Some plants and animals with many cells can also reproduce without using sex cells. Tiny new plants or animals may grow on the body of the parent. These then split off to make new plants or animals.

A plant called **Bryophyllum** can reproduce without sex cells.

These new plants all have exactly the same genes as each other and exactly the same genes as the parent plant. We say they are **clones** of the parent plant.

26 (a) Where, on a Bryophyllum plant, do the new plants grow?
(b) How do they become separate plants?

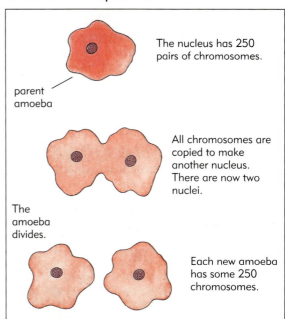

How an amoeba reproduces.

parent amoeba

The nucleus has 250 pairs of chromosomes.

All chromosomes are copied to make another nucleus. There are now two nuclei.

The amoeba divides.

Each new amoeba has some 250 chromosomes.

leaf

plantlets (these fall to the ground)

young plant growing

How a Bryophyllum plant reproduces.

How strawberry plants reproduce

New strawberry plants can grow from runners. The diagram shows how.

27 How many parents does a strawberry runner have?

28 The young strawberry plants in the drawing are clones of the parent plant. What does this tell you about the young plants?

Strawberry plants can also be grown from seeds. They then have two parents.

29 Strawberry plants grown from seeds will all be slightly different from each other. Explain why.

30 Imagine you are a gardener. You have a strawberry plant which grows large and tasty strawberries. You want to grow more plants exactly like it. How would you grow them? Explain your answer.

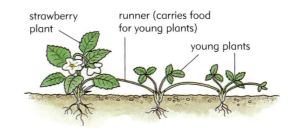

strawberry plant

runner (carries food for young plants)

young plants

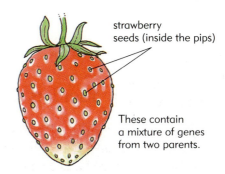

strawberry seeds (inside the pips)

These contain a mixture of genes from two parents.

Some human genes

Genes control many of your characteristics

Each of your 23 pairs of chromosomes is made of lots of genes. Chromosomes are made of long molecules of a substance called **DNA**. So each gene is a section of a DNA molecule.

1 Some people compare the genes on a chromosome to beads on a necklace. Explain why.

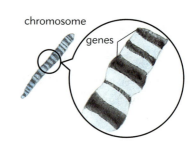

chromosome

genes

Chromosomes come in pairs, so genes come in pairs. Sometimes only one pair of genes controls a characteristic. But that gene may have more than one form. The DNA in each form is slightly different, so each produces a different effect.

One gene controls whether you can roll your tongue or not. It has two different forms or **alleles**. A **recessive** allele only shows up when the **dominant** allele is not present. The girl in the photograph has muscles that let her roll her tongue. She has an allele (R) for tongue rolling and an allele (r) for non-rolling.

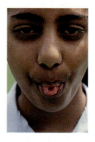

2 What do we call different forms of a gene?

3 Is the allele for tongue-rolling dominant or recessive? Explain your answer.

4 Look at the table.

Which person is:
(a) a tongue-roller?
(b) a non-roller?

Name	Alleles
John	RR
Carl	Rr
Sam	rr

Some human disorders are inherited

Some inherited disorders cause serious problems or disability. Others do not.

Huntington's disorder

Michael and his father both have a disease called **Huntington's disorder**. Michael did not <u>catch</u> the disorder from his father. He <u>inherited</u> it.

Huntington's disorder is a disease which damages your nervous system. The symptoms appear as you get older, usually when you are about 35 years old.

People with the disorder cannot control their muscles. Their bodies may jerk suddenly. As the disorder gets worse they cannot think clearly.

There is no cure for Huntington's disorder and people die from it in middle age.

5 Michael feels healthy now. How will the disorder affect him as he gets older?

How do people get Huntington's disorder?

A faulty allele causes Huntington's disorder. Just **one** faulty allele, from **one** of your parents, is enough to give you this disorder. We say that the allele is **dominant**.

6 Which parent passed on the faulty allele to Michael?

7 Why can't people be carriers of Huntington's disorder?

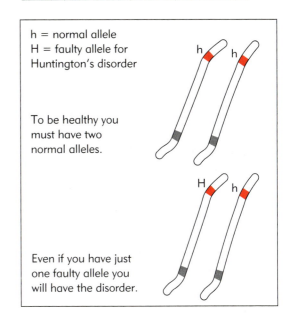

h = normal allele
H = faulty allele for Huntington's disorder

To be healthy you must have two normal alleles.

Even if you have just one faulty allele you will have the disorder.

Higher

More about Huntington's disorder

The allele that causes this condition is dominant. Everyone who has this allele develops the disorder, but it doesn't develop until a person is 30 to 40 years old. So, the allele may already have been passed on to their children.

Look at Michael's family tree. Michael has the disorder. His brother, Wayne, has had a test. He has only normal alleles. Paul, aged 21, can't decide whether to have the test or not.

8 Why does everyone with the Huntington's disorder allele always develop symptoms?

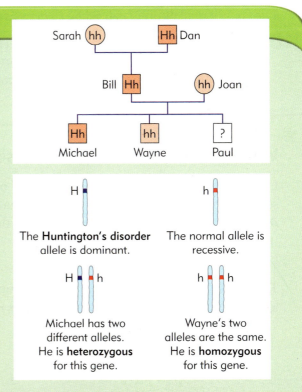

The **Huntington's disorder** allele is dominant.

The normal allele is recessive.

Michael has two different alleles. He is **heterozygous** for this gene.

Wayne's two alleles are the same. He is **homozygous** for this gene.

9 The diagram shows the allele combinations possible in Bill and Joan's children. What is the probability that Paul has only normal alleles?

10 For Paul, write down the advantages and disadvantages of having a test.

11 Explain what these words mean:
(a) dominant;
(b) recessive;
(c) homozygous;
(d) heterozygous.

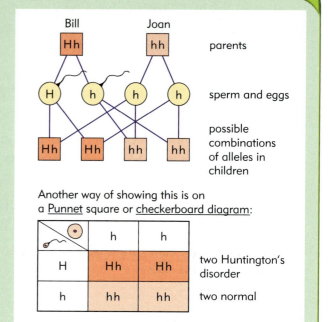

Another way of showing this is on a Punnet square or checkerboard diagram:

Cystic fibrosis

You can catch a cold or measles. Microorganisms cause these diseases. You cannot catch **cystic fibrosis**. It is caused by genes passed on to children by their parents. It is an inherited disorder caused by a **recessive** allele.

12 Imagine you have cystic fibrosis. You are in hospital for a check-up. A friend is worried about coming to see you. Write her a note explaining why she cannot catch the disorder from you.

What is it like to have cystic fibrosis?

Laura has cystic fibrosis. It is a disorder of cell membranes, so it affects Laura in many ways. She has to have physiotherapy every day to help her get rid of the very thick and sticky mucus that blocks the tubes to her lungs.

This mucus affects her breathing. Also her lungs are easily infected as microorganisms are trapped in this mucus.

Another problem is that Laura's digestive glands, such as her pancreas, do not work properly. She has to take enzymes every time she has a meal. These enzymes help her to digest her food. If she forgets to take them, she cannot digest her food very well.

13 Laura cannot play sports. Why is this?

14 Laura is smaller than most other girls of her age. Write down <u>one</u> possible reason for this.

How do people get cystic fibrosis?

Faulty alleles cause cystic fibrosis. The allele is recessive so you can get cystic fibrosis only if you get a faulty allele from **both** parents.

People who have one faulty allele but don't have the disorder are called **carriers**.

15 Explain why you need a faulty allele from both parents to get cystic fibrosis.

16 A person with two healthy parents could still get cystic fibrosis. Explain how.

If you have two normal alleles you are healthy.

If you have one normal allele and one faulty allele you are healthy.

If you have two faulty alleles you have the disorder.

Higher

More about cystic fibrosis

One in 20 people carries the cystic fibrosis allele. Until Laura was born, her parents had no idea that they were carriers of cystic fibrosis. After Laura was diagnosed with the disease, they were worried about the risk of having another child with the disorder. Their doctor sent them to a genetic counsellor.

17 Look at the counsellor's charts.
   (a) Write down the alleles of Laura and her parents.
   (b) Why do we call her parents 'carriers'?
   (c) Imagine that you are the counsellor. Explain why there is a probability of 1 in 4 of Laura's parents having another affected child.

18 Some people think that anyone wanting to start a family should have a test for cystic fibrosis. Explain why.

Cystic fibrosis
Chromosomes

	normal	carrier
No symptoms	C C	C c
Suffers symptoms		cystic fibrosis c c

If you are both carriers:

sperm \ eggs	C	c
C	CC	Cc
c	cC	cc

Sickle cell anaemia

Jomo has **sickle cell disorder**. Sometimes he feels well. At other times he feels tired. He catches infections easily. Normal red blood cells are disc-shaped. Jomo's red blood cells can go out of shape and block capillaries, causing painful swellings. Sometimes he has to go into hospital.

19 Describe <u>two</u> differences between the red blood cells of people with sickle cell and normal red blood cells.

20 Explain why Jomo feels tired more often than he should.

The sickle cell allele must be present in both of Jomo's parents, just like the allele for cystic fibrosis. The normal allele is dominant so that one recessive sickle cell allele produces no symptoms. We call people with one recessive allele **carriers**. Even when both parents are carriers they don't always pass on the disease.

Jomo's red blood cells. Sickle cells don't carry oxygen as well as normal cells do.

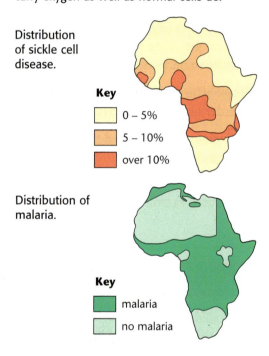

Distribution of sickle cell disease.

Key
- 0 – 5%
- 5 – 10%
- over 10%

The sickle cell allele can be useful

Malaria is a disease caused by a parasite that lives in red blood cells.

Doctors think that people with one sickle cell allele hardly ever catch malaria.

21 The maps support this idea. Explain how.

22 (a) Why can it be useful to have one sickle cell allele?
(b) Why is it harmful to have two copies?

Distribution of malaria.

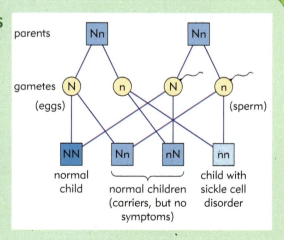

Key
- malaria
- no malaria

Higher

The chances of inheriting two sickle cell alleles

The allele for sickle cell disorder is recessive.

23 Draw diagrams, similar to the ones shown, for an nn mother and an Nn father.

eggs \ sperm	N	n
N	NN normal	Nn carrier
n	nN carrier	nn sickle cell disorder

parents: Nn   Nn

gametes: N   n   N   n
(eggs)   (sperm)

NN — normal child
Nn, nN — normal children (carriers, but no symptoms)
nn — child with sickle cell disorder

You need to be able to construct genetic diagrams and to predict and explain the results of crosses.

Selective breeding and cloning

> **Remember from Key Stage 3**
>
> Farmers and other breeders of plants and animals have produced new varieties by selective breeding. For example, new varieties produce increased yields of wheat, eggs and milk.

As well as increasing yields, people often set out to breed plants or animals with other features. At other times, a new feature of a living thing suddenly appears.

In the 18th century, a farmer noticed that one of his lambs was born with short legs. He decided to breed from this lamb to produce more short-legged sheep.

1 Why did the farmer decide that short legs are useful?

We call breeding living things in this way **selective** breeding. Another name for this is **artificial selection**.

Our best friend

A dog may be our best friend, but sometimes we are not very friendly to dogs. We have bred many different kinds of dogs. But some of the features we have selected cause problems for the dogs.

2 Write down the names of <u>three</u> different breeds of dog.

3 (a) Write down <u>two</u> features which people selected when they bred basset hounds.
(b) What problems do these features cause?

4 Why is a basset hound more likely to have back problems than a bulldog?

Other breeds of dog have other problems.

5 Dogs with short noses have breathing problems. Which breed of dog might find it hard to breathe?

6 Bulldogs have narrow hips. What problem does this cause for female bulldogs?

7 Some people think that we should change some of the features of pedigree dogs that we select for. Why do you think this is?

8 The diagram shows how the cocker spaniel breed was produced. Describe what the breeders did.

Droopy eyelids – they often get eye infections.   Long ears – they sometimes trip up over them.

very long back

Basset hound

Narrow hips – this makes it harder for females to give birth to their pups.

French bulldog

Italian spaniels

They chose the dogs with the longest ears and longest backs for breeding.
↓

Norfolk spaniels
↓

They chose the dogs with the shortest tails and shortest legs for breeding.
↓

cocker spaniels

Breeding a cocker spaniel.

When is a cauliflower not a cabbage?

Cabbages, cauliflowers, Brussels sprouts and broccoli were all bred from the same ancestor plant. Farmers bred cauliflowers by choosing plants with the largest flower heads. They then used these plants to produce the next lot of seeds. They did this again and again for many years. Eventually they produced cauliflowers.

Brussels sprouts are large side buds.

9 Tell the story of how Brussels sprouts were produced from the ancestor plant.

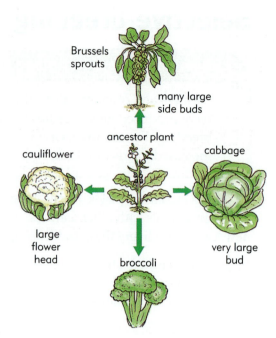

Growing identical plants

Plant and animal breeders can produce new varieties by sexual reproduction. New plants don't always grow from sex cells. Reproduction that doesn't use sex cells is called **asexual** or **non-sexual** reproduction. New plants sometimes grow from the ordinary cells of plants. Plants sometimes grow this way by themselves. If they don't do this, we can produce more plants from parts of older plants. We call these parts **cuttings**.

When we grow plants from seeds, we do not know what the plants will be like. Cuttings, on the other hand, have the same characteristics as the plant we take them from.

10 Why do plants from cuttings look exactly like the plant they came from?

11 Making new plants from cuttings is called **asexual** reproduction. Explain why.

Taking cuttings from plant shoots

Dawn wants to take cuttings from a geranium plant. The diagram shows what she needs to do.

Cuttings from the same plant have exactly the same genes. We say they are **genetically identical**.
We call them **clones**.

12 Write down the following stages in the right order. Use the diagram to help you.

- Plant the cutting in compost.

- Dip the cut end of the shoot into rooting hormone.

- Cut a young shoot from the parent plant.

- Cover the cutting with a polythene bag. (Cuttings need to be in a damp atmosphere until the roots grow.)

- Take off some of the lower leaves.

13 What does rooting hormone do?

14 What <u>two</u> things should Dawn do so the plant loses less water?

Plants will wilt and may die if they don't have enough water.

Using other parts of a plant for cuttings

We take cuttings because it helps us to produce many plants quickly and cheaply. We can also take cuttings from leaves and roots.

Taking a leaf cutting.

African violets are usually grown from **leaf cuttings**.

15 Write down <u>three</u> reasons why plant nurseries usually grow African violets from cuttings rather than by sowing seeds.

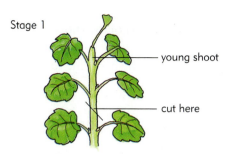
Stage 1
young shoot
cut here

Stage 2
Take off the lower leaves so that the cutting loses less water.

Stage 3
Dip the cut end into rooting hormone. This makes roots grow better.
rooting hormone

Stage 4
polythene bag
Plant the cutting in compost and water it. Cover it with a polythene bag.

Stage 5 (a few weeks later)
Roots can now take in all the water the plant needs.

Taking geranium shoot cuttings.

The diagrams show how you can make new rhubarb plants.

16 (a) Which part of the rhubarb plant is used to
    make cuttings?

  (b) When is the best time to take these cuttings?

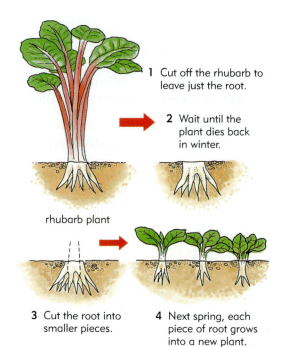

1 Cut off the rhubarb to leave just the root.

2 Wait until the plant dies back in winter.

rhubarb plant

3 Cut the root into smaller pieces.

4 Next spring, each piece of root grows into a new plant.

Making new rhubarb plants.

Reappearing dandelions

Cuttings can sometimes grow where you do not want
them to. When you dig up dandelions from a flower bed or
lawn they often grow again.

17 Look at the diagrams. Why did the dandelions reappear
    after two weeks?

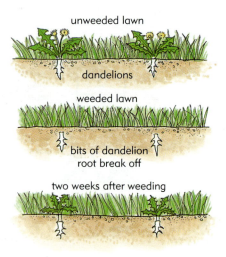

unweeded lawn

dandelions

weeded lawn

bits of dandelion root break off

two weeks after weeding

Another way of cloning plants

A way of growing a large number of identical plants even
more quickly is by using **tissue culture**. Scientists can
sterilise the surface of cells without damaging them.
Then they grow the cells on sterile jelly called **agar**
so that there aren't any bacteria or fungi.
The cells divide to produce a mass of identical cells.
These can be divided to produce lots of new plants.

At first, scientists grow them on agar containing nutrients
and hormones that make shoots grow. Then they transfer
them to agar with nutrients and hormones that make
roots grow.

18 Write down <u>two</u> reasons why bacteria and fungi don't
    grow in the culture tube.

19 Scientists add different chemicals to the agar at
    each stage. What do they add and why?

New varieties of potatoes

The parts of potato plants that we eat are underground stems called **tubers**. They are produced by asexual reproduction, so they are clones. It is these clones that a farmer plants to grow more potatoes. So, if he plants a type of potato called Desiree, his whole crop will be Desiree potatoes.

Potato plants also have flowers which produce seeds. So, scientists produce <u>new</u> types called **varieties** of potatoes by sexual reproduction. They breed from plants with the characteristics they want. They grow new plants from the seeds and choose the best plants to breed from. This is called **selective** breeding or **artificial selection**. When they find a useful variety, they clone it to produce potato tubers for the farmers to plant.

20 Look at the picture. Write down <u>two</u> advantages of growing a potato crop by cloning.

21 Describe how breeders produce new varieties of potatoes.

Remember

- In asexual reproduction, a parent and its offspring have exactly the same alleles. They are clones.

- In sexual reproduction, the alleles from the parents are mixed up. So the offspring show a lot of variation.

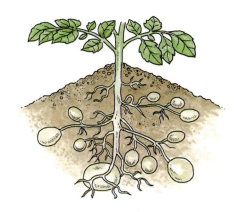

Cloning produces a large number of genetically identical tubers fairly quickly.

Animals can be cloned too

Angora goats produce valuable wool. So, they are expensive. If a breeder buys only a few goats, it will be a long time before he has bred a big herd. The diagram shows how scientists can use ordinary goats as the 'mothers' of Angora goats.

22 (a) Describe <u>one</u> way of cloning an animal.
(b) Explain why cloning is useful.

Cloning produces genetically identical animals or plants. So, in a population of cloned organisms, the genetic variation decreases. The number of different alleles in the population falls.

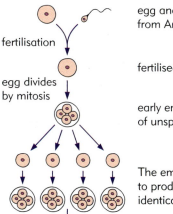

egg and sperm from Angora goats

fertilisation

fertilised egg

egg divides by mitosis

early embryo – a ball of unspecialised cells

The embryo is divided to produce four identical embryos.

Each **embryo** is transplanted into the womb (uterus) of an ordinary goat.

Problems with cloning

If conditions change, plants and animals that we have
selected for farming may not grow as well as they did.
Different alleles may make them better able to survive in
the new conditions. But, if these alleles have been bred out
of the population, they are lost forever. This is why
scientists save seeds from old or wild varieties of plants
and why some farmers keep rare breeds of farm animals.

23 (a) Explain why it is useful to have a lot of different
alleles in a population.
(b) Wild potato plants produce poor crops of tiny
potatoes. Explain, as fully as you can, why plant
breeders are interested in them.

Genetic engineering

Finding and transferring genes

All your body cell nuclei contain 23 pairs of
chromosomes. Each chromosome is made of a large
number of genes. Scientists have made **chromosome
maps** to show where all the human genes are.
This is the **Human Genome Project**. It should help
scientists to detect and treat some inherited diseases.

1 (a) Where exactly are genes?
(b) What are they made of?
(c) What does a gene do?

2 Why did scientists want to map human chromosomes?

When scientists find out where a gene is, they can 'cut'
it out and make copies. They use **enzymes** to make the
'cuts'. They use different enzymes to make the copies.
They can then transfer the genes into the cells of other
living things. These cells may then make proteins
that they wouldn't normally make. We call this
genetic engineering. Making copies of genes is
called **gene cloning**.

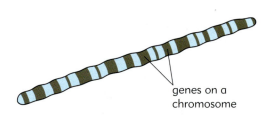

genes on a
chromosome

Chromosomes are made of long molecules of
DNA. A gene is a section of this DNA.
It is a code that controls the order of amino
acids in a protein.

The hormone **insulin** is a protein. Scientists have transferred the human insulin gene to bacteria. Now, many people with diabetes use this 'human' insulin. In the past, they used insulin from pigs or cattle.

3 (a) Describe how scientists produced the bacteria that make insulin.
(b) What do they do to make a batch of insulin?

4 Some people think that it is wrong to use insulin from animals. Think of some reasons for this and write them down.

More gene transfers

Some scientists have transferred genes for different proteins into fertilised sheep eggs. These cells make copies of the gene when their nuclei divide. So, all the new nuclei of the embryo contain the new gene. Examples of human genes transferred to sheep are those producing:

- human insulin;
- a protein needed to treat emphysema;
- factor VIII, a protein needed for blood clotting.

Doctors hope to find a way to transfer genes into human cells. For example, if normal genes could be transferred into enough cells of cystic fibrosis patients, they could live healthy lives. We call this **gene therapy**.

Genes can be transferred into plant cells too. For example, scientists transferred a gene for herbicide resistance from a bacterium into sugar beet.

5 Describe <u>one</u> example of gene transfer in an animal and <u>one</u> in a plant.

6 Many farmers save money by planting seeds from a previous year's crop. Some charities think that herbicide resistant crops are bad for farmers in poor countries. Why is this?

You may be asked to consider economic, social and ethical issues concerning cloning and genetic engineering.

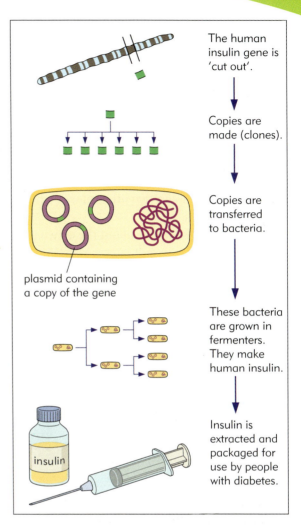

The human insulin gene is 'cut out'.

Copies are made (clones).

Copies are transferred to bacteria.

plasmid containing a copy of the gene

These bacteria are grown in fermenters. They make human insulin.

Insulin is extracted and packaged for use by people with diabetes.

insulin

This sugar beet will not be killed when the farmer sprays the field with herbicide (weedkiller). But the farmer has to buy new seed and the matching weedkiller each year. Saved seeds do not grow.

The mystery of fossils

The fossils in the picture were found near Whitby on the Yorkshire coast. The people of Whitby used to say that an ammonite was a snake turned into stone by St Hilda. They called another fossil a devil's toenail.

Fossils are found in rocks. That is why cliffs at the seaside are good places to look for them.

ammonite

devil's toenail (gryphaea)

1 (a) Write down <u>two</u> other places where you could look for fossils.

(b) What is similar about these two places?

Quarrying.

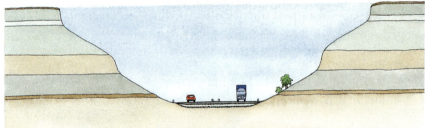

Road cutting.

What do scientists think fossils are?

Scientists think that fossils are the remains of dead plants and animals from millions of years ago. The devil's toenail and the ammonite both lived in the sea more than 180 million years ago.

The diagram shows how ammonites became fossils.

2 How long ago did ammonites live?

3 Write down the following sentences in the right order. Use the diagrams to help you.

● Ammonite fossils were then surrounded by rock.

● Layers of sand and mud covered the ammonites.

● Ammonites died and fell to the bottom of the sea.

● The sand and mud turned into rock.

Ammonites lived in the sea about 180 million years ago.

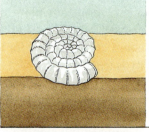

When they died they fell to the bottom and were covered with sand and mud.

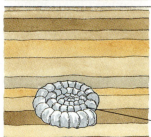

The sand and mud slowly changed into rock.

fossil ammonite

What sort of remains are found?

Hard parts such as shells and bones are more likely to become fossils than other parts of plants and animals.

Soft parts of dead animals and plants may be eaten by animals. They also rot or decay easily. They do not last long enough to become fossils. Most fossils are formed from hard parts, which do not decay very quickly. Sometimes only traces remain. We may find signs of plant rootlets or animal burrows and footprints.

The photograph shows a dinosaur's footprint. The diagrams show how this footprint has lasted for millions of years.

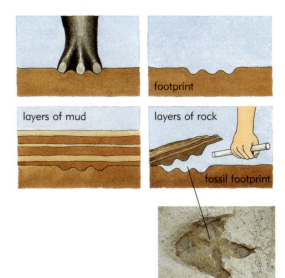

4 What is similar about the parts of plants and animals which are common fossils?

5 Explain how the dinosaur's footprint came to be in the rock.

Where did the plants and animals live?

About 90 per cent of all fossils found are the remains of sea creatures. They were buried in the mud or sand on the sea bed.

6 Why are the remains of land animals and plants uncommon?

Ammonites are similar to a sea creature called a pearly nautilus which still lives in tropical seas.
Look at its tentacles.

7 The fossil ammonite has no tentacles. Write down two things which could have happened to them.

Pearly nautilus.   Ammonite.

Sometimes hard parts stay the same

A shell is made of a mineral called **calcite**. The diagrams show what happens when a shell is buried in bits of other shells.

8 Look at the diagrams. What happens to:
(a) the shell?
(b) the bits of shells?

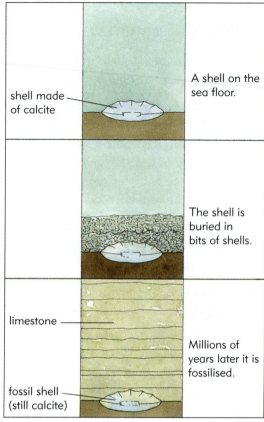

A shell on the sea floor.

shell made of calcite

The shell is buried in bits of shells.

limestone

Millions of years later it is fossilised.

fossil shell (still calcite)

Sometimes hard parts are turned into stone

When hard parts are turned to stone we say they are **petrified**.

Sometimes a shell is buried in layers of sand. Sand contains a mineral called **silica**.

The diagrams show what can then happen.

9 Describe what happened to the shell buried in sand.

The photograph shows a slice across a fossil tree.
The wood has been turned into stone.

Silica has replaced the wood a bit at a time.

10 How do we know the fossil was once a tree?

11 You can still see the detail of what the wood was like. Explain why.

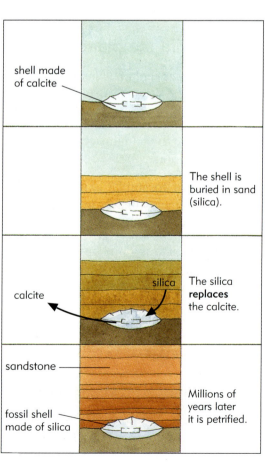

shell made of calcite

The shell is buried in sand (silica).

calcite   silica

The silica **replaces** the calcite.

sandstone

Millions of years later it is petrified.

fossil shell made of silica

Are soft parts of plants and animals ever preserved?

Soft parts of plants and animals usually **decay** quite quickly. This is because microorganisms feed on them. Microorganisms need oxygen, moisture and warmth to feed and grow.

Without all of these conditions, microorganisms cannot cause decay. The plants and animals are **preserved**.

We often find fossil insects in amber. Sometimes soft parts of plants and animals can be preserved for long enough in other ways to form fossils.

This baby mammoth was found frozen in ice.

Spider in amber. Resin from a tree set hard around this spider. Oxygen could not get to it.

12 The animals in the pictures were preserved because one of the conditions for decay was missing. In each case, give <u>one</u> reason why the animal did not decay.

Three billion years of life

Scientists think that the Earth was formed about 4600 million years ago. They think that life on Earth began over 3000 million (three billion) years ago.

13 For about what fraction of its history has there been life on Earth? Choose from:
(a) half;
(b) more than half;
(c) less than half.

Rocks that are more than about 600 million years old contain very few fossils. But this does not mean that there weren't many living things before then.

14 Look at the picture of a fossil which has been found in rocks more than 600 million years old.
Explain why this type of fossil is rare.

This shows a fossilised jellyfish. Jellyfish have no hard parts.

New forms of life

Fossils can also tell us how life on Earth has changed. Most scientists think that all species of living things on Earth today came from earlier, simpler species. We say that living things have **evolved**.

15 What do scientists think the first life forms were like and what do they think happened to them?

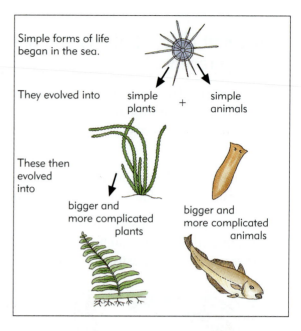

Simple forms of life began in the sea.

They evolved into simple plants + simple animals

These then evolved into bigger and more complicated plants

bigger and more complicated animals

Some ancient species

We know from fossils that some plants and animals are very like their ancestors. They have changed very little.

16 Show which fossils were the ancestors of the modern plants and animals in the diagram.
For example:

A ⟶ ginkgo

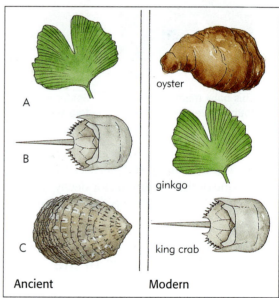

A

B

C

oyster

ginkgo

king crab

Ancient   Modern

Which plants and animals lived when?

Scientists can use fossils to tell them which plants and animals lived when. The diagram opposite shows some of the things scientists have found out.

You can see from the diagram that graptolites were very common between 400 and 500 million years ago. Then they died out. We say they became **extinct**.

17 When did trilobites first appear?

18 When did trilobites become extinct?

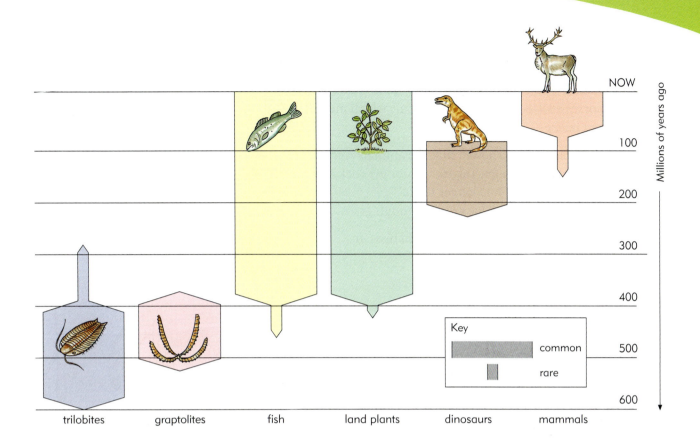

Key
common
rare

trilobites   graptolites   fish   land plants   dinosaurs   mammals

Millions of years ago

NOW
100
200
300
400
500
600

19 Which group of animals first appeared about 220 million years ago?

20 Name a group of animals which became common between 300 and 400 million years ago.

21 How long ago did the first mammals appear?

Fossil detective stories

Fossils can tell us a lot about living things in the past and how they changed. They tell us how living things **evolved**.

There are lots of things we don't know about past life because there is no fossil evidence. We say that there are <u>gaps</u> in the fossil record.

Some changes were big

Scientists have studied fossil horses. They think that today's horses evolved from small mammals which ate soft leaves from bushes.

The fossils in older rocks showed that the animals were small and had teeth with few ridges. In younger rocks the animals were bigger and had longer teeth.
The teeth had strong ridges for grinding tough grass.

22   As horses developed, what were the changes in:
(a) the size?
(b) the teeth?

23   Write down how the toes changed:
(a) between horses 1 and 2;
(b) between horses 2 and 3;
(c) between horses 3 and 4.

Scientists think that these changes took place because the environment changed. They think that grassy plains took the place of trees and bushes.

The horses which were best at chewing tough grass and at running fast survived. They bred and passed on their genes, so the young horses they produced were also good at running and chewing tough grass.

24   Why is speed more important on grassy plains than where there are bushes?

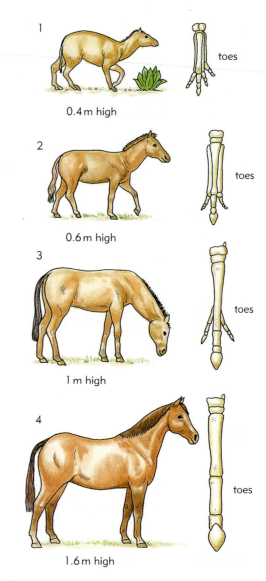

1   0.4 m high   toes

2   0.6 m high   toes

3   1 m high   toes

4   1.6 m high   toes

Changes to the horse over many years.

Other changes were smaller

A scientist worked out how an oyster called **Ostrea** evolved into a different oyster called **Gryphaea**.
The chart shows what he found.

25   The Gryphaea shell has more turns than the Ostrea shell. About how many turns has
(a) the oldest shell?
(b) the youngest shell?

The most curved modern oysters live in the muddiest seas. The curve keeps the opening of the oyster above the sea bed. This stops the shell filling up with mud.

26   Ostrea probably evolved into Gryphaea because of a change in the environment 180 million years ago.
What could this change have been?

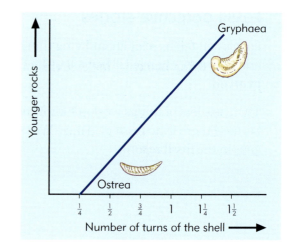

Younger rocks

Gryphaea

Ostrea

$\frac{1}{4}$ $\frac{1}{2}$ $\frac{3}{4}$ 1 $1\frac{1}{4}$ $1\frac{1}{2}$

Number of turns of the shell

Why did dinosaurs become extinct?

If the environment changes quickly but an animal or plant cannot change quickly, it will die out. It will become **extinct**. Dinosaurs became extinct about 65 million years ago. Scientists argue about the reason. Look at some of their ideas.

27 Choose <u>one</u> of the ideas which you think is <u>not</u> very sensible. Explain why you think it is probably wrong. (Use the 'Information box' to help you.)

28 Explain, as fully as you can, how a meteorite could have made dinosaurs extinct.

Ideas that could explain the extinction of dinosaurs

● New predators killed them.

● A very big meteorite hit the Earth. Dust blocked out the sun.

● A new disease killed them.

● New competitors took over their food.

Information to help you

● Plants cannot grow without the Sun.

● The biggest dinosaurs were bigger than other animals at that time.

● Most dinosaurs ate plants.

● Dinosaurs probably couldn't survive very cold weather.

You should be able to explain how fossil evidence supports the theory of evolution.

Mutation and change

Do you ever make a mistake when you copy a sentence? A small mistake can make a big difference.

1 Sentence B is a copy a student made of sentence A. How many letters were copied wrongly?

2 What difference did it make to the reindeer?

Before it divides, a nucleus makes a new copy of its chromosomes. Sometimes there are mistakes in these copies. We call these mistakes **mutations**. They are often harmful.

Remember

The genes that control your features are on your chromosomes. Different forms of the same gene are called alleles.

A Your nose is not red.   **B** Your nose is now red.

Sometimes alleles are faulty

Joanna has inherited the same faulty allele from both of her parents. She has sickle cell disease.

When Joanna's blood cells go out of shape they block tiny blood vessels. This makes her joints, hands and feet swell up and hurt. She also gets pains in her stomach and back. Sometimes she has to go into hospital.

3 Explain how the changed genes affect Joanna's red blood cells.

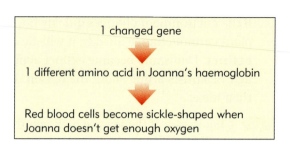

1 changed gene

⬇

1 different amino acid in Joanna's haemoglobin

⬇

Red blood cells become sickle-shaped when Joanna doesn't get enough oxygen

Chromosomes can also be faulty

Look at the pictures of the chromosomes of two brothers. Mark has **Down's syndrome**. He has mental and physical problems.

4 Write down <u>one</u> difference between the chromosomes of the two boys.

This one difference caused Mark's problems.

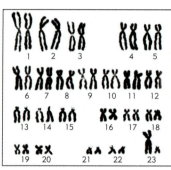

Ken's chromosomes

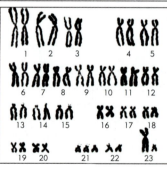

Mark's chromosomes

What causes mutations?

Mutations happen by chance. However, some things increase the chance of mutation. One of these is radiation from radioactive substances. We call this **ionising radiation**.

The more radiation you receive, the more chance there is of mutations.

5 What is the biggest source of the radiation each of us receives?

6 What percentage of radiation we receive comes from fallout from weapons tests?

7 How much radiation does a person receive, on average, for medical reasons?

X-rays, ultra-violet rays and some chemicals, e.g. mustard gas, also increase mutation rates.

8 Write down why it is against the Geneva Convention to use mustard gas as a weapon.

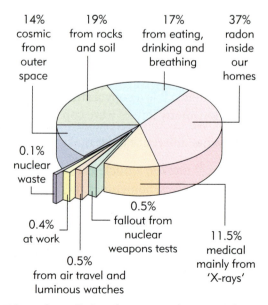

14% cosmic from outer space

19% from rocks and soil

17% from eating, drinking and breathing

37% radon inside our homes

0.1% nuclear waste

0.4% at work

0.5% from air travel and luminous watches

0.5% fallout from nuclear weapons tests

11.5% medical mainly from 'X-rays'

Where the radiation that we receive comes from.

Not all mutations are harmful

Scientists use radiation to cause mutations in seeds such as wheat. They see the effects when they grow the seeds. The picture shows some of these effects.

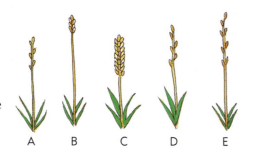

A B C D E

9 The scientists are very pleased with plant C. Why is this?

Looking for explanations of change

In nature, plants and animals usually produce more offspring than there is space or food for. So, only a few of the offspring survive long enough to breed.

10 (a) Describe the two forms of moth in the picture.
 (b) Which one is the more likely to die before it can breed? Explain your answer.

So, the moths that breed are the ones that are best suited to that particular **environment**. They pass on the alleles. It <u>looks as if</u> the environment has 'selected' moths with the useful characteristics. So we call it **natural selection**.

11 (a) What changes did the Industrial Revolution cause in the moths' environment?
 (b) In what way did the changes affect the moth population?

A more worrying change

We use **antibiotics** to kill the bacteria that cause disease. Sometimes mutations in bacteria stop an antibiotic killing them. We say the bacteria have become **antibiotic resistant**.

In 1972, there was an epidemic of typhoid in Mexico. Normally an antibiotic called **chloramphenicol** cured it. This time it didn't work. Over 14 000 people died. Eventually they found an antibiotic that worked.

12 Explain, as fully as you can, why chloramphenicol didn't control this epidemic.

You need to be able to explain how over-use of antibiotics can lead to the evolution of antibiotic resistant bacteria.

Remember

We use artificial selection to produce plants and animals with the characteristics we want. Many don't survive long enough to breed. So they don't pass on their genes to the next generation.

Two peppered moths resting on a tree trunk during the day. Only the most well-camouflaged moths escape being eaten by birds.

Before the Industrial Revolution, tree trunks were light coloured and covered in lichen.
Most peppered moths were light then.
Once the Industrial Revolution started, many trees had no lichen and their bark was covered in soot. Most peppered moths were then dark.

Chloramphenicol was used in people without prescription or supervision. It was over-used.

↓

A few bacteria developed resistance to chloramphenicol. They were mutants.

↓

The mutant forms multiplied.

↓

Only mutant forms survived. Chloramphenicol was no use against them.

Explaining change

In the nineteenth century, Charles Darwin collected a lot of evidence about changes in plants and animals. He was not the first scientist to suggest the idea of change or **evolution**. But he and Alfred Russel Wallace were the first to explain <u>how</u> it could happen. They called it evolution by **natural selection**. Many people mocked the idea. It was not what they had been taught and it went against what the Bible said. Now many people do accept it, but not everyone.

13 Why did it take so long for Darwin's idea to be accepted?

Causes of natural selection

You have seen that some animals are better at avoiding predators than others.

14 Write down <u>two</u> other reasons why some plants and animals are better able to survive than others.

Thousands of years ago, the animals that evolved into giraffes were not as tall as giraffes are now. Over a long period of time, the legs and necks of giraffes got longer and longer. They could reach leaves high in the trees, as well as reaching down for water.

15 Draw a diagram, similar to the one shown, to show how natural selection produced giraffes with such long legs and necks.

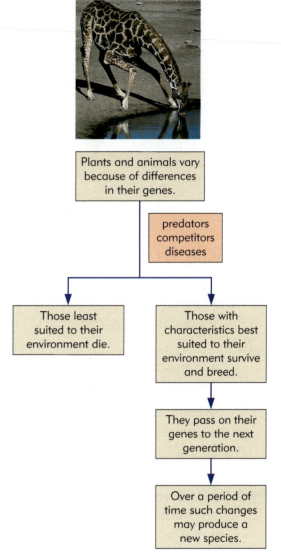

Plants and animals vary because of differences in their genes.

predators
competitors
diseases

Those least suited to their environment die.

Those with characteristics best suited to their environment survive and breed.

They pass on their genes to the next generation.

Over a period of time such changes may produce a new species.

How natural selection can produce a new species. (Darwin's idea of natural selection was very similar, except that he didn't know about genes.)

You need to be able to suggest why Darwin's theory of natural selection was only gradually accepted.

Higher

Ideas about evolution

In the eighteenth and nineteenth centuries scientists found out more and more about fossils. Many accepted that fossils were the remains of ancient plants and animals. So they wanted to know why some new ones appeared and why some became extinct. The Church taught that species did not change. A French scientist called Georges Cuvier suggested that a series of Noah's floods and creations explained the fossil record.

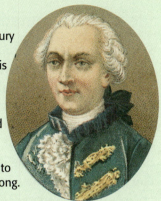

An eighteenth-century French naturalist called Georges Louis de Buffon thought that the Earth was much older than the Bible suggested and that species had changed. The Church forced him to say that he was wrong.

16 (a) What was Cuvier's explanation of new species and extinctions?
(b) Why weren't people willing to suggest other explanations at the time?

Lamarck observed artificial selection. He thought that something similar might happen in nature. He thought that an organ would change when it was necessary. He would have said that giraffes developed long necks because they needed them to reach up into the trees for food. He also thought that characteristics which an animal developed during its life, such as stronger muscles, would be passed on to its offspring. This is not true. So it was easy to challenge his other ideas.

Darwin and Wallace's explanation for change was harder to challenge. This was evolution by natural selection. Another name for it is 'the survival of the fittest'.

17 Lamarck and Darwin both suggested theories of evolution, but they explained how it happened in different ways. Write down the differences in their explanations.

Darwin thought that the animals or plants best suited to their environments would survive, breed and pass on their characteristics.

18 (a) Why do you think that so many people had ideas about evolution in the eighteenth and nineteenth centuries?
(b) Why were people's ideas different?

You need to be able to identify differences between different theories of evolution and to suggest reasons for different theories.

Ideas about mutation

Darwin knew about variation but not the causes. He didn't know about genes, alleles and mutations. Mutations produce new alleles. A few of these new alleles are useful, most are harmful and others seem to make no difference. But they do add to the range of alleles (the **gene pool**). They increase the number of variations on which natural selection can act. They may turn out to be useful if the environment changes. Useful mutations are ones that increase the chances of survival.

> **Remember**
>
> Mutations are changes in genes and chromosomes. Most mutations are harmful. For example, a mutation produces the sickle cell allele.

19 (a) Describe <u>two</u> examples of harmful effects of mutations in reproductive cells.
   (b) Explain why a neutral mutation might become useful.

Mutations of genes in body cells can also be harmful. Some of our genes control cell division. Some are like 'on' switches and make a cell start to divide. Others are like 'off' switches and stop cell division. Factors in the environment such as cigarette smoke and radiation can affect these **switch genes**. If the switch stays on, cells do not stop dividing when they should. They are out of control. This is **cancer**.

20 (a) Write down <u>one</u> difference between division of cancer cells and normal cells.
   (b) What do we call the change that makes a normal cell change into a cancer cell?

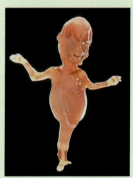

This fetus died because it was abnormal. A mutation in a reproductive cell caused this abnormality.

Discoveries about DNA

In 1859 a scientist called Friedrich Miescher was able to separate DNA from the cell nucleus. He called it **nuclein**. Then, in 1944, other scientists showed that this nuclein is our genetic material and that it controls what we are like. In 1953 Francis Crick and James Watson worked out the chemical structure of DNA. They based their work on discoveries made by chemists, physicists, biologists and mathematicians.

> **Information box**
>
> Chargaff found a pattern in the amounts of the 4 bases in the DNA molecule.
>
> amount of base A = amount of base T
> amount of base C = amount of base G
>
>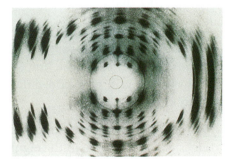
>
> Maurice Wilkins and Rosalind Franklin found that there was a repeating pattern in the structure of DNA.

Higher

21 Look at the information box on page 170. Write down <u>two</u> patterns that helped Crick and Watson to work out the structure of DNA and build a model of it.

22 Who discovered each of these patterns?

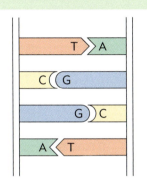

Think of a DNA molecule as a bit like a ladder. The base pairs are like the rungs of the ladder. They are either A–T or C–G.

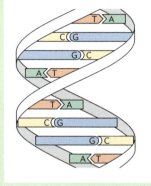

The shape of the DNA molecule turned out to be a double helix. So think of the ladder twisted into a spiral.

23 DNA is made of two strands linked together.
    (a) What is the shape of the molecule?
    (b) What links the two strands together?

There was still the problem of how DNA did its job. No one knew how DNA coded for the genetic information.

24 What discoveries did these people make about DNA:
    (a) Wilkins and Franklin?
    (b) Crick and Watson?

Cracking the genetic code

There are over 20 different amino acids. Cells combine them in different numbers and different orders to make different proteins. So we can make lots of different proteins. Our cells make proteins in tiny structures in the cytoplasm. They get their instructions in code from the DNA in the nucleus. Scientists know that one gene codes for one protein. What they needed to find out was the codes for the amino acids which make up proteins.

In 1961, Crick worked out the code.

25 It is possible to make hundreds of different proteins from just 20 amino acids. Explain why.

Higher

How does the code work?

- If <u>one</u> base codes for one amino acid, four bases (A,T,C,G) can only code for four amino acids.

- If <u>two</u> bases code for one amino acid, 16 different amino acid codes are possible.

- But there are more than 20 different amino acids.

- So a <u>three</u>-base code is needed. Each group of three bases is the code for one amino acid.

26 The diagram shows only 10 three-base codes. Work out at least 10 more for yourself.

If you wanted to, you could work out 64 different codes.

The order of these bases on the DNA molecule controls the sequence in which the amino acids are joined to make a protein. A change in the order of bases causes a change in the protein produced. This is a mutation.

27 This is the code for a small part of a protein molecule.

GTT ATG TGG TTT GTT

Write down the sequence of amino acids that this codes for.

28 (a) Write down the sequence of amino acids for this DNA: GTT/TGG/ATG/G
(b) One letter is lost. So it is now: TTT/GGA/TGG. Write down the new sequence of amino acids.
(c) What do we call this sort of change in our DNA?

If one base codes for one amino acid, four bases can only code for four amino acids.

If two bases code for each amino acid, sixteen different amino acid codes are possible.

	A	T	G	C
A	AA	AT	AG	AC
T	TA	TT	TG	TC
G	GA	GT	GG	GC
C	CA	CT	CG	CC

If three bases code for each amino acid, there are more possible codes than there are amino acids.

GCA	GCG	GTA	CAT	ATA
AAT	TAG	TTT	CAA	AAG

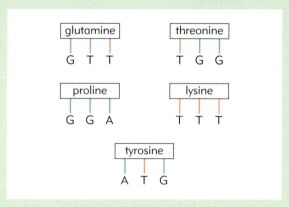

These are the codes for five amino acids.

You do not need to remember this story or the codes for different amino acids.
You do need to remember that DNA contains <u>four</u> different bases and that <u>three</u> bases code for one amino acid.

Drawing diagrams of chemical apparatus

You will find the information on these pages useful throughout Chapters 5 to 8 in this book.

When we do an experiment, we often need to draw a diagram. Adding a diagram to our written notes makes it easier to show what happened in the experiment.

Drawing diagrams of containers

The simplest container we can use is a **test tube**. We can heat solids or liquids in a test tube. For bigger volumes of liquid we use a **beaker** or a **flask**. We can use round flasks or conical flasks. We can use an **evaporating basin** to grow crystals from a solution.

Measuring the liquids we need

Most beakers have a scale marked on the outside. This gives a rough idea of the volume of liquid inside. We can use a **measuring cylinder** if we need to measure liquids more carefully. A **burette** lets us measure liquids very accurately.

Special apparatus that we use with gases

We can collect many kinds of gas in a **gas jar**. If we want to measure an amount of gas, we can use a **gas syringe** which has a scale on it.

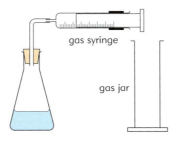

gas syringe

gas jar

Filtering mixtures to separate them

We need a **filter funnel** and some **filter paper** to separate sand from a solution. The solution goes through the filter paper but the sand does not. The job of the filter funnel is to support the filter paper.

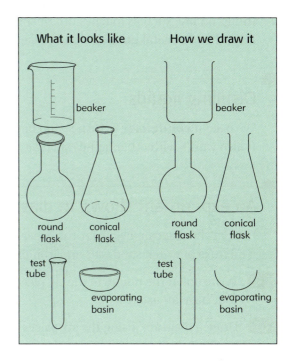

What it looks like   How we draw it

beaker   beaker

round flask   conical flask   round flask   conical flask

test tube   evaporating basin   test tube   evaporating basin

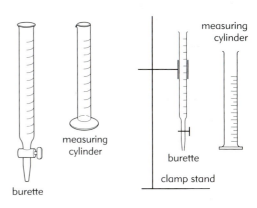

measuring cylinder

measuring cylinder

burette

clamp stand

burette

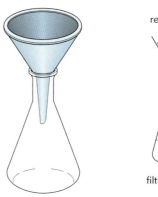
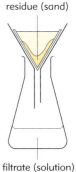

residue (sand)

filtrate (solution)

Making things hotter

Many experiments need a **Bunsen burner** to make them work. We can use the gas burner to heat test tubes or to set things on fire, such as magnesium ribbon. We also need a **tripod** and a metal **gauze** to hold beakers or flasks.

Distilling liquids

We need a **condenser** to turn steam back into water. It also works with other liquids such as alcohol.

An experiment shown in diagrams

Look at the diagram below for the experiment to separate a mixture of water, sand, salt and iron filings.

1 Write down the names of the <u>six</u> pieces of apparatus that have an asterisk (*).

2 Draw a diagram, using the six pieces of apparatus from question 1, to show how you could produce salt and water from a salt solution that also contained sand.

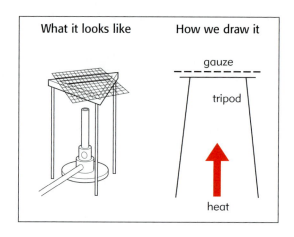

What it looks like | How we draw it
gauze
tripod
heat

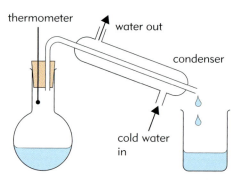

thermometer
water out
condenser
cold water in

This is how we draw the distilling apparatus.

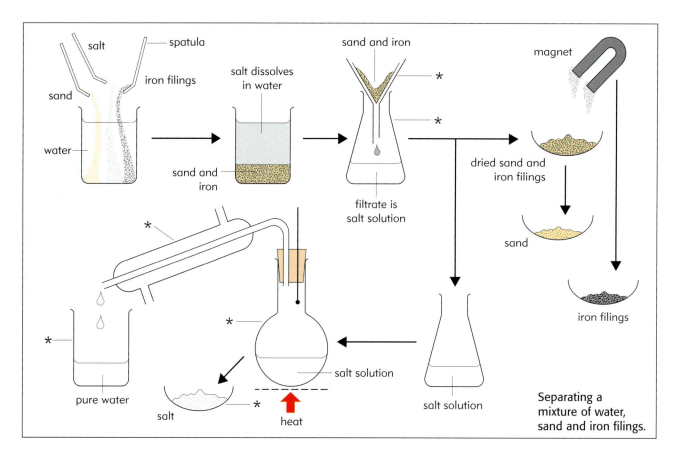

salt — spatula
iron filings
sand
water

salt dissolves in water
sand and iron

sand and iron *
*
filtrate is salt solution

magnet

dried sand and iron filings

sand

iron filings

*
pure water
*
salt solution
*
salt
heat

salt solution

Separating a mixture of water, sand and iron filings.

Knowing when to be careful

We have to be careful when we use some substances. This is because the substances may harm us in some way. When we use harmful substances, we need to look for warnings on the labels. The labels usually have **hazard symbols**. There are different hazard symbols for substances that harm us in different ways.

You need to be able to handle chemicals safely throughout your Science course. Your knowledge of hazard symbols will be assessed with your work in Chapter 8.

Starting a fire

Some substances catch fire very easily. We call these **highly flammable** substances.

1. The label on a bottle of methylated spirits says:

   Keep container tightly closed. No smoking.

   Why does the label say this?

highly
flammable

This symbol appears on labels for substances like methylated spirits.

Helping it burn more fiercely

When things burn they use up oxygen from the air. Some substances contain oxygen, which lets other materials burn even better. We say that substances like this are **oxidising** substances.

2. Sodium chlorate is a strong weedkiller. It can kill all the weeds on a garden path.

   A gardener might store sodium chlorate in a garden shed near dry sacks and wood. Why is this a bad idea?

oxidising

This symbol appears on labels for substances like sodium chlorate.

Don't take chances with these materials

Pirate flags used to have a skull and crossbones. Pirates were dangerous and could kill you! Substances that can kill you are called **toxic** substances.

We use <u>tiny</u> amounts of chlorine to kill dangerous bacteria in our drinking water. We also use chlorine to treat the water in swimming pools. But if you breathe in a <u>lot</u> of chlorine it can kill you.

3. Why do we add a little chlorine to the water in a swimming pool?

toxic

This symbol appears on labels for substances like chlorine.

Still bad, but at least it won't kill you

Some substances are **harmful**, but they are not as dangerous as toxic materials.

Copper sulphate forms beautiful blue crystals, but if you swallow solid copper sulphate or some of its solution, it is harmful.

4 A student used copper sulphate to grow crystals. Why should she wash her hands before eating food?

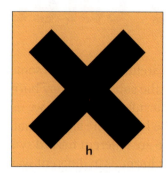

harmful

This symbol appears on labels for substances like copper sulphate.

There is more than one way to get burned

If you touch the inside of a hot oven you will burn yourself. There are chemicals that can destroy your skin and damage your eyes. We call these **corrosive** materials. Corrosive materials can give skin burns.

5 Sulphuric acid is used in experiments and to fill up car batteries. Why should you use safety glasses when using sulphuric acid?

corrosive

This symbol appears on labels for substances like sulphuric acid.

Some substances are irritants

Some substances can make your skin go red or form blisters. If the substance is a dry powder it may make you cough. We call substances like this **irritants**. They are less dangerous than corrosive materials but you must still take care.

Copper carbonate is a beautiful green colour. We can use copper carbonate to make copper metal. If you spill the green powder and breathe it in, it can make you cough. We call the powder an irritant.

irritant

This symbol appears on labels for substances like copper carbonate.

Acids, alkalis and neutralisations

Remember from Key Stage 3

Acids

All **acids** dissolve in water to give colourless and corrosive solutions. Strong acids will harm most living tissue and dissolve or corrode most metals and rocks.

1 Car batteries contain sulphuric acid. This is a very strong acid. Write down <u>two</u> reasons why you must take care not to spill it.

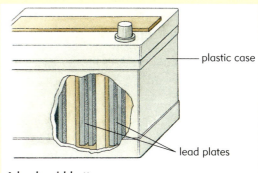

A lead–acid battery.

Alkalis – the opposites of acids

<u>Alkalis</u> also dissolve in water to form colourless solutions.

Like acids, strong alkalis attack living tissue. But unlike acids, they would turn bits of you to soap. Many powerful oven cleaners contain strong alkalis.

This warning sign tells you that a substance is corrosive.

2 Why is it important to wear safety spectacles or goggles when using alkalis?

3 What other protection do you need when using an oven cleaner?

Not an acid, not an alkali

Water is not an acid and not an alkali. Water is <u>neutral</u>.

When a substance dissolves in water, it makes an <u>aqueous</u> solution. Aqueous means 'watery'. We have seen that we can have aqueous solutions of acids and alkalis. Salt, sugar, alcohol and many other substances dissolve in water to give solutions that are not acidic or alkaline. They are neutral.

4 Look at the three liquids. Explain why you cannot tell which is which just by looking.

The labels have fallen off these three bottles.

Indicators

Indicators are dyes that change colour with acids and alkalis. They tell us whether the solution is <u>acidic</u> or <u>alkaline</u> or neutral.

5 What colour does litmus give with:
(a) an acid?
(b) an alkali?

6 What colour does litmus give with a neutral liquid like water?

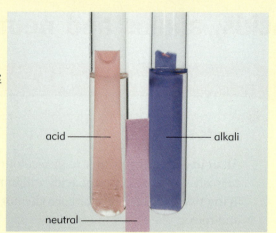

Using litmus as an indicator.

How strong is my acid or alkali?

Strong acids have a pH of 0–1

Neutral solutions have a pH of 7

Strong alkalis have a maximum pH of 14

0 1 2 3 4 5 6 7 8 9 10 11 12 13 14

← increasingly acidic ——————— increasingly alkaline →

We use a scale of numbers called the <u>pH</u> **scale** (the 'pee-aitch' scale) to tell us how strong an acid or alkali is.

7 Which substance in the table has:
(a) the highest pH number?
(b) the lowest pH number?

Substance	pH
ammonia cleaning liquid	11.5
blood	7.5
coffee	5
liquid X	7
liquid Y	8.5
liquid Z	4
orange juice	3
oven cleaner	14
stomach acid	1.5
urine	6

Universal indicator – a chemical rainbow

Universal indicator is a mixture of dyes. Each dye changes colour at a different pH so the mixture gives us different colours as we go through the pH range. We add the indicator in drops, so the chemical we test must be colourless for us to see the proper colour change.

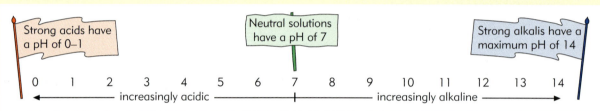

Colour	red		orange		yellow		green	blue		navy blue			purple	
pH	0	1	2	3	4	5	6	7	8	9	10	11	12	13	14

increasingly acidic ←   neutral   → increasingly alkaline

8 What colour would universal indicator turn if you put it into:
(a) ammonia cleaning liquid?
(b) liquid Y?
(c) stomach acid?

Adding acid to alkali

The diagram shows what happens as you add more and more <u>acid</u> to an <u>alkali</u>.

9 What colour is the indicator in flask B?

10 Is the solution in flask B acidic, alkaline or neutral?

If you add just the right amount of acid to an alkali you get a solution that is neutral. We say that the acid and alkali **neutralise** each other. We call the reaction between an acid and an alkali **neutralisation**.

11 The solution in flask C is acidic. Explain why.

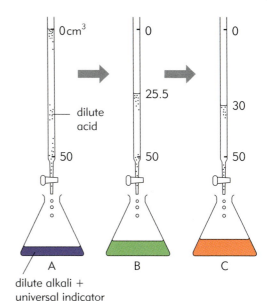

A   B   C

dilute alkali + universal indicator

What happens to the acid and the alkali during neutralisation?

When you mix some acid with just the right amount of alkali you get a neutral solution. But all the particles from the acid and alkali are still there. They have reacted to make new substances. The diagram shows how you can obtain some of the new substance made in a neutralisation reaction. The reaction also produces more water.

12 Write a word equation for the reaction between sodium hydroxide and hydrochloric acid.

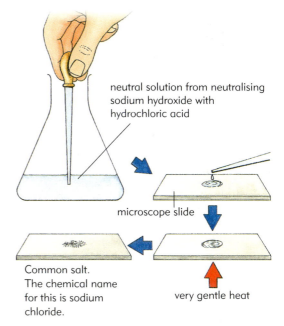

neutral solution from neutralising sodium hydroxide with hydrochloric acid

microscope slide

Common salt. The chemical name for this is sodium chloride.

very gentle heat

How do you make different kinds of salt?

Sodium chloride is the salt you put on your food. But it isn't the only kind of salt. Whenever you neutralise any acid with any alkali you get a **salt** and water.

When you neutralise hydrochloric acid, the salt you make is always a **chloride**. The salt takes the first part of its name from the metal in the alkali you use.
So neutralising sodium hydroxide with hydrochloric acid gives you **sodium** chloride.

13 What salt do you get if you neutralise potassium hydroxide with hydrochloric acid?

The salts of nitric acid are **nitrates**.
The salts of sulphuric acid are **sulphates**.

14 Write a word equation for the reaction between potassium hydroxide and sulphuric acid.

15 Write a word equation for a reaction that makes sodium nitrate and water.

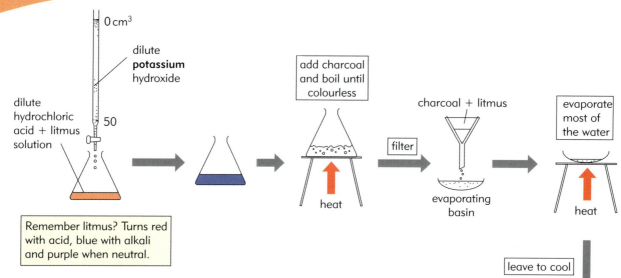

0 cm³

dilute **potassium** hydroxide

dilute hydrochloric acid + litmus solution

50

add charcoal and boil until colourless

charcoal + litmus

evaporate most of the water

filter

heat

evaporating basin

heat

Remember litmus? Turns red with acid, blue with alkali and purple when neutral.

leave to cool

pure potassium chloride

16 The diagram shows how to make potassium chloride.
(a) Why is litmus added to the acid?
(b) Why is the neutral solution boiled with charcoal and then filtered?

Higher

What makes a solution acidic or alkaline?

Acidic solutions are acidic because they contain hydrogen atoms that have a positive (+) electrical charge. These are called **hydrogen ions**. The symbol for a hydrogen ion is H^+.

Alkaline solutions are alkaline because they contain **hydroxide ions**. These have a negative (−) electrical charge. The symbol for a hydroxide ion is OH^-.

When an acidic solution and an alkaline solution are mixed, each hydrogen ion joins up with a hydroxide ion to form a **water** molecule.

In acid and alkali neutralisation reactions the hydrogen ions from the acid are turned into water by the hydroxide ions from the alkali. Showing only the ions that combine, you can write it like this:

$$H^+(aq) + OH^-(aq) \longrightarrow H_2O(l)$$

17 What is a hydrogen ion and what is the symbol for it?

What is a hydroxide ion and what is the symbol for it?

18 Use the idea of hydrogen ions and hydroxide ions to suggest an explanation for the difference:

19 (a) between solutions with pH 1 and pH 5;
(b) between solutions with pH 8 and pH 13.

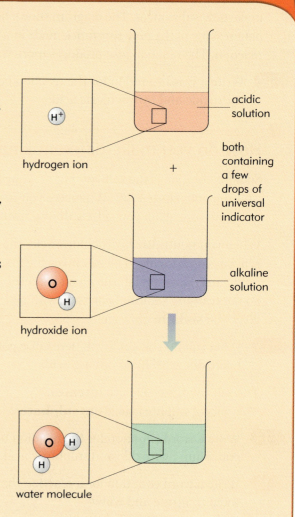

acidic solution

H^+

hydrogen ion

both containing a few drops of universal indicator

+

O −
H

hydroxide ion

alkaline solution

O H
H

water molecule

What's special about metals?

We use **metals** for lots of different things.
Our lives wouldn't be the same without them.
It is not always easy to tell whether something is a
metal or not. We have to look at lots of its <u>properties</u>.

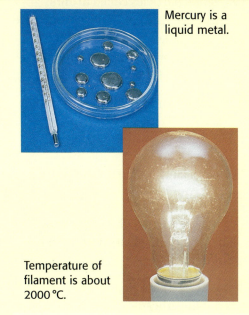

Mercury is a liquid metal.

Metals don't usually melt easily

Metals are usually <u>solids</u>. Only one metal is a liquid at
room temperature, but we can melt all metals if we heat
them enough. Metals usually have <u>high melting points</u>.
Look at the photographs and the table.

1. (a) Which metal is a liquid at room temperature?
   (b) Write down <u>one</u> use for this liquid.

2. Which metal is used to make lamp filaments?

3. Which metal has a low melting point, but is a solid at
   room temperature (20 °C)?

Temperature of filament is about 2000 °C.

Metal	Melting point in °C
mercury	−39
gold	1063
iron	1535
sodium	98
tungsten	3410

Metals let heat and electricity pass through them

Both <u>heat</u> and <u>electricity</u> flow easily through all metals.
We say that metals are good <u>conductors</u> of heat and
electricity. This is a good way to tell the difference
between metals and non-metals. If we know that a
substance conducts electricity, then we are almost sure
that it is a metal. The only non-metal that conducts
electricity well is a type of carbon called **graphite**.

4. Is the substance in the diagram a metal?
   Give a reason for your answer.

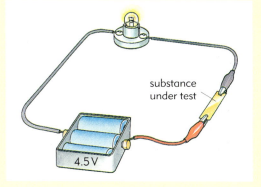

substance under test

4.5 V

Metals are usually strong and tough

Most metals are <u>strong</u>. They can hold large weights
without snapping. Metals are <u>tough</u>. They do not shatter
easily. They do not crack when we hit them or squeeze
them. But we can force a metal to bend into a different
<u>shape</u>. This is easier if the metal is thin.

5. Why do tall buildings have steel frameworks?

6. Look at the diagram. Why is steel useful for making
   car body parts?

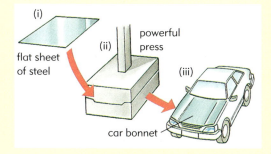

(i) flat sheet of steel

(ii) powerful press

(iii)

car bonnet

Some metals are elements

Metals, like everything else, are made of very small particles called **atoms**.
A substance that contains just one type of atom is called an **element**.
For example, iron is an element as it contains only iron atoms.

7 Write down the names of <u>two</u> other metals that are elements.

How many metal elements are there?

The table shows all of the different elements we find in the natural world around us.

8 (a) How many elements are there altogether?
(b) How many of these elements are metals?
(c) Would you say that about a quarter, about a half or about three-quarters of the elements are metals?

A table of elements set out in this way is called the **Periodic Table**.

9 What do you notice about where the non-metals and metals are in this table?

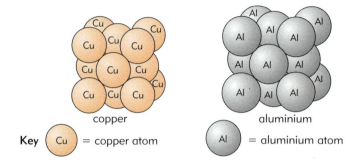

copper   aluminium

Key (Cu) = copper atom   (Al) = aluminium atom

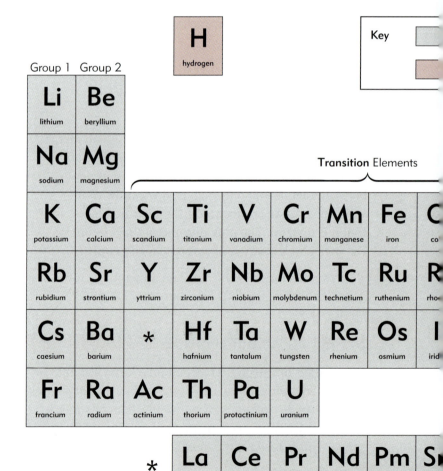

The Periodic Table

Alloys – mixtures of metals

We don't use just the metals that are elements. We also mix metals together to get the properties we want to use. These mixtures of metals are called **alloys**.

Alloys are usually harder and stronger than the metals from which they are made. For example, aluminium is often alloyed with magnesium.

10 How is the aluminium alloy different from aluminium?

						Group 0
						He helium
Group 3	Group 4	Group 5	Group 6	Group 7		
B boron	C carbon	N nitrogen	O oxygen	F fluorine		Ne neon
Al aluminium	Si silicon	P phosphorus	S sulphur	Cl chlorine		Ar argon

Ni nickel	Cu copper	Zn zinc	Ga gallium	Ge germanium	As arsenic	Se selenium	Br bromine	Kr krypton
Pd palladium	Ag silver	Cd cadmium	In indium	Sn tin	Sb antimony	Te tellurium	I iodine	Xe xenon
Pt platinum	Au gold	Hg mercury	Tl thallium	Pb lead	Bi bismuth	Po polonium	At astatine	Rn radon

metals

non-metals

Eu europium	Gd gadolinium	Tb terbium	Dy dysprosium	Ho holmium	Er erbium	Tm thulium	Yb ytterbium	Lu lutetium

Using metals

Most of the metals that you meet in everyday life are **transition elements**. Two transition metals that we use a lot are iron and copper. Iron is usually turned into steel before we use it.

Using copper

Here are some facts about copper:

- copper is easy to shape into pipes and wires;
- copper pipes and wires are easy to bend;
- copper is a better conductor of heat and electricity than most other metals;
- copper does not corrode as quickly as iron or steel;
- copper is a fairly expensive metal.

1 (a) Write down <u>two</u> uses for copper in everyday life.
(b) In each case give a reason for using copper.

Some of the ways we use copper.

Electrical cables.

Water pipes.

Using steel

We use more steel than any other metal. Millions of tonnes are used every year. Steel is strong, tough and easily shaped. It is also cheaper than most other metals. Many structures, including vehicle bodies, are made of steel.

2 Look at the picture. Write down <u>five</u> things in the picture that are usually made from steel.

3 Why is steel used so much as a structural material?

steel supports

steel railway lines

Using aluminium

Another metal that we use a lot of is aluminium. It is not a transition metal, but it is still really useful. Aluminium is a very lightweight metal, and it does not easily corrode. These two properties make aluminium an important metal.

But pure aluminium is weak and soft and easy to bend. This means that we can't use it for many jobs. We can mix aluminium with another light metal called magnesium. We get an alloy that is stronger, harder and stiffer than aluminium. Aluminium alloy is used in aircraft construction.

Aluminium is a very good conductor of electricity. It isn't quite as good as copper but it is much lighter than copper. So it is used for overhead power cables. Steel is much stronger than aluminium, but doesn't conduct electricity so well.

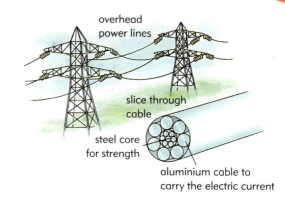

overhead power lines
slice through cable
steel core for strength
aluminium cable to carry the electric current

4 What are overhead power cables made from
(a) on the outside?
(b) in the centre?

5 Why are the cables made like this?

The alkali metals

The metals lithium, sodium and potassium are all very much alike. So we say that these elements are all part of the same **chemical family**. We call this family the **alkali metals**.

The diagram shows part of the Periodic Table of elements.

1 The alkali metals are all in the same Group of the Periodic Table. Which Group is it?

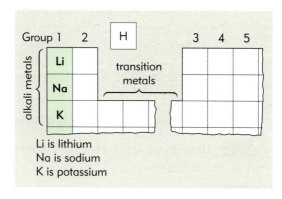

Group 1 2 H 3 4 5
alkali metals
Li
Na
K
transition metals

Li is lithium
Na is sodium
K is potassium

What are the alkali metals like?

Alkali metals are like other metals in many ways. There are also some differences.

2 Write down <u>two</u> ways in which alkali metals are the same as other metals.

3 Write down <u>three</u> ways in which alkali metals are different from other metals.

Alkali metals are also very **reactive**. They are so reactive that we have to store them under oil, away from the air and water. The properties of alkali metals mean that they are unsuitable to use for making things like pans, cars and bridges.

4 Write down <u>two</u> properties of alkali metals that make them unsuitable as structural materials.

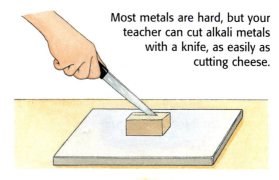

Most metals are hard, but your teacher can cut alkali metals with a knife, as easily as cutting cheese.

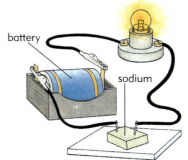

battery
sodium

Like other metals, alkali metals conduct electricity and heat, but they melt more easily than most other metals.

Potassium, like lithium and sodium, is lighter (less dense) than other metals. It is so light that it floats on water. The potassium darts about as it reacts with the water, making it fizz.

Why do we call them alkali metals?

Alkali metals all react very fast with cold water.
They fizz and move around on the water as they react.
A gas called hydrogen is produced.

This is the word equation for the reaction of potassium
with water:

$$\text{potassium} + \text{water} \longrightarrow \text{hydrogen} + \text{potassium hydroxide}$$

The potassium hydroxide dissolves in the water as
the potassium reacts. Potassium hydroxide solution
is alkaline.

The diagrams show what happens when sodium reacts
with water.

5 Why does the water fizz as it reacts with the sodium?

6 How do we know that the colourless solution
is alkaline?

7 Sodium reacts with cold water in the same sort of way
as potassium. Write a word equation for this reaction.

8 Sodium gives an alkali with water.
What is the name of this alkali?

All alkali metal hydroxides dissolve in water to give
alkaline solutions. This is why we call these metals the
'alkali metals'.

The sodium moves
about on top of the
water, making it fizz.
A colourless solution
is left behind.

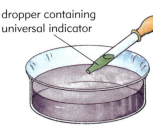

dropper containing
universal indicator

The indicator turns
purple, which shows
that the solution is
alkaline.

Do alkali metals react with other elements?

Not all of the elements are metals. Oxygen is an example of
a **non-metal**. Some non-metals will react with metals to
make compounds.

Hot sodium metal reacts violently with oxygen gas.
This is the word equation for the reaction:

$$\text{sodium} + \text{oxygen} \longrightarrow \text{sodium oxide}$$

The other alkali metals react with oxygen in the same way.

9 Write down the word equation for the reaction between
lithium and oxygen.

Alkali metals also react with the family of non-metal
elements called the **halogens** in Group 7 of the
Periodic Table.

The transition metals

Looking at transition metal compounds

Like alkali metals, nearly all of the transition metals will react with non-metal elements to form compounds.

For example, they react with oxygen to form **oxides** and with chlorine to form **chlorides**. The diagram shows some alkali metal compounds and some transition metal compounds.

1 Write down the differences that you notice between alkali metal compounds and transition metal compounds.

Alkali metal compounds

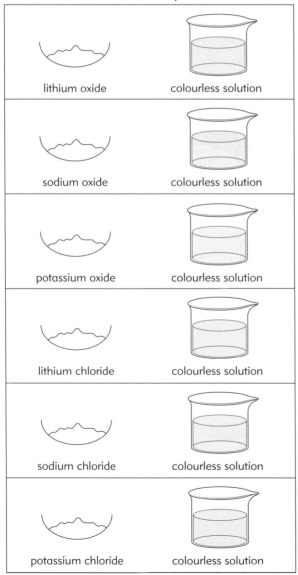

lithium oxide	colourless solution
sodium oxide	colourless solution
potassium oxide	colourless solution
lithium chloride	colourless solution
sodium chloride	colourless solution
potassium chloride	colourless solution

Transition metal compounds

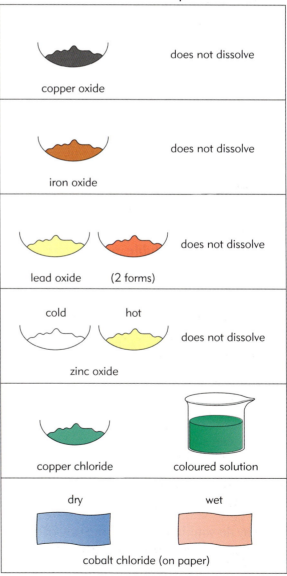

copper oxide — does not dissolve

iron oxide — does not dissolve

lead oxide (2 forms) — does not dissolve

cold   hot
zinc oxide — does not dissolve

copper chloride — coloured solution

dry   wet
cobalt chloride (on paper)

Using transition metal compounds

Because they are coloured, we use transition metal compounds to colour glass. We also use them to produce coloured glazes on pottery.

When copper is used on the roof of a building it eventually goes green. This happens because of the copper compounds produced when copper reacts with substances in the air.

2 Why are transition metals used as glazes for pottery?

3 Why do the roofs of some buildings turn green?

The copper compound that gives this roof its green colour is called verdigris.

Another use for transition metals

Some transition metals can also be used to speed up chemical reactions. Substances that are used in this way are called **catalysts**.

The Haber process for making ammonia from nitrogen and hydrogen uses an iron catalyst. To make nitric acid from the ammonia, it is first reacted with oxygen using a platinum catalyst.

4 What two transition metals are used as catalysts in the production of nitric acid?

Another way of making salts

Transition metal oxides and hydroxides do not dissolve in water. They are **insoluble** in water. So they can't be used to make alkaline solutions. This means that you can't use acid + alkali reactions to make salts of transition metals.

> ### Remember
>
> You can make an alkali metal salt using a neutralisation reaction:
>
> acid + alkali ⟶ salt + water
>
> You can use an indicator to tell when an acid and an alkali have neutralised each other.

How can you make transition metal salts?

Even though they do not dissolve in water, transition metal oxides (or hydroxides) can still neutralise acids to make salts. The diagram shows a reaction of this kind.

1 Why can copper oxide not be used to make an alkaline solution?

2 How do you know that the copper oxide reacts with the hydrochloric acid?

3 Why do you think that the mixture of copper oxide and hydrochloric acid is heated and stirred?

4 What two new substances are produced in this reaction?

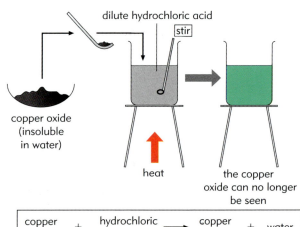

dilute hydrochloric acid

stir

copper oxide (insoluble in water)

heat

the copper oxide can no longer be seen

| copper oxide | + | hydrochloric acid | ⟶ | copper chloride | + | water |

A new name for anti-acids

All substances that react with acids to produce a salt and water are called **bases**.

So we can say:

acid + base ⟶ salt + water

Alkalis are bases that dissolve in water. They are **soluble** bases.

5   The diagram shows another reaction between an acid and a base.

Write down a word equation for this reaction.

Remember

Neutralising hydrochloric acid ⟶ a chloride
Neutralising nitric acid ⟶ a nitrate
Neutralising sulphuric acid ⟶ a sulphate

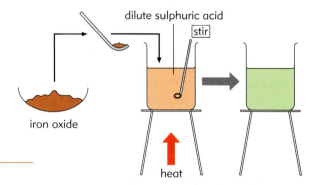

dilute sulphuric acid
stir
iron oxide
heat

How can we tell when all the acid has been neutralised?

When we neutralise an acid with an alkali, we use an **indicator**. This tells us when all the acid has been neutralised so that we add just the right amount of alkali.

When we neutralise an acid with an insoluble base, we don't need an indicator. The diagram explains why.

6   (a) How do you know when all the acid has been neutralised?
     (b) What do you then do to separate the salt solution from any insoluble metal oxide that is left over?
     (c) How can you get <u>solid</u> salt from the salt solution?
     (d) What is the name of the salt produced from the acid and base used in the reaction shown in the diagram?

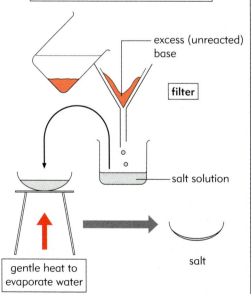

lead oxide (insoluble base)
stir
nitric acid
heat

Keep adding the insoluble base until no more will **react**, this tells you that all of the acid has been neutralised.

excess (unreacted) base
filter
salt solution
gentle heat to evaporate water
salt

Alkalis and salts that don't contain metals

Ammonia is a gas made from nitrogen and hydrogen. It dissolves in water to produce an alkaline solution called **ammonium hydroxide**.

ammonia + water ⟶ ammonium hydroxide
(gas)   (liquid)   (solution)

Ammonium hydroxide can be used to neutralise an acid and produce an **ammonium salt**.

7   The word equation shows how you can make ammonium chloride.

Write down a word equation to show how you can make ammonium nitrate.

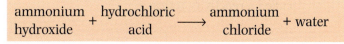

ammonium hydroxide + hydrochloric acid ⟶ ammonium chloride + water

Comparing the reactivities of metals

Burning metals

Many metals burn. But some metals burn more easily than others.

Magnesium burns quickly in the oxygen in the air with a brilliant white flame. The picture shows how we can use this white light.

1 Write the word equation for the reaction where magnesium burns in oxygen.

2 When you burn magnesium, a white powder is left behind. What is it?

Zinc doesn't burn as easily as magnesium. The photographs show how you can make it burn.

We can heat up iron filings in the same way as the zinc powder and put them into oxygen. They may glow red-hot and produce a few sparks. Iron oxide powder is produced. Iron burns much less easily than zinc.

If we do the same thing with copper powder there is very little reaction. It just glows a little. Afterwards we see that the surface of the copper has changed from brown to black copper oxide.

Magnesium is used in flares.

zinc powder

zinc oxide is produced

oxygen

Making zinc oxide.

Putting the metals into order

Magnesium burns easily. We say that it is very **reactive**. Zinc does not react so easily with oxygen. It is less reactive than magnesium.

3 Write down the metals copper, iron, magnesium and zinc in order. Start with the most reactive and end with the least reactive.

A list of metals in order of their reactivity is called a reactivity series.

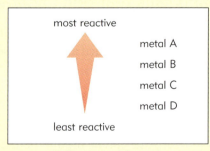

most reactive

metal A
metal B
metal C
metal D

least reactive

How to make a reactivity series.

Reacting metals with water

The diagrams show calcium and magnesium reacting with water.

4 Which metal reacts faster with water?

If you put some zinc into cold water you cannot see any reaction.

The reaction between sodium and water is so fast that it is dangerous. Potassium reacts even more quickly with water. (See pages 185–186.)

5 Using the information above, add calcium, sodium and potassium to your reactivity series.

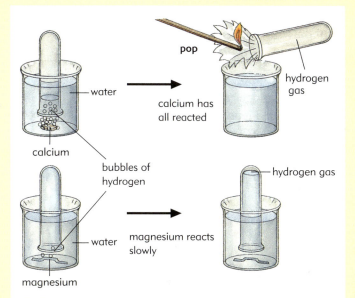

Reacting metals with dilute acids

The diagrams show how four different metals react with dilute acid. The metals are all in powder form.

Silver is less reactive than copper and gold is less reactive still.

6 Using the above information, add silver and gold to your reactivity series.

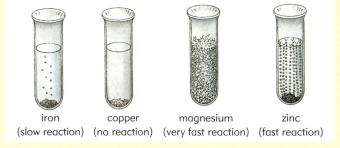

iron
(slow reaction)

copper
(no reaction)

magnesium
(very fast reaction)

zinc
(fast reaction)

Word equations

Instead of describing a chemical reaction in sentences, you can use a **word equation**.

Here is the word equation for the reaction between magnesium and hydrochloric acid:

magnesium + hydrochloric acid $\longrightarrow$ magnesium chloride + hydrogen

Another way of writing chemical equations

Instead of writing the names of all the elements in chemical compounds we can use their symbols. We then get a symbol equation. For example:

$$Mg(s) \quad + \quad 2HCl(aq) \quad \longrightarrow \quad MgCl_2(aq) \quad + \quad H_2(g)$$

| magnesium (a solid) | hydrochloric acid (aqueous – this means 'dissolved in water') | magnesium chloride (aqueous) | hydrogen (a gas) |

(Don't worry, for now, about what the numbers mean.)

Competing metals

The photographs show what happens when you put an iron nail into copper sulphate solution.

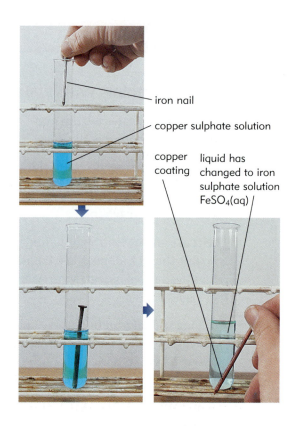

iron nail

copper sulphate solution

copper coating

liquid has changed to iron sulphate solution $FeSO_4(aq)$

7 Look at the photographs and answer these questions.
(a) What happens to the iron nail when it is dipped into copper sulphate solution?
(b) What happens to the copper sulphate solution when an iron nail is dipped into it?
(c) How can you tell that the copper sulphate solution has changed?

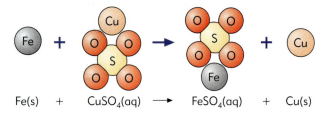

$$Fe(s) \quad + \quad CuSO_4(aq) \quad \longrightarrow \quad FeSO_4(aq) \quad + \quad Cu(s)$$

8 Write down a word equation for this reaction.

Iron is more reactive than copper. The iron pushes the copper out of the copper sulphate solution. We say that the iron has **displaced** the copper.

Which metals compete best?

A more reactive metal usually displaces a less reactive metal from a solution of one of its compounds.

9 Look at the photographs.
Which is the more reactive metal?

10 What word is used to say what magnesium does to the copper from the copper sulphate solution?

11 Write a word equation to describe this reaction.

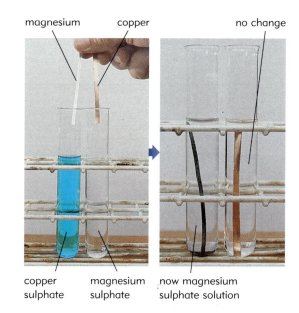

magnesium   copper   no change

copper sulphate   magnesium sulphate   now magnesium sulphate solution

Using the reactivity series

You can use the reactivity series to predict whether a displacement reaction will happen.

12 Copy the table. Use the short reactivity series to decide whether a displacement reaction will happen. Fill in the missing ticks and crosses.

Put a tick if you think a reaction will happen and a cross if you think it won't.

most reactive

magnesium	Mg
iron	Fe
copper	Cu

least reactive

	Magnesium sulphate solution	Iron sulphate solution	Copper sulphate solution
Mg			✓
Fe			✓
Cu	✗		

Pushing a metal out of its solid compound

Aluminium is more reactive than iron. So if you heat up aluminium powder with iron oxide there is a reaction.

$$2Al(s) + Fe_2O_3(s) \longrightarrow Al_2O_3(s) + 2Fe(l) + energy$$
$$[l = liquid]$$

13 Write a word equation for this reaction.

When the reaction starts, it makes so much heat that the iron is melted. This reaction is used to weld together railway lines. The welders build a small mould around the gap between the railway lines. The molten iron runs down into the gap and welds the two sections together.

14 The welders who carry out this reaction stand back at a distance from the reaction and wear thick heatproof gloves. Why is this?

A non-metal that can push out metals

Carbon isn't a metal, but if we heat it up, it can displace some metals from their oxides. Some metals can also be displaced from their heated oxides by passing **hydrogen** over them.

Even though carbon and hydrogen are not metals, they will displace some metals from their oxides. So we can put them into the reactivity series of metals. Carbon and hydrogen will only displace metals that are below them in the reactivity series.

15 Will carbon displace:
  (a) aluminium from aluminium oxide?
  (b) iron from iron oxide?
  (c) copper from copper oxide?

16 Which of the metals in question 15 will hydrogen displace from its oxide?

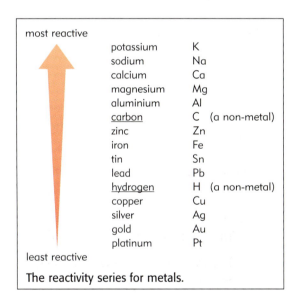

most reactive

potassium	K
sodium	Na
calcium	Ca
magnesium	Mg
aluminium	Al
carbon	C (a non-metal)
zinc	Zn
iron	Fe
tin	Sn
lead	Pb
hydrogen	H (a non-metal)
copper	Cu
silver	Ag
gold	Au
platinum	Pt

least reactive

The reactivity series for metals.

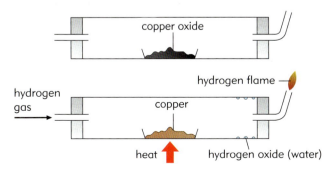

Using hydrogen to displace a metal.

Where do metals come from?

Where do we find metals?

We find metals mixed with rocks in the Earth's **crust**. We find gold in the Earth's crust as the metal itself. The pieces of gold in rocks contain just gold and nothing else. You can collect lots of small pieces of gold, heat them until they melt and then pour the molten gold into a mould. The gold sets hard as it cools. Gold is a very rare metal. Many other metals are much more common than gold.

1 Which are the two most common metals in the Earth's crust?

2 Why don't we show gold in the pie chart?

We find most metals, including iron and aluminium, as metal **ores**. In the ore, the metal is joined with other elements. Metals are often joined with oxygen in compounds we call metal **oxides**. For example, most iron ores contain iron oxide. Metals may also be joined with sulphur in compounds we call metal **sulphides**.

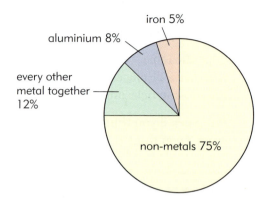

Elements in the Earth's crust. Gold makes up only three parts in every billion (thousand million) parts of the Earth's crust.

Looking at ores

Three different metal ores are called haematite, galena and malachite. The table shows some information about these ores.

3 Which ore contains the elements lead and sulphur?

4 Which ore can be used to produce:
(a) iron metal?
(b) copper metal?

Name of the ore	Main substances in the ore	Main elements in the ore
Haematite	Iron oxide	Iron, Oxygen
Galena	Lead sulphide	Lead, Sulphur
Malachite	Copper carbonate	Copper, Carbon, Oxygen

How much metal is there in metal ores?

Metal ores contain rock as well as the valuable metal compounds. Different ores contain different amounts of rock.

5 Look at the charts about iron ore, nickel ore and copper ore. How much metal compound is there in 100 g of each of these ores?

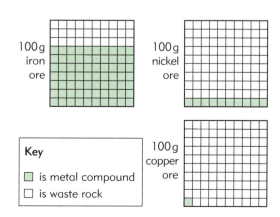

Many rocks contain metals, usually in the form of metal compounds. But most rocks are no use as ores. If a rock does not contain enough metal it costs too much to extract the metal. We say that it is **uneconomic** to use the rock as an ore.

An ore that only contains a small amount of metal may still be worth mining if the metal is valuable enough. The more valuable a metal is, the less there needs to be in rocks to make them worth using as ores.

Gold is worth mining even if there are only 10 parts in every million. This is about the same amount as two wedding rings in a truck-load of rock.

Some ores contain only a small amount of metal compound. This needs to be separated from the rest of the rock before the metal can be extracted from it. The diagram shows one way of doing this.

The metal can then be extracted from the **concentrated** ore.

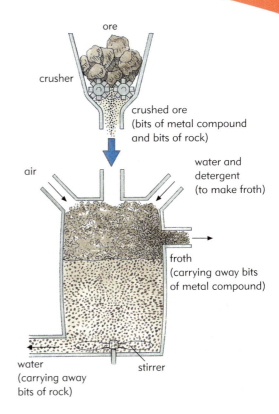

ore

crusher

crushed ore (bits of metal compound and bits of rock)

water and detergent (to make froth)

air

froth (carrying away bits of metal compound)

water (carrying away bits of rock)

stirrer

How can we extract metals from their ores?

To get pure metals from ores you must split up the metal compound in the ore. You can release, or **extract**, some metals by heating the metal oxide with carbon. We can extract copper by heating copper oxide with charcoal, a form of carbon. The charcoal reacts with the oxygen in the copper oxide. This leaves copper metal.

6 (a) What other substance is produced?
(b) Write a word equation for this reaction.

Removing oxygen from a metal oxide is called **reduction**. So carbon has **reduced** the copper oxide. Reducing iron oxide with carbon needs a much higher temperature than for copper. Aluminium oxide cannot be split using carbon. Aluminium oxide can be reduced to aluminium metal only by using electricity.

7 Use the reactivity series to explain why:
(a) aluminium oxide cannot be reduced using carbon;
(b) gold is found as the metal itself in the Earth's crust.

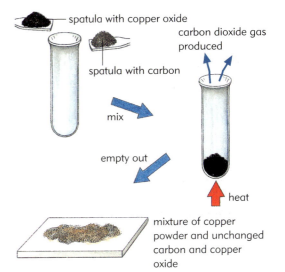

spatula with copper oxide

carbon dioxide gas produced

spatula with carbon

mix

empty out

heat

mixture of copper powder and unchanged carbon and copper oxide

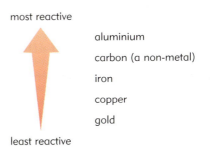

most reactive

aluminium

carbon (a non-metal)

iron

copper

gold

least reactive

Where does steel come from?

Steel is a strong and tough material. It is also cheap to make. Steel is mostly iron, so to make steel, we must first extract iron from iron ore. The commonest iron ore is called **haematite**.

How can you get iron from iron ore?

We extract iron from iron ore in a **blast furnace**.

8 Look at the diagram of the blast furnace.
   (a) Write down the <u>three</u> things that go into the top of the blast furnace.
   (b) What goes into the blast furnace near to the bottom?

Why is coke needed in the blast furnace?

First the coke reacts with oxygen from the air to make carbon dioxide gas. This reaction releases lots of heat.

$$C(s) + O_2(g) \longrightarrow CO_2(g) + energy$$

9 Write down a word equation for this reaction.

Next the carbon dioxide reacts with hot coke to give carbon monoxide gas.

$$CO_2(g) + C(s) \longrightarrow 2CO(g)$$

10 Write down a word equation for this reaction.

11 What are the <u>two</u> different jobs that coke does inside the blast furnace?

The carbon monoxide removes oxygen from the haematite (iron oxide) to give iron metal. Carbon dioxide gas is also made at the same time.

carbon monoxide + iron oxide → iron + carbon dioxide

The iron in the iron oxide <u>loses</u> oxygen. So we say that it is **reduced**.
The carbon in the carbon monoxide <u>gains</u> oxygen. So we say that it is **oxidised**.

coke
+
limestone
+
haematite (iron oxide)

waste carbon dioxide plus nitrogen (gases)

conveyor belt

molten slag floats on top of the molten iron

molten iron comes out here

hot air blast

hot air blast

molten slag comes out here

molten iron

slag ladle

iron ladle

A blast furnace.

What happens to the air that is blasted into the furnace?

The hot air that is blasted into the furnace contains both oxygen and nitrogen gases. Nitrogen is not a reactive gas, so it goes through the furnace without changing. The oxygen from the air ends up in carbon dioxide.

12 What are the <u>two</u> main gases in the waste that comes out of the top of the furnace?

Why is limestone needed in the blast furnace?

Iron ore contains solid waste, such as sand, as well as the useful iron oxide. This waste would make the iron weak so it must be removed. Limestone reacts with the solid waste to produce **slag**. The blast furnace is so hot that both the iron and the slag melt and trickle down to the base where they collect.

13 How are the molten iron and the slag removed from the base of the furnace?

14 Why is it easy to keep the iron separate from the slag?

15 Iron made in the blast furnace is often turned into steel. Why do you think that mild steel is a more useful material than iron from the blast furnace?

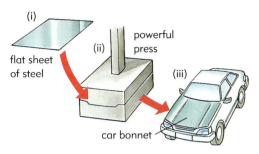

Steel can be rolled into thin sheets then pressed into the shapes we want.

Material	% carbon	Properties
iron from the blast furnace	4.0	brittle
mild steel	0.4	tough

Higher

Oxidising and reducing

When oxygen <u>combines</u> with elements or compounds, converting them into other substances, it <u>oxidises</u> them. The process is called **oxidation**. The diagram shows one example.

When some or all of the oxygen is <u>taken away</u> from a compound, this is **reduction**. For example:

copper oxide + hydrogen → copper + water
$$CuO + H_2 \rightarrow Cu + H_2O$$

We say that hydrogen is the <u>reducing agent</u> here.

16 The box shows how iron oxide is reduced to iron in a blast furnace.
(a) What is the name given to the process in the blast furnace where oxygen is removed from iron oxide to leave iron?
(b) When the carbon monoxide is oxidised, what does it change into?
(c) What is the reducing agent in this reaction?

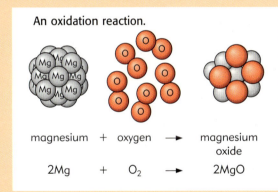

An oxidation reaction.

magnesium + oxygen → magnesium oxide

$$2Mg + O_2 \rightarrow 2MgO$$

Reducing iron oxide to iron

The following reaction occurs in a blast furnace.

iron oxide + carbon monoxide → iron + carbon dioxide

$$Fe_2O_3 + 3CO \rightarrow 2Fe + 3CO_2$$

Redox reactions

In the blast furnace, iron oxide is reduced and carbon monoxide is oxidised. This is just one of many examples of reactions in which one substance is oxidised and another is reduced. We call them redox reactions.

REDuction + OXidation = REDOX

17 (a) Explain why the reaction between copper oxide and carbon is a redox reaction.
(b) What is the reducing agent in this reaction?

When zinc reacts with copper oxide, zinc oxide and copper are produced. The box on the right gives more details of what happens in this reaction.

18 (a) How can oxidation be described in terms of electrons?
(b) How can reduction be described in terms of electrons?

Reducing copper oxide to copper

$$\text{copper oxide} + \text{carbon} \rightarrow \text{copper} + \text{carbon dioxide}$$

$$2CuO + C \rightarrow 2Cu + CO_2$$

Using zinc to displace copper

$$\text{zinc} + \text{copper oxide} \rightarrow \text{zinc oxide} + \text{copper}$$

$$Zn + CuO \rightarrow ZnO + Cu$$
$$[Cu^{2+}O^{2-}] \rightarrow [Zn^{2+}O^{2-}]$$

The zinc atoms combine with oxygen.
So they are oxidised.
But zinc atoms also become zinc ions.
To do this they lose two electrons (e^-).
So oxidation is <u>losing</u> electrons.

Oxygen is taken from the copper ions.
So they are reduced.
But copper ions also become copper atoms.
To do this they gain two electrons.
So reduction is <u>gaining</u> electrons.

Oxidation without oxygen

Oxidation and reduction involve **electron transfer**, i.e. atoms losing and gaining electrons. So the terms 'oxidation' and 'reduction' are also used in other reactions, even if oxygen itself does not take part.

19 In the reaction between iron and copper sulphate, which substance is oxidised and which substance is reduced? Give reasons for your answers.

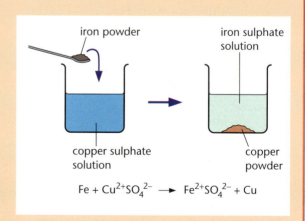

$$Fe + Cu^{2+}SO_4^{2-} \longrightarrow Fe^{2+}SO_4^{2-} + Cu$$

To help you remember
OIL RIG   **O**xidation **I**s **L**oss (of electrons).
Reduction **I**s **G**ain (of electrons).

Using electricity to split up metal compounds

We can get some metals from their ores by heating them with carbon. Another way to get metals from their compounds is to pass electricity through the compound.

Solid metal <u>compounds</u> such as metal ores do not conduct electricity. To make a metal compound conduct electricity we must either melt it or dissolve it.

1 What is the problem with trying to pass electricity through metal compounds?

Getting copper from copper chloride

Copper chloride is a metal compound that dissolves in water. The diagram shows what happens when an electric current passes through copper chloride solution.

2 Write a word equation for the reaction that happens when electricity passes through copper chloride solution.

3 (a) What substance is produced at the negative electrode when electricity is passed through copper chloride solution?
   (b) What gas is produced at the positive electrode when electricity is passed through copper chloride solution?

The electricity has split up the copper chloride. We say that the copper chloride has been **decomposed**.

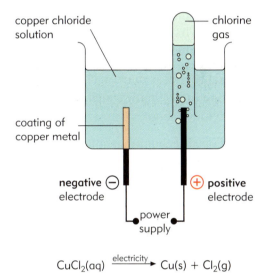

copper chloride solution — chlorine gas

coating of copper metal

negative ⊖ electrode   ⊕ **positive** electrode

power supply

$$CuCl_2(aq) \xrightarrow{electricity} Cu(s) + Cl_2(g)$$

Getting lead from lead bromide

Lead bromide is another metal compound, but it does not dissolve in water. To make electricity pass through lead bromide you must first melt it.

The diagram shows what happens when you melt lead bromide and pass an electric current through it.

4 Write down a word equation for this reaction.

Splitting up a compound by passing an electric current through it is called **electrolysis**.

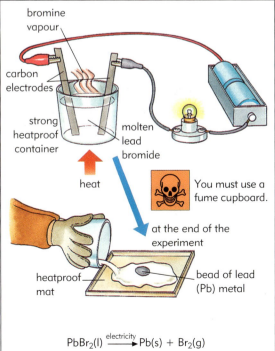

bromine vapour

carbon electrodes

strong heatproof container

molten lead bromide

heat

You must use a fume cupboard.

at the end of the experiment

heatproof mat

bead of lead (Pb) metal

$$PbBr_2(l) \xrightarrow{electricity} Pb(s) + Br_2(g)$$

How does electrolysis work?

In copper chloride solution:

- the copper atoms have a positive (+) charge;
- the chlorine atoms have a negative (−) charge.

Electrically charged atoms are called **ions**.
When we dissolve solid copper chloride in water, the ions can move about.

5 Look at the diagram.
   (a) What charge do copper ions have?
   (b) Which electrode do copper ions move to?
   (c) What charge do chloride ions have?
   (d) Which electrode do chloride ions move to?

Lead bromide is made of charged ions. We call it an **ionic** compound. When we melt solid lead bromide, the ions can then move about.

6 Copy the diagram showing lead bromide.
   Mark on the diagram the way that the ions move during electrolysis.

Metal compounds are ionic. Look at the table of ions.

7 Which electrode do metals form at during electrolysis?

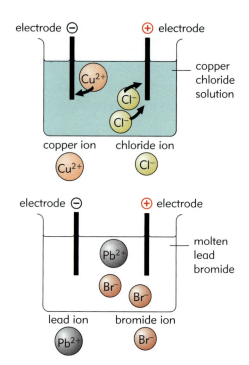

The ions move to the electrode with the opposite charge. Opposite charges attract.

Metal ions	Non-metal ions
sodium Na^+	chloride Cl^-
copper Cu^{2+}	bromide Br^-
lead Pb^{2+}	oxide O^{2-}
aluminium Al^{3+}	

Metal ions always have a positive charge.

Where does aluminium come from?

Aluminium is the most common metal in the Earth's crust. But you never find pieces of natural aluminium. This is because aluminium is a very reactive metal. Aluminium combines with other elements to form compounds.

We get aluminium from an iron ore called **bauxite**. Bauxite is mainly aluminium oxide.

8 What is the name of natural aluminium ore?

9 Which aluminium compound do you find in aluminium ore?

How do we get aluminium from its ore?

We use carbon to extract iron from iron oxide, but we cannot use carbon to extract aluminium from aluminium oxide. It doesn't work.

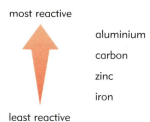

most reactive

aluminium

carbon

zinc

iron

least reactive

Look at the reactivity series.

10 Why can't carbon push aluminium out of aluminium oxide?

We must extract aluminium in a different way. We need to use electricity to extract aluminium from its oxide. We must make the aluminium ore conduct electricity to do this.

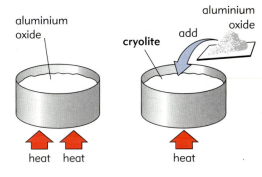

To melt the aluminium oxide you need to heat it to a high temperature, more than 2000 °C. This is much hotter than a Bunsen burner flame.

A mixture of cryolite and aluminium oxide melts at a lower temperature of 950 °C.

How to make aluminium oxide conduct electricity

Aluminium oxide does not dissolve in water to give a solution, so to make it conduct we have to melt it.

Look at the diagrams.

11 What is the problem with melting aluminium oxide?

12 How can we solve the problem?

When we have melted aluminium oxide we can use electricity to split it up. This is called **electrolysis**.

Using electrolysis to split up aluminium oxide

The diagram shows how aluminium is made from melted aluminium oxide.

13 (a) What two materials does aluminium oxide give when it splits up?
(b) Write down where each of these materials is formed.

14 (a) What do we use to make the electrodes?
(b) The positive electrode burns away and we must replace it with a new one. Write down why it burns away.

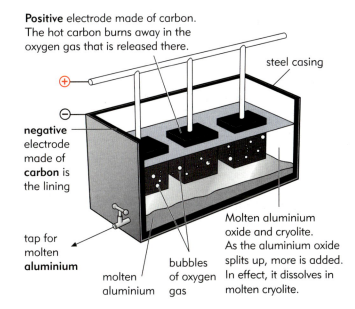

Positive electrode made of carbon. The hot carbon burns away in the oxygen gas that is released there.

steel casing

negative electrode made of **carbon** is the lining

tap for molten **aluminium**

molten aluminium

bubbles of oxygen gas

Molten aluminium oxide and cryolite. As the aluminium oxide splits up, more is added. In effect, it dissolves in molten cryolite.

Why does electrolysis work?

In aluminium oxide both the aluminium and the oxygen have electrical charges. We call them aluminium ions and oxide ions. If you melt aluminium oxide, the ions can move about in the liquid.

15 (a) What charge do aluminium ions have?
(b) Which electrode do aluminium ions move to?
(c) What charge do oxide ions have?
(d) Which electrode do oxide ions move to?

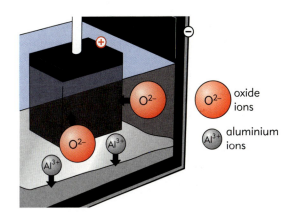

O^{2-} oxide ions

Al^{3+} aluminium ions

More about electrolysis

Metal compounds, when molten or in solution, will conduct electricity. We call the liquid an **electrolyte**. When we split up a compound by passing electricity through it, we call this **electrolysis**.

In the electrolysis of copper chloride solution:

- at the negative electrode, copper ions gain electrons to form copper atoms:
  $$Cu^{2+} + 2e^- \rightarrow Cu$$

- at the positive electrode, chloride ions lose electrons to form chlorine molecules:
  $$2Cl^- - 2e^- \rightarrow Cl_2$$

16 (a) In electrolysis, positively charged ions gain electrons at the negative electrode. Why can this be referred to as reduction?
(b) What happens at the positive electrode in electrolysis? Why can this be described as oxidation?

17 What is meant by a 'redox reaction'? Use an example to explain your answer.

18 Aluminium is made by the electrolysis of molten aluminium oxide. These are the reactions that take place at the electrodes:

- at the positive electrode: $2O^{2-} - 4e^- \rightarrow O_2$
- at the negative electrode: $Al^{3+} + 3e^- \rightarrow Al$

Describe this reaction in terms of oxidation and reduction.

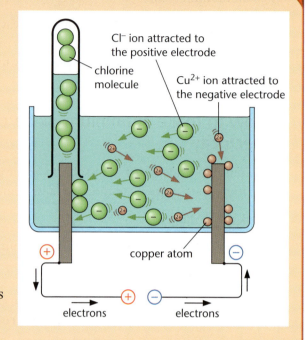

Cl⁻ ion attracted to the positive electrode

chlorine molecule

Cu²⁺ ion attracted to the negative electrode

copper atom

electrons   electrons

Making very pure copper

We all use many different electrical appliances such as computers, CD players, TVs and electric kettles. We need copper to make these appliances.

The copper that is produced from copper ore in a furnace is 98–99% pure. But this is not pure enough to make cables or electric circuits. The copper that we use for these jobs must be very pure indeed.

The impure copper from the furnace is made into 99.98% pure copper by a process called **electrolysis**. The diagrams show how this is done.

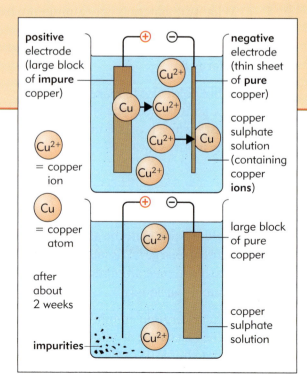

positive electrode (large block of **impure** copper)

negative electrode (thin sheet of **pure** copper)

copper sulphate solution (containing copper **ions**)

Cu^{2+} = copper ion

Cu = copper atom

after about 2 weeks

impurities

large block of pure copper

copper sulphate solution

19 (a) What happens to copper atoms at the positive electrode?
(b) What happens to copper ions at the negative electrode?

20 (a) What happens, over a period of two weeks, to all of the
copper from the positive electrode?
(b) What happens to the impurities?

In this electrolysis reaction, the copper sulphate is <u>not</u> decomposed.
It remains unchanged. Copper ions enter the solution at the positive
electrode and leave the solution at the negative electrode.

Higher

Electron transfers when purifying copper

Oxidation and reduction can be understood in
terms of **electron transfer** (see Box).

We can also use the idea of electron transfer to understand
what is happening when copper is purified.

When an electric current flows through a circuit connected
by wires to a battery or power supply:

- electrons flow <u>into</u> the positive (+) terminal;
- electrons flow <u>out of</u> the negative (−) terminal.

In the electrolysis that is used to purify copper, the flow of
electrons has the following effects:

- At the positive electrode, copper atoms lose two
  electrons (e^-) and become copper ions (Cu^{2+}).
  These go into the solution (electrolyte).
- At the negative electrode, copper ions gain two electrons
  and become copper atoms.
  These are deposited on the electrode.

21 Use the ideas of oxidation and reduction to describe
what happens at each electrode.

> **Remember**
>
> Oxidation Is Loss (of electrons).
> Reduction Is Gain (of electrons).

The process that is used to purify copper can also be used to give
objects made from a cheaper metal a thin coating of copper,
chromium or silver. The object is **plated** with one of these metals.

In silver plating, for example:

- the positive electrode is made of silver;
- the object being plated is the negative electrode;
- the electrolyte is a solution containing silver ions (Ag^+).

Sports cups are usually silver
plated rather than solid silver.

22 Use the ideas of electron transfer, oxidation and reduction to describe what
happens during silver plating. Use a labelled diagram as part of your answer.

Preventing corrosion

How do other metals affect the corrosion of iron?

Iron and steel (which is mainly iron) **corrode** more quickly than most other transition metals. The diagrams show how the rate of corrosion is affected when the iron or steel is connected to other metals.

Key = rust

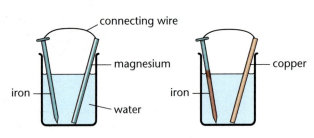 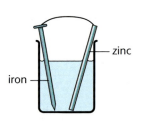 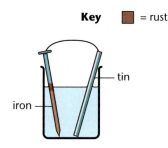

1 How is the rate of corrosion of iron related to the reactivity of the metals to which it is attached?

most reactive

magnesium
zinc
iron
tin
copper

least reactive

When a more reactive metal is in contact with iron, this metal corrodes instead of the iron.

The more reactive metal is 'sacrificed' to protect the iron. This method of protection is called **sacrificial protection**.

2 Describe and explain how:
(a) gas and oil pipelines, made from steel, are protected against corrosion in the ground;
(b) the steel hulls of ships are protected against rusting.

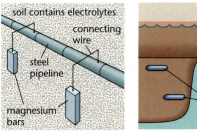 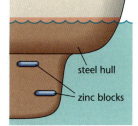

Examples of sacrificial protection.

What is stainless steel?

Stainless steel is a special type of steel that does not go rusty. We can make stainless steel (an alloy) by mixing iron with two other metals called **nickel** and **chromium**. Stainless steel does not rust but it is expensive.

3 Write down <u>three</u> things that we make from stainless steel.

4 Car exhausts made from stainless steel last a lot longer than car exhausts made from ordinary (mild) steel.

Why do you think all exhausts are not made from stainless steel?

Stainless steel in the kitchen.

Why aluminium doesn't corrode

Metals corrode by reacting with oxygen or water (or both). Aluminium is a much more reactive metal than iron so you would expect it to corrode more quickly. The diagram shows why this doesn't happen.

5 How does a layer of aluminium oxide protect the aluminium metal underneath it?

Sometimes things made from aluminium are deliberately given a layer of aluminium oxide. This is called **anodising**.

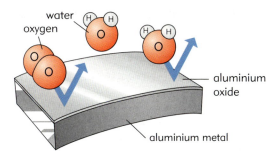

A tough, thin layer of aluminium oxide protects the metal underneath. Water and oxygen cannot get through this layer.

Higher

Explaining sacrificial protection

It is easy to understand why a metal that is more reactive than iron might corrode more quickly than iron. But it isn't so easy to understand why connecting a more reactive metal to iron should slow down the corrosion of iron.

This happens because when you put two different metals, such as iron and zinc, into a solution containing ions you make a simple **electric cell**.

An electric cell is basically an electron 'pump'. The diagram shows how a zinc–iron electric cell can prevent the iron from rusting.

6 Draw and label a similar diagram to explain why connecting iron to a less reactive metal such as copper makes the iron rust faster.

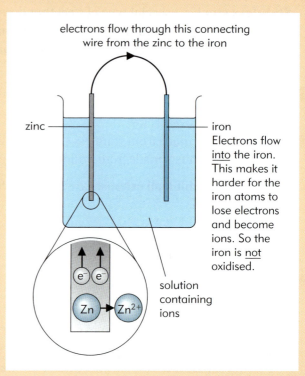

electrons flow through this connecting wire from the zinc to the iron

zinc

iron
Electrons flow into the iron. This makes it harder for the iron atoms to lose electrons and become ions. So the iron is not oxidised.

solution containing ions

7 You can measure the potential difference (voltage) across the two different metals in a simple cell using a voltmeter.

Metals in cell	Potential difference
zinc + iron	0.4 volts
magnesium + iron	2.0 volts

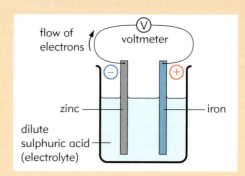

flow of electrons

voltmeter

zinc

iron

dilute sulphuric acid (electrolyte)

(a) Which metal, zinc or magnesium, would be better at preventing iron from rusting?
(b) Which of these two metals would itself corrode more quickly when connected to iron?
(c) Use the idea of electron transfers to suggest an explanation for each of your answers.

6 Earth materials

Using limestone

The ground under our feet is made of **rock** but you don't always see it. This is because the rock is often covered with **soil**. We also cover the ground with roads, pavements and buildings. But if you dig down far enough you always reach solid rock. There are many different kinds of rock. One common rock is called **limestone**.

Limestone is not a very hard rock so we can cut limestone into blocks and slabs quite easily. This makes limestone very useful for buildings. But there is a problem with using limestone for buildings, as the pictures show.

1 Why is limestone a useful building material?

2 What is the problem when we use limestone for buildings?

3 Why is this problem worse today than it was hundreds of years ago?

Weather changes limestone. Acid rain makes it change even faster.

Heating limestone

We can use limestone to make other materials. If we make limestone really hot we can change it into **quicklime**. We use a lime kiln to do this.

4 A lime kiln is heated in <u>two</u> ways. Write them down.

A word equation for the reaction is:

limestone + energy ⟶ quicklime + carbon dioxide

The chemical name for limestone is calcium carbonate. The chemical name for quicklime is calcium oxide.

5 Write down the word equation using the chemical names.

A reaction that uses heat or thermal energy to break down a substance into different substances is called **thermal decomposition**.

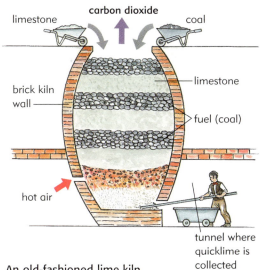

An old-fashioned lime kiln.

What can we do with quicklime?

If you heat a piece of limestone strongly, it changes into a new material called quicklime.

6 (a) What is the chemical name for quicklime?
(b) What other substance is produced when you heat limestone to make quicklime?

Many other **carbonates** also split up (**decompose**) when you heat them.

7 What <u>two</u> substances are produced when you heat copper carbonate?

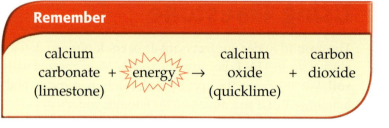

Remember

calcium carbonate (limestone) + energy → calcium oxide (quicklime) + carbon dioxide

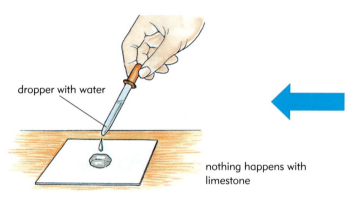

dropper with water

nothing happens with limestone

pieces of limestone (calcium carbonate)

Quicklime looks almost the same as limestone, but when you add a few drops of water you can see the difference.

8 What happens when you add a few drops of water to limestone?

9 What happens when you add a few drops of water to quicklime?

The quicklime reacts with the water to form a new material.

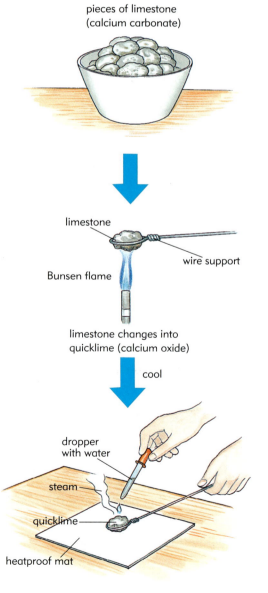

limestone

wire support

Bunsen flame

limestone changes into quicklime (calcium oxide)

cool

dropper with water

steam

quicklime

heatproof mat

What is the new material?

The new material formed from quicklime is called **slaked lime**.

10 Write a word equation for what happens when water is added to quicklime.

The chemical name for slaked lime is calcium hydroxide.

11 Write down the word equation using the chemical names for quicklime and slaked lime.

What is so special about slaked lime?

Slaked lime dissolves just a little in water.
We call this solution **lime water**.

paper stays yellow

water

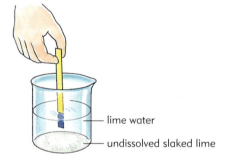

lime water

undissolved slaked lime

12 What happens when we test lime water with indicator paper?

13 What does this tell you about lime water?

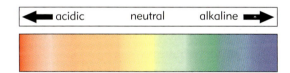

acidic   neutral   alkaline

Using slaked lime

Soil in fields and gardens may be too acidic for some plants. We can use an alkali to **neutralise** the acidity. We need to add just the right amount of alkali.

Look at the photograph.

14 What is the farmer spreading on the field?

15 Write down <u>two</u> reasons why lime helps the soil.

16 Why must the farmer be careful not to add too much lime to the field?

Fish don't like acidic water. So we can also use lime to neutralise the acidity of some lakes.

Spreading slaked lime. The lime neutralises soil acid. It also makes clay soil less sticky.

Other useful materials made from limestone

Many of the things we build today are made from **concrete**. When wet concrete sets, it becomes as hard as stone. When we mix concrete it can be poured into moulds. This is how we make concrete into lots of different shapes.

These objects were made using concrete.

17 Write down <u>two</u> things we can make using concrete.

18 Write down <u>two</u> reasons why concrete is useful for making these things.

To make concrete you need **cement**. Cement is made from limestone.

Making cement

We need to use two materials from the ground to make cement.
These are the **raw materials**.

19 What <u>two</u> raw materials do you need to make cement?

20 What do you have to do to these raw materials to turn them into cement?

21 Write down <u>two</u> reasons why the kiln rotates all the time.

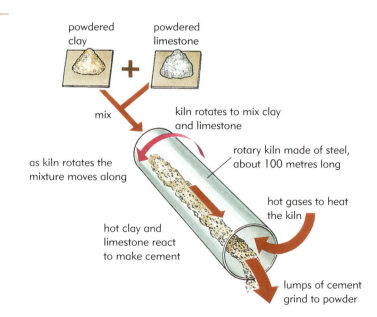

powdered clay

powdered limestone

mix

kiln rotates to mix clay and limestone

rotary kiln made of steel, about 100 metres long

as kiln rotates the mixture moves along

hot gases to heat the kiln

hot clay and limestone react to make cement

lumps of cement grind to powder

Making concrete

The diagram shows how you can make concrete.

22 What <u>four</u> things must you mix together to make concrete?

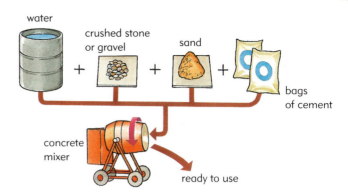

Using concrete

Once you have mixed some concrete you need to make it the right shape. The diagram shows how you can do this. The water reacts slowly with the cement to make the concrete set hard as stone. This can take a few days.

23 How can you keep the sides of the new concrete step straight?

24 Why should you wait a few days before removing the wooden frame?

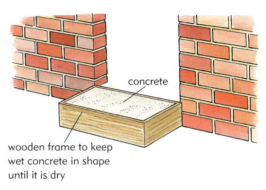

Making a concrete step for a house.

Making glass

Glass is another very useful material that we make using limestone.

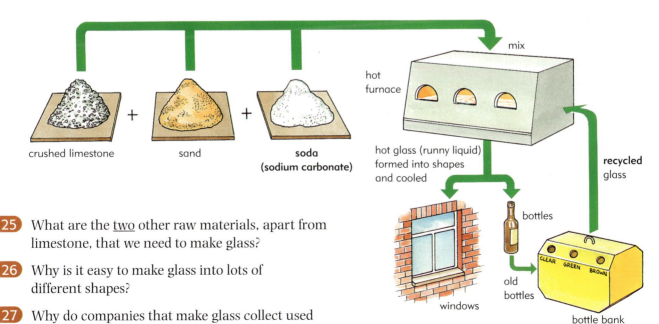

25 What are the <u>two</u> other raw materials, apart from limestone, that we need to make glass?

26 Why is it easy to make glass into lots of different shapes?

27 Why do companies that make glass collect used glass from bottle banks?

Using crude oil

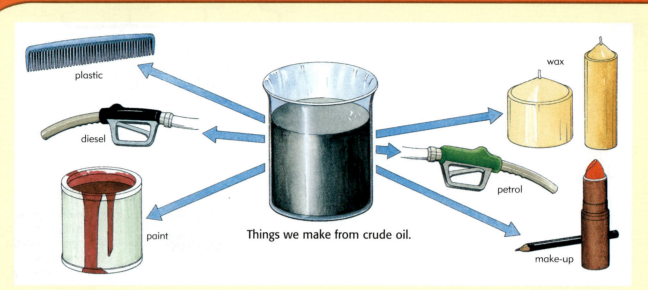

plastic

diesel

paint

Things we make from crude oil.

wax

petrol

make-up

We find many useful substances in the Earth's crust.
One of these substances is oil.
We call the oil that comes out of the ground **crude oil**.

1 Write down some of the things that we can make
from crude oil.

How did the crude oil get there?

The story of oil starts millions of years ago in the sea.
Large numbers of animals and plants died and fell to the
sea bed. If the remains were buried quickly, other animals
could not eat them. If there was no oxygen, they did not
decompose fully. The organic remains turned into oil.

2 Put these sentences into the right order to explain
how oil was made.
The first sentence has been put in place for you.

- Millions of tiny plants and animals lived in the seas.

- The pressure and heat of the rocks turned the
remains of the dead plants and animals into oil.

- As the dead plants and animals decayed they were
covered with layers of mud and other sediments.

- When the plants and animals died their bodies fell
to the bottom of the sea.

- The layers of mud and other sediments slowly
changed into layers of sedimentary rock.

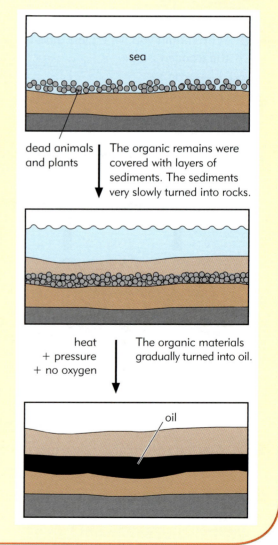

sea

dead animals
and plants

The organic remains were
covered with layers of
sediments. The sediments
very slowly turned into rocks.

heat
+ pressure
+ no oxygen

The organic materials
gradually turned into oil.

oil

How do we get oil from the ground?

Oil forms in small drops which are spread through lots and lots of rock. Luckily for us, the oil doesn't stay like this. A lot of oil often collects together in one place, as you can see in the diagrams.

3 (a) Explain why oil floats on top of the permeable rocks.

(b) Describe how oil gets trapped.

4 What must you do to collect the trapped oil?

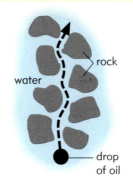

Many rocks have lots of tiny spaces. Oil and water can move through these spaces. We say the rocks are permeable. Oil floats on water. So oil rises to the top of the permeable rocks.

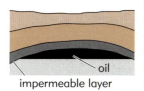

When the oil reaches an impermeable layer it can't rise any further. The oil is trapped under the rock. (Impermeable means not permeable.)

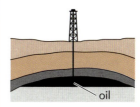

To get the oil you have to drill down through the impermeable layer.

What are fossil fuels?

Fossils are the remains of plants and animals that we find in sedimentary rocks. Oil is also the remains of plants and animals from millions of years ago. So we call it a fossil fuel.

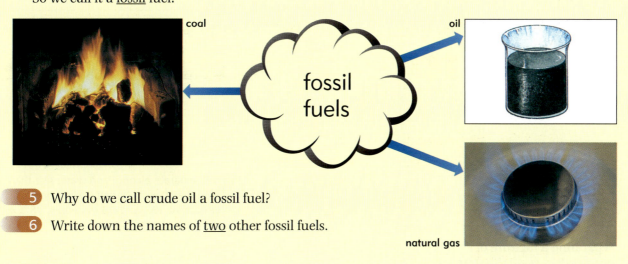

coal

oil

fossil fuels

natural gas

5 Why do we call crude oil a fossil fuel?

6 Write down the names of two other fossil fuels.

Crude oil is a mixture

Crude oil is a mixture of <u>lots</u> of different liquids. These liquids are very useful, but we can't use them until we have separated them from each other.

7   Write down the <u>two</u> main uses for the liquids in crude oil.

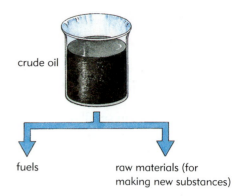

crude oil

fuels

raw materials (for making new substances)

How to separate a mixture of liquids

If you heat up a liquid it changes to a vapour. We say it **evaporates**. If you make the liquid hot enough, it boils. A boiling liquid evaporates very quickly.

If you cool a vapour it changes back into a liquid. We say it **condenses**.

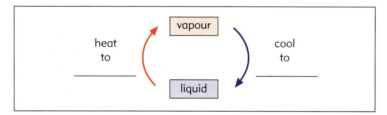

heat
to

vapour

cool
to

liquid

Evaporating a liquid and then condensing it again is called distillation.

8   Copy the diagram, then complete it.

We can use this idea to separate a mixture of liquids.

Wine is a mixture of water, alcohol and small amounts of other chemicals. The diagram shows how brandy is made from wine.

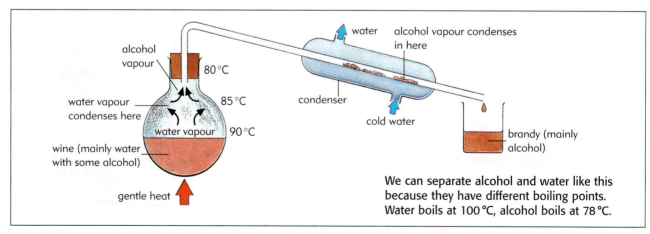

alcohol vapour

80 °C

water   alcohol vapour condenses in here

water vapour condenses here

85 °C

condenser

water vapour   90 °C

cold water

wine (mainly water with some alcohol)

gentle heat

brandy (mainly alcohol)

We can separate alcohol and water like this because they have different boiling points. Water boils at 100 °C, alcohol boils at 78 °C.

Separating a mixture of liquids into parts or fractions like this is called **fractional distillation**.

Separating crude oil into fractions

We use fractional distillation to separate crude oil into different parts or **fractions**. The different fractions boil at different temperatures.

9 (a) Which fraction of crude oil has the highest boiling point?
(b) Which fraction has the lowest boiling point?

10 Explain why crude oil can be separated by fractional distillation.

11 Why is separating crude oil into fractions more difficult than making brandy?

Fraction of crude oil	Boiling points in °C
dissolved gases	below 0
petrol	around 65
naphtha	around 130
kerosene	around 200
diesel oil	around 300
bitumen	over 400

An oil fractionating tower

In Britain, 250 000 tonnes of oil are produced every day! To separate all of this oil into its fractions we use enormous **fractionating towers**.
The diagram shows one of these.

12 Explain why bitumen falls to the bottom of the tower and methane rises to the top.

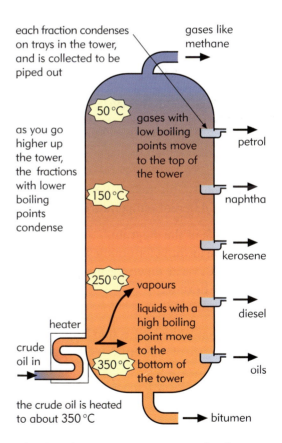

each fraction condenses on trays in the tower, and is collected to be piped out

gases like methane

50°C

as you go higher up the tower, the fractions with lower boiling points condense

gases with low boiling points move to the top of the tower

petrol

150°C

naphtha

kerosene

250°C vapours

liquids with a high boiling point move to the bottom of the tower

diesel

heater

crude oil in

350°C

oils

the crude oil is heated to about 350°C

bitumen

A fractionating tower to separate crude oil.

What are the chemicals in crude oil?

In nature, there are about 90 different kinds of atom, which we call the **elements**. Substances that contain more than one kind of atom joined together are called **compounds**. Most of the substances in crude oil are compounds that are made from just two kinds of atom. The smallest part of each compound is a **molecule**. The diagram shows two of the molecules you find in crude oil.

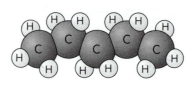

This molecule has 5 carbon atoms and 12 hydrogen atoms. We write this C_5H_{12}. This is the formula of the compound.

13 Which <u>two</u> kinds of atom do these molecules contain?

14 What is the difference between the two molecules?

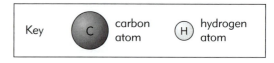

Key — C carbon atom — H hydrogen atom

15 Write down the formula of:
(a) the smaller molecule;
(b) the larger molecule.

Molecules made only of hydrogen atoms and carbon atoms are called **hydrocarbons**. Most of the molecules in crude oil are hydrocarbons.

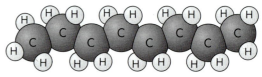

This molecule has 8 carbon atoms and 18 hydrogen atoms.

Differences between hydrocarbons

Crude oil is a **mixture** of many different hydrocarbons. The hydrocarbon molecules are all different sizes and masses. The boiling point of a molecule depends on its size and mass. This means that the hydrocarbon molecules all boil at different temperatures.

16 Look at the diagram.
The formula for butane is C_4H_{10}. Its boiling point is 0 °C. Write down the formula and boiling point for :
(a) hexane;
(b) decane.

17 What is the connection between the boiling point and the size of molecule?

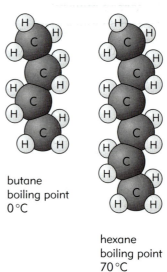

butane
boiling point
0 °C

hexane
boiling point
70 °C

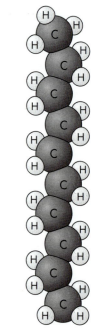

decane
boiling point
175 °C

Which hydrocarbons are in which fractions?

Hydrocarbons with the largest molecules have the highest boiling points. These large molecules condense to liquids lower down the fractionating tower.

18 Which fraction of crude oil has the smallest molecules?

19 Which fraction of crude oil has the largest molecules?

A simple way to say how big hydrocarbon molecules are is to count how many carbon atoms they have.
The diagram below shows the hydrocarbons in some of the oil fractions.

20 How many carbon atoms are there in:
(a) the hydrocarbon molecules in petrol?
(b) the hydrocarbon molecules in diesel?
(c) the hydrocarbon molecules in bitumen?

21 How many carbon atoms would you expect to find in the hydrocarbon gases?

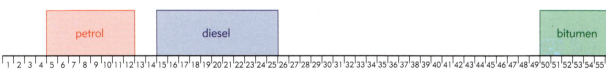

Number of carbon atoms in a molecule.

The crude oil fractions are different in more ways than their different boiling points. This is why we can use them for different jobs.

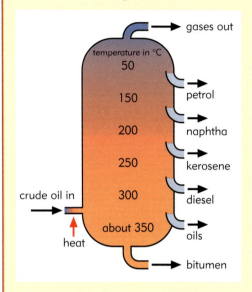

Remember

The different fractions in crude oil condense to liquids at different temperatures.

Each fraction contains more than one hydrocarbon compound. The hydrocarbons in a fraction have a similar number of carbon atoms and similar boiling points.

Different hydrocarbons for different jobs

Different hydrocarbons have molecules of different sizes. This gives them different properties. One example is their boiling points. Hydrocarbons with smaller molecules have lower boiling points. The properties of different hydrocarbons make them useful for different jobs.

> **Remember**
>
> The compounds in crude oil contain hydrogen and carbon atoms. They are hydrocarbons.

Oil fraction	petrol	diesel	lubricating oil	bitumen
Number of carbon atoms	5 to 12	15 to 25	26 to 50	more than 50
Size of the molecules	small	fairly small	big	very big
Boiling point	low	fairly low	fairly high	high
Appearance				
A few drops left in the open air.	Quickly changes to a vapour. We say it is very volatile.	Slowly changes to vapour.	Very slowly changes to vapour.	Hardly changes to vapour at all.
A few drops soaked into glass wool.	Catches fire very easily. We say it is very flammable.	Catches fire quite easily.	Hard to light.	Hard to light.
How easy is it to pour?	Easy to pour.	Easy to pour.	Not easy to pour, it sticks to the sides.	Almost solid, very slow to pour. We say it is viscous.

1. Petrol and diesel are both used as fuels in engines.
   Which properties make them useful for this job?
   Give reasons for your answers.

2. Lubricating oil is used to make engines run smoothly.
   The oil reduces friction between moving parts.
   Which properties make it useful for this job?
   Explain why.

3. Bitumen is used to make the tarmac on roads.
   Which properties make it useful for this job?
   Give reasons for your answer.

Making long hydrocarbons more useful

We use more crude oil for **fuel** than for anything else. However, there are lots of long hydrocarbon molecules in crude oil and these are not very useful as fuels.

4 Write down <u>two</u> reasons why long hydrocarbon molecules do not make good fuels.

We can make long hydrocarbon molecules more useful if we split them up into smaller molecules. We call this **cracking**. We make most of our petrol this way.

5 Write a word equation for what happens when decane is split up by cracking into octane and ethene.

A simpler way to write down this chemical reaction is to use the formula of each compound.

6 Rewrite the word equation for what happens when decane is split up, by cracking, into octane and ethene using the formulas for decane, octane and ethene.

Cracking is another example of a thermal decomposition reaction.
We use a hot catalyst to speed up this reaction.

7 Explain why cracking is a thermal decomposition reaction.

Remember

When thermal energy (heat) is used to break down or decompose a substance, we call the process thermal decomposition.

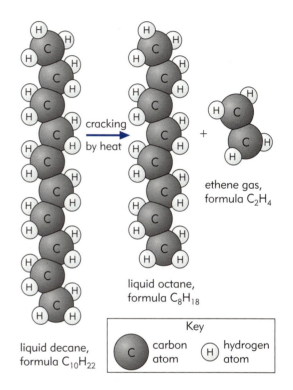

ethene gas, formula C_2H_4

liquid octane, formula C_8H_{18}

liquid decane, formula $C_{10}H_{22}$

Key			
C	carbon atom	H	hydrogen atom

Higher

More about hydrocarbons

The strong bonds that hold atoms together in molecules of hydrocarbons are formed by shared pairs of electrons. We call them **covalent** bonds.
The diagram shows two different hydrocarbon molecules.

Strong bonds hold the atoms together in the molecule, but there are only weak bonds <u>between</u> molecules. This means that many hydrocarbons are gases or liquids, or solids with low melting points.

Two families of hydrocarbons are **alkanes** and **alkenes**.

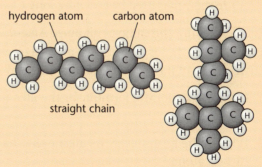

hydrogen atom   carbon atom

straight chain

branched chain

Hydrocarbon molecules may be in the form of branched or straight chains, with carbon atoms forming the spine.

The alkanes

Crude oil is a mixture of hydrocarbons, most of which are alkanes. The smallest alkane is methane: CH_4.

Sometimes it is useful to represent a substance with a **structural formula**. A structural formula shows how atoms are bonded in a molecule.

In all alkanes, the carbon atoms are joined by **single** bonds (that is, a single shared pair of electrons). Single bonds are shown as — in structural formulas. The structural formula for an alkane shows that:

- each carbon atom can form four bonds;
- each hydrogen atom can form one bond.

8 Draw the structural formula for the alkane butane, C_4H_{10}.

Alkanes are **saturated** hydrocarbons. This means that every carbon atom has used up all of its four bonds to link to other atoms.

methane (CH_4)

structural formula for methane

Name of alkane	Molecular formula	Structural formula
ethane	C_2H_6	H H | | H—C—C—H | | H H
propane	C_3H_8	H H H | | | H—C—C—C—H | | | H H H

The general formula for alkanes is C_nH_{2n+2}, where n is any number.

Alkanes are not very reactive because a lot of energy is needed to break C–H bonds. Reactions with other substances tend to be slow.

The alkenes

The alkenes are only found in small quantities in nature. Most are manufactured by cracking long chain alkanes.

Alkenes are not the same as alkanes. Alkenes have a **double** bond (that is, two shared pairs of electrons) between two of the carbon atoms in the chain.

9 Draw the structural formula for the alkene propene, C_3H_6.

10 What is the formula for the alkene that has five carbon atoms?

The alkenes are much more reactive than the alkanes because the double carbon–carbon bonds can 'open up' to provide extra bonds. Extra atoms of other elements can be attached to these bonds. So we say that alkenes are **unsaturated**.

Structural formula for ethene (the simplest alkene).

H   H
 \   /
  C = C
 /   \
H   H

double bond

The general formula for alkenes is C_nH_{2n}.

How do we test for alkenes?

Bromine is one of the elements that can join with the extra bonds formed when the double bond in an alkene opens up. Bromine water is yellow–brown. It becomes colourless as bromine reacts with ethene.
So we can use bromine water as a **test** for alkenes.

Alkanes do not react with bromine.

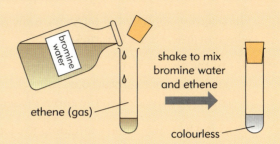

shake to mix bromine water and ethene

ethene (gas)

colourless

11 Write down what you would see when bromine water is shaken with:
(a) an alkene;
(b) an alkane.

Alkenes joining together

Because of the double bond, alkene molecules can also react with each other. For example, when heated under pressure with a catalyst, ethene molecules will join together to form very long chain molecules.
The box shows what happens.

12 Explain why alkenes are more useful than alkanes.

13 This formula represents the polymer poly(styrene).
Draw a formula to represent <u>one</u> molecule of the monomer from which the polymer was formed.

$$\left(\begin{array}{cc} H & H \\ | & | \\ -C & -C- \\ | & | \\ H & C_6H_5 \end{array}\right)_n$$

14 This formula shows a vinyl chloride monomer. Write down a representation of the reaction where n vinyl chloride monomers produce the polymer poly(vinylchloride).

$$\left(\begin{array}{cc} H & H \\ | & | \\ C & =C \\ | & | \\ H & Cl \end{array}\right)$$

Addition polymerisation

'Mono' means 'one'; 'poly' means 'many'.

many small molecules of ethene (**monomers**) ⟶ long chain molecule (**polymer**) called poly(ethene)

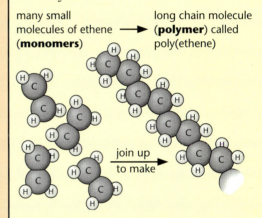

join up to make

In addition polymerisation, the polymer is the only product. The reaction can be represented like this:

$$n\left(\begin{array}{cc} H & H \\ | & | \\ C & =C \\ | & | \\ H & H \end{array}\right) \longrightarrow \left(\begin{array}{cc} H & H \\ | & | \\ -C & -C- \\ | & | \\ H & H \end{array}\right)_n$$

monomer ethene molecules

poly(ethene)

where n is a large number.

Using products from oil

Cracking hydrocarbons at a refinery

The diagram shows what happens in the part of a refinery where hydrocarbons are cracked.

1 Put the sentences in the right order to explain how we crack hydrocarbons.
The first sentence is in the correct place.

- We heat the liquid containing the long hydrocarbon molecules.

- The long molecules split up into a mixture of smaller ones.

- We pass the vapour over a hot catalyst.

- We separate the different small molecules.

- The long hydrocarbons form a vapour. We say they evaporate.

2 Why do we use a catalyst in cracking?

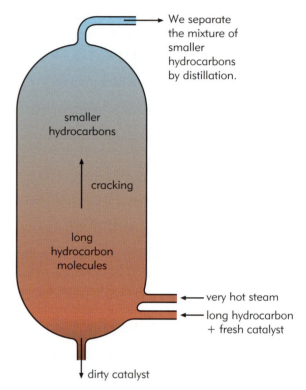

We separate the mixture of smaller hydrocarbons by distillation.

smaller hydrocarbons

cracking

long hydrocarbon molecules

very hot steam

long hydrocarbon + fresh catalyst

dirty catalyst

The catalyst makes the reaction happen
- faster
- at a lower temperature.

Cracking a hydrocarbon at school

Liquid paraffin is a hydrocarbon. It contains fairly large molecules. So it is not very runny.

The diagram shows how you can crack liquid paraffin.

3 Which <u>two</u> things show that you have cracked some of the liquid paraffin?

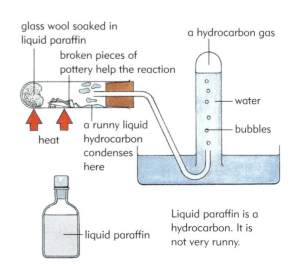

glass wool soaked in liquid paraffin

broken pieces of pottery help the reaction

a hydrocarbon gas

water

bubbles

heat

a runny liquid hydrocarbon condenses here

liquid paraffin

Liquid paraffin is a hydrocarbon. It is not very runny.

How we use the cracked hydrocarbons

Cracking produces the smaller molecules that we need as **fuels**. This is how we get most of the octane in petrol.

4 Look at the table.
Why is octane better than decane for petrol?

Hydrocarbon	Formula	Boiling point (°C)
hexane	C_6H_{14}	70
octane	C_8H_{18}	126
decane	$C_{10}H_{22}$	175

Fuels with lower boiling points catch fire or ignite more easily.

We join up some of the small molecules to make new materials such as **plastics**. For example, we use ethene to make the plastic **poly(ethene)**. 'Poly' means many. So poly(ethene) means many ethenes. It is a very large molecule.

As well as plastics, many other large molecules are made of lots of small molecules joined together. We call them **polymers**.

5 What does the 'poly' in polymer mean?

6 Poly(propene) is a plastic. What smaller molecules is it made from? Explain your answer.

We make plastics by joining together lots of smaller molecules. It's a bit like joining lots of paper clips or daisies to form chains.

Plastics from oil

We can crack long hydrocarbon molecules to give smaller molecules. We can join up short hydrocarbon molecules to give larger molecules. We call the large molecules **polymers**.

Using ethene to make a plastic

One of the small molecules we get by cracking hydrocarbon molecules is called **ethene**. If we join many ethene molecules together, we get a very long molecule that is a useful plastic.

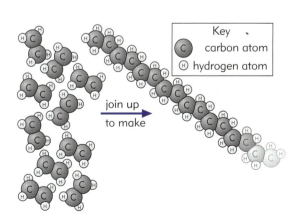

Lots of small ethene molecules join together to give a long molecule of poly(ethene). 'Poly' means 'many'.

7 (a) What is the name of the plastic made from ethene?
(b) Why does the plastic have this name?

People often call this plastic 'polythene'.

What do we use poly(ethene) for?

The picture shows some of the things we can make using poly(ethene). The box shows some of the properties of poly(ethene).

Some properties of poly(ethene)

It is soft – you can scratch it easily and it wears away.

It is tough – even if you drop it, it doesn't break.

It isn't very clear – you cannot see clearly through it unless it is very thin.

It is strong – it is hard to tear.

It is flexible – you can bend it easily.

It melts easily – but can stand boiling water.

It is waterproof – liquids cannot soak through it.

8 Why wouldn't we use poly(ethene) to make these things:

    (a) saucepans?   (b) shoes?   (c) a car windscreen?

More 'polys'

We can join up other hydrocarbon molecules to make polymers too. Different polymers have different properties. So we use them for different things.

9 Write down the name of the polymer made from propene.

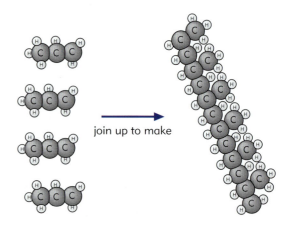

join up to make

Lots of small propene molecules join up to form poly(propene).

What do we use poly(propene) for?

We use poly(propene) to make milk crates and fibres for ropes and carpets. We used to make all ropes from natural fibres such as hemp. We often use poly(propene) and other synthetic polymers now because they:

- are stronger;
- are more hard-wearing;
- do not shrink;
- do not rot.

We choose different polymers for different jobs. For example, poly(propene) is a better polymer than poly(ethene) for making ropes because it doesn't stretch as easily.

10 Poly(propene) is also strong and hard-wearing. Explain why these two properties are important in a rope.

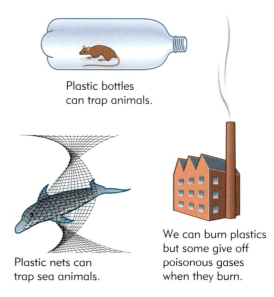

Plastic bottles can trap animals.

Plastic nets can trap sea animals.

We can burn plastics but some give off poisonous gases when they burn.

Plastics stay in the environment for a long time.

What happens to waste plastics?

Microorganisms break down natural fibres such as cotton and wool. So when we bury them in landfill sites, they rot. We say that they are **biodegradable**. Most plastics don't rot. They are not biodegradable.

11   What happens to plastics in landfill sites? Explain your answer.

12   Describe two other problems caused by plastics in the environment.

13   Scientists are developing some biodegradable plastics. Why do you think they are doing this?

14   Write down one problem of recycling plastics.

PE   HDPE   PET

We can recycle many plastics, but we have to sort them into their different kinds. Look out for recycling codes like these on plastics.

Burning fuels – where do they go?

We get gases, petrol and diesel from crude oil. When we burn these fuels, energy is released. New substances are also produced.

Look at the diagram.

1   What reacts with petrol to make it burn?

2   What happens to the new substances that are produced?

3   Write down a word equation for the reaction between petrol and oxygen.

All the fuels we get from crude oil produce the same new substances when they burn.

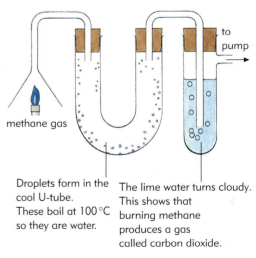

waste **gases** from exhaust go into the air

oxygen from the air

petrol

engine   petrol tank

What new substances are made when fuels burn?

To find out what new substances are made when fuels burn, you need to trap them.

The diagram shows how you can do this.

4   What two substances are made when methane burns?

5   Write down a word equation for the reaction between methane and oxygen.

methane gas   to pump

Droplets form in the cool U-tube. These boil at 100 °C so they are water.

The lime water turns cloudy. This shows that burning methane produces a gas called carbon dioxide.

Trapping the new substances made when methane burns.

anhydrous copper sulphate (anhydrous means 'without water')

hydrated copper sulphate (hydrated means 'with water')

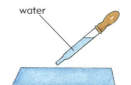

dry cobalt chloride paper

wet cobalt chloride paper

Two more tests for water.

6 Write down <u>three</u> different ways of showing that the droplets of liquid that collect in the U-tube are water.

7 How can you tell that the gas produced is carbon dioxide?

What happens to molecules when methane burns?

The diagram shows what happens to a methane molecule when it burns.

8 Look at the diagram.
(a) Describe what happens to the carbon atom in a methane molecule when methane burns.
(b) Describe what happens to the hydrogen atoms in a methane molecule when methane burns.

When anything burns, its atoms join with oxygen atoms to make **oxides**. Water is hydrogen oxide.

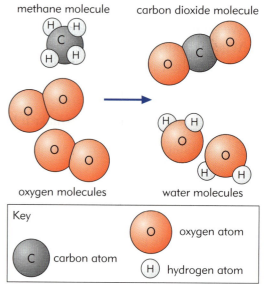

methane molecule   carbon dioxide molecule

oxygen molecules   water molecules

Key

O oxygen atom

C carbon atom

H hydrogen atom

We can write this: $CH_4 + 2O_2 \longrightarrow CO_2 + 2H_2O$

Burning other fuels

Fuels from crude oil are all hydrocarbons.
The diagrams show two different hydrocarbon molecules.

9 Burning hydrocarbons always makes water and carbon dioxide. Why is this?

Many fuels contain sulphur as well as carbon compounds. When sulphur burns it makes a gas called **sulphur dioxide**. This gas is an important cause of acid rain.

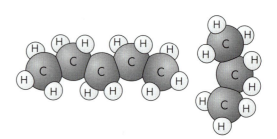

It's raining acid

Acids are dangerous substances.
We know that they can 'eat away' at some things.

10 (a) What has happened to the statue in
       the photograph?
   (b) What has caused this to happen to the statue?

Acid rain can kill trees and the fish in lakes.

Acid rain is a serious problem in many countries
including Britain. As well as damaging buildings,
acid rain can harm animals and plants.

11 Write down <u>two</u> ways in which acid rain can
    harm living things.

We need to prevent acid rain from forming.
To do this we have to understand what causes it.

What turns our rain into acid?

When fuels burn they react with oxygen.
Atoms in the fuel join with oxygen atoms in the air.
New substances called **oxides** are made.

Most fuels contain carbon atoms.

12 What new substance do the carbon atoms make when
    a fuel burns?

Many fuels also contain some sulphur atoms.

13 (a) What new substance do the sulphur atoms make
       when the fuel burns?
   (b) Write down a word equation for this reaction.

Sulphur dioxide is a gas that can turn rain into acid.

carbon atom	oxygen molecule		carbon dioxide molecule
sulphur atom	oxygen molecule		sulphur dioxide molecule

We can also write these reactions like this:

$$C(s) + O_2(g) \longrightarrow CO_2(g)$$
$$S(s) + O_2(g) \longrightarrow SO_2(g)$$

(s) = solid, (g) = gas

How sulphur dioxide makes acid rain

The diagram shows what happens when fuels containing
sulphur burn. Burning the fuels releases **sulphur dioxide**
gas into the air. This gas combines with oxygen in the
air and then dissolves in water to produce drops of
sulphuric acid.

14 What type of fuels are the cause of acid rain?

Acid rain does not usually fall where it is made.
Winds can blow the 'acid clouds' for hundreds of
kilometres before they fall as rain.

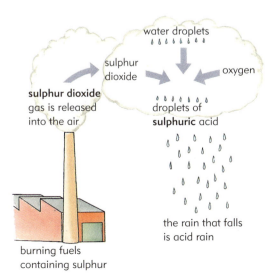

water droplets

sulphur dioxide   oxygen

sulphur dioxide
gas is released   droplets of
into the air   **sulphuric** acid

the rain that falls
is acid rain

burning fuels
containing sulphur

How rain turns into acid.

Don't just blame sulphur

It's not just the <u>sulphur</u> in fuels that causes acid rain.
When we burn things at high temperatures the nitrogen
in the air can join up with oxygen.
This happens inside car engines.

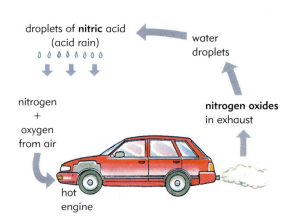

15 What new substances are made when the nitrogen
atoms join up with oxygen atoms?

16 What substance is produced when nitrogen oxides
dissolve in droplets of water?

The Earth's changing atmosphere

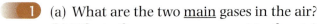

The atmosphere today

The Earth's **atmosphere** is a layer of gases above the
surface. The air pressure gets lower and lower as you go
up. At 150 km above the surface the pressure is so low that
scientists sometimes call this the top of the atmosphere.
The pie chart shows the main gases in the atmosphere.
It has been like this for about 200 million years.

1 (a) What are the two <u>main</u> gases in the air?
(b) What other gases are present in the air?

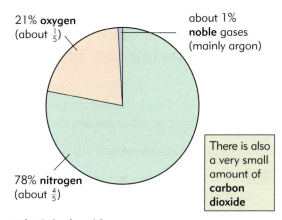

There is also a very small amount of **carbon dioxide**

What's in the air?

From earliest times

The Earth was formed about 4600 million years ago.
It was so hot that it was molten for millions of years.
Then, as it cooled, a solid **crust** formed. There were lots of
volcanoes. They poured out lava and a mixture of gases.
These gases formed the Earth's atmosphere.
It was a bit like the atmosphere on Venus.

2 Write down <u>two</u> differences between the Earth's early
atmosphere and the atmosphere today.

3 The Earth's early atmosphere was not suitable for
humans and other animals. Explain why.

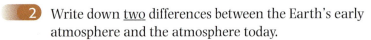
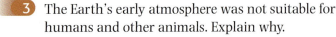

> **The atmosphere 4000 million years ago**
> * little or no oxygen
> * mainly carbon dioxide
> * some nitrogen and water vapour
> * small amounts of methane and ammonia

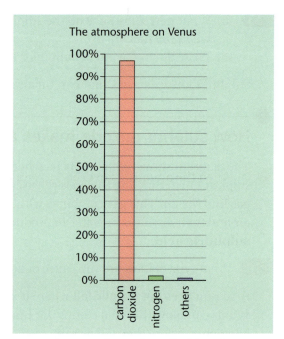

We need oxygen, but carbon dioxide poisons us.
So we couldn't live on Venus.

Changes soon began

The Earth continued to cool. Water vapour condensed and fell as rain. At first it was so hot that the water evaporated straight away. Later, parts of the surface were cool enough for some water to be liquid. By about 3800 million years ago, water was collecting in hollows on the surface. The first lakes and oceans were forming.

4 Write down the temperature at which water vapour can condense to form liquid water.

Carbon dioxide dissolves in water. So the amount of carbon dioxide in the atmosphere began to fall as the gas dissolved in the oceans and lakes.

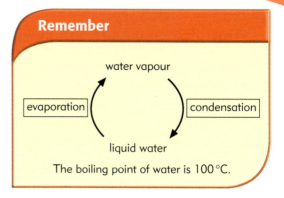

Remember

water vapour

evaporation | condensation

liquid water

The boiling point of water is 100 °C.

**The atmosphere
3800 million years ago**
- most of the water vapour lost
- carbon dioxide levels beginning to fall

Living things cause changes

Scientists are not sure about how life began or what it was like. The first living things were probably microorganisms that could use carbon dioxide in photosynthesis to produce food (carbohydrate), just like plants do.

carbon dioxide + water ⟶ carbohydrate + oxygen

These microorganisms took carbon dioxide from the atmosphere and released oxygen.

5 Write down the name of the gas that photosynthesis adds to the atmosphere.

The first bacteria and other microorganisms lived in an atmosphere that had little or no oxygen. But photosynthesis began to 'pollute' the atmosphere with oxygen. The oxygen was poisonous to some of the microorganisms. We say that they could not **tolerate** oxygen. So, gradually the number of habitats suitable for these microorganisms went down.

6 When the amount of oxygen increased, what do you think happened to the microorganisms that couldn't live where there was oxygen?

When oxygen was released, it reacted with the ammonia and methane in the atmosphere.
So the atmosphere changed even more.

methane + oxygen ⟶ carbon dioxide + water

ammonia + oxygen ⟶ nitrogen + water

7 Which gases took the place of methane and ammonia?

**The atmosphere
2200 million years ago**
- slow increase in oxygen levels
- nitrogen level increasing
- carbon dioxide level falling slowly

By 2200 million years ago, oxygen levels were high enough to oxidise iron; banded red ironstone rocks are evidence of this.

**The atmosphere
1000 million years ago**
- increasing levels of carbon dioxide, nitrogen and oxygen
- oxygen forms about 1% of the atmosphere

From the oceans to the land

As the atmosphere changed, tiny plants and then larger plants evolved. They all took in carbon dioxide and gave out oxygen. After millions of years, plants grew on the land too. Plants grew where there was light, water and a suitable temperature. So, the amount of oxygen in the atmosphere continued to increase.

8 (a) What was the percentage of oxygen in the atmosphere 400 million years ago?
(b) Why was the amount of carbon dioxide going down?

By 200 million years ago, the amounts of oxygen and carbon dioxide in the atmosphere became steady. Microorganisms and animals were using oxygen and producing carbon dioxide at the same rate that plants were doing the opposite. The atmosphere had become nearly the same as it is today.

> **The atmosphere
> 400 million years ago**
> ● oxygen level rising to 2% of the atmosphere
> ● carbon dioxide level falling

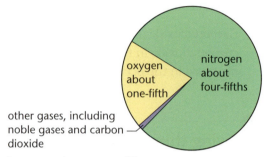

The atmosphere 200 million years ago.

Keeping carbon dioxide out of the atmosphere

Between 600 and 400 million years ago, many animals with shells evolved. The first ones were microscopic. Later, there were large animals such as corals and crinoids too. Most of these animals had hard parts made of calcium carbonate. When these animals died and sank to the bottom of the sea, their shells formed **carbonate rocks** such as limestone and chalk. These are **sedimentary** rocks and carbon can stay 'locked up' in them for millions of years.

9 (a) Write down the name of <u>one</u> carbonate rock.
(b) What is the main carbon compound in this rock?

Plants and animals break down the carbon compounds that plants make in photosynthesis. They release carbon dioxide back into the air.

However, this doesn't always happen. Oil is the remains of microorganisms that were not broken down. So we call it a **fossil fuel**. The carbon compounds in oil and other fossil fuels have been 'locked up' for millions of years.

10 Look at the picture. Where does the carbon 'locked up' in coal come from?

The tropical forests of the Carboniferous period formed much of the world's coal.

Higher

The amount of nitrogen increases

Today, about $\frac{4}{5}$ of the Earth's atmosphere is nitrogen. The Earth's early atmosphere contained only small amounts of nitrogen. Some of the nitrogen was the element itself. The rest of it was in a compound called **ammonia**. Ammonia dissolves really well in water. So most of the ammonia in the early atmosphere dissolved in the rain and went into the seas. Some plants and bacteria used this ammonia. They released most of the nitrogen from ammonia.

11 Explain how ammonia got into the seas.

12 Describe a chemical reaction that releases nitrogen from ammonia.

13 Chemical reactions that release nitrogen also take place in living things. What kind of bacteria release nitrogen into the atmosphere?

The ozone layer forms

As the amount of oxygen increased, **ozone** formed. It collected in a layer in the upper atmosphere.

The **ozone layer** is important for living things. It filters out ultra-violet radiation from the Sun. This radiation can harm living things. Too much ultra-violet radiation can cause disorders such as skin cancer and cataract. Cataract is a clouding of the lens of the eye. This is why doctors say that it harms us to get sun-burned. Some scientists think that it was the ozone layer that let the first land plants evolve. Once plants managed to survive on the land, animals went there too.

14 Look at the diagram. Where is the ozone layer?

15 Write down <u>two</u> reasons why the ozone layer is important for living things.

Remember

Photosynthesis adds oxygen to the atmosphere.

ammonia + oxygen → nitrogen + water

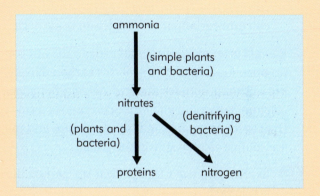

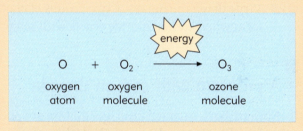

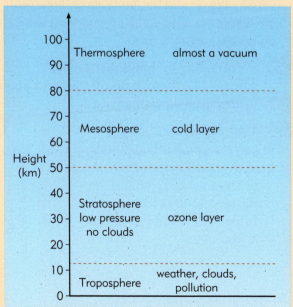

Layers of the atmosphere. Scientists usually call 150 km the top of the atmosphere.

'Locked-up' carbon dioxide is released

The carbon in fossil fuels and carbonate rocks is not 'locked-up' forever. Sometimes Earth movements bury carbonate rocks so deeply in the Earth that they decompose and carbon dioxide is released. It gets back into the atmosphere through volcanoes.

$$\text{calcium carbonate} \xrightarrow{\text{heat}} \text{calcium oxide} + \text{carbon dioxide}$$

We add even more carbon dioxide to the atmosphere when we burn fossil fuels. Some of this carbon dioxide ends up in the sea as carbonates and hydrogencarbonates. Insoluble calcium carbonate and other carbonates form sediments. Hydrogencarbonates, mainly of calcium and magnesium, are soluble so they stay in the water.

16 Explain <u>two</u> ways that 'locked-up' carbon dioxide is released.

17 Write down the names of <u>two</u> groups of chemicals that form when carbon dioxide reacts with sea water.

A problem with carbon dioxide

Now, we burn so much fossil fuel that we are putting more carbon dioxide into the atmosphere than plants and the sea are taking out. This carbon dioxide acts like a blanket and makes the Earth warmer.

18 Write down <u>two</u> things that we can do to stop the amount of carbon dioxide in the atmosphere increasing further. Explain your answers.

19 The graph shows how the percentage of carbon dioxide in the Earth's atmosphere is believed to have changed over the past 4500 million years. Explain these changes as fully as you can.

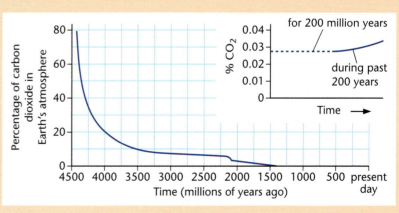

The changing Earth

The Earth's crust formed when molten material cooled and became solid. New rocks still form like this today. We call them **igneous** rocks.

We call molten rock <u>magma</u>.

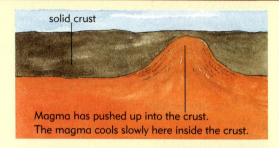

solid crust

Magma has pushed up into the crust.
The magma cools slowly here inside the crust.

Two kinds of igneous rock

Basalt forms from molten rock that cools down quickly outside the Earth's crust. The diagram shows how another igneous rock forms. This rock is **granite**, and it forms from molten rock that cools down slowly inside the Earth's crust.

Basalt is called an **extrusive** igneous rock because it forms outside the Earth's crust. Granite is called an **intrusive** igneous rock because it forms inside the Earth's crust.

Granite has large crystals.

Basalt has small crystals.

Rocks under the microscope. You can see crystals of a variety of minerals.

1. Write down <u>two</u> differences in the ways that basalt and granite form.

2. Look at the pictures of thin slices of basalt and granite.
What is the difference in the crystal size between the two rocks? Give a reason why there is a difference in crystal size.

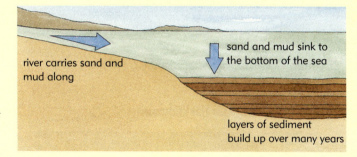

river carries sand and mud along

sand and mud sink to the bottom of the sea

layers of sediment build up over many years

New rocks from old

Old rocks break down and become the raw materials for making new rocks.

Rocks in the Earth's crust
↓ weathering
Bits of rock (e.g. sand and mud)
↓ erosion and deposition
Sediments
↓ cementing
Sedimentary rocks

Sandstone, mudstone and limestone are examples of sedimentary rocks.

3. Sand from a sandstone in a hillside can become a new sedimentary rock.
Write a few sentences about how this happens.

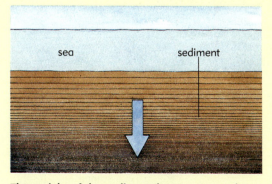

sea

sediment

The weight of the sediment layers presses down on the layers below. This squeezes out any water. Natural chemicals stick the bits of sand and mud together. We say they are cemented.

Changed rocks

Earth movements bury rocks deep inside the Earth's crust. These rocks get squeezed very hard and become very hot. The high pressure and high temperature can change rocks into new kinds of rocks called **metamorphic** rocks. This can happen to sedimentary, igneous and metamorphic rocks. Marble is a hard rock made of crystals. It can be polished to give a shiny surface. It is a metamorphic rock. Marble is made from limestone by heat and pressure. Limestone is a sedimentary rock. It is a fairly soft rock that contains shells.

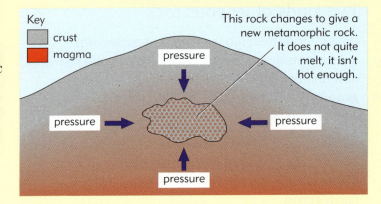

sediments ⟶ sedimentary rocks

cooling
molten magma ⟶ igneous rocks

The rock cycle

The materials that rocks are made from are being **recycled** all the time.

4 Copy and complete the diagram.

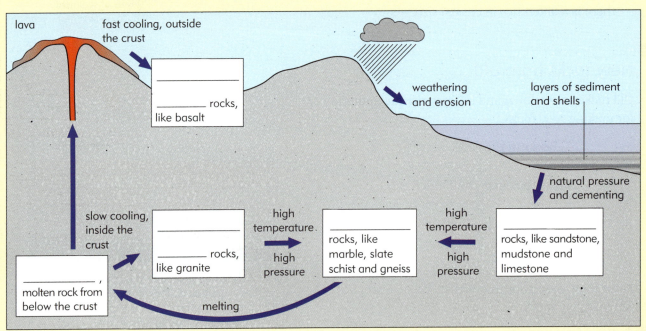

What shape is the Earth's surface?

The Earth is nearly spherical. This means that it is round like a ball, although it is not perfectly round.

Look at the picture.

5 Write down <u>two</u> things that mean the Earth is not perfectly round.

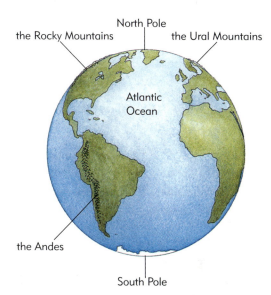

The surface of the Earth is bumpy. The Earth is slightly flattened at the North and South Poles.

Inside the Earth

Even the deepest drill cannot make a hole through the Earth's crust. However, scientists think that the Earth is made of different layers. They have found out a bit about these layers by studying the way vibrations from explosions and earthquakes travel through the Earth. The diagram shows what these layers are probably like.

Sometimes part of the lower crust or the outer mantle melts. We call the molten rock **magma**.

The Earth is much heavier than if it were only rock inside. We say that it has a higher **density**.

6 Why is the whole Earth much denser than the rocks that we find in the Earth's crust?

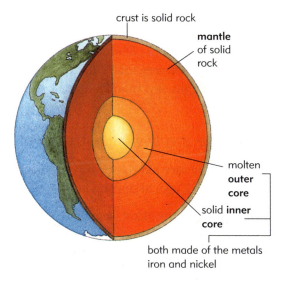

Nickel and iron are much heavier materials than rock. We say that they are denser than rock. The mantle is made of denser rock than the crust. This rock is solid, but the upper part is so near to its melting point that it is hot enough to flow very slowly.

Changes to the Earth's crust

The Earth's crust is changing all the time. Some of the changes happen quickly, but other changes are slow and can take millions of years. Look at the two diagrams and the two text boxes.

7 Write down <u>two</u> things that can change the Earth's crust quickly.

New rocks can form when things happen to the Earth's crust. Some of the changes that happen to the Earth's crust wear away the hills and mountains. Other changes help to form new hills and mountains.

8 Write down <u>two</u> changes that help to make new mountains.

When new mountains are made, rocks can be pushed up thousands of metres. This needs huge forces.

Layers of rock

Sediments that form sedimentary rock sometimes stop forming and then start again. This means that sedimentary rocks often have **layers** in them. Sometimes the layers are very thick, up to several metres. Other layers are very thin, less than a millimetre.

Where are the oldest layers?

Rocks are very old compared to people.
But in rock age, 80 million years is really quite young.

9 Where is the oldest layer of sedimentary rock in a cliff made of horizontal layers? Give a reason for your answer.

Even if rocks have moved we can often tell which ones are the oldest.

10 Look at the rock layers A, B, C, D.
(a) Which is the youngest rock?
(b) Which is the oldest rock?

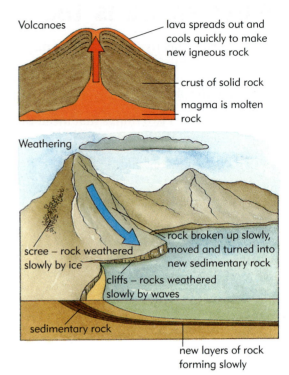

Volcanoes — lava spreads out and cools quickly to make new igneous rock
crust of solid rock
magma is molten rock

Weathering
scree – rock weathered slowly by ice
cliffs – rocks weathered slowly by waves
rock broken up slowly, moved and turned into new sedimentary rock
sedimentary rock
new layers of rock forming slowly

Earthquakes	Rocks folding
The crust can suddenly move or break in an earthquake.	Large forces can make rocks fold slowly. When rocks are folded upwards, they can form new hills and mountains. Some rocks can be changed into metamorphic rocks.

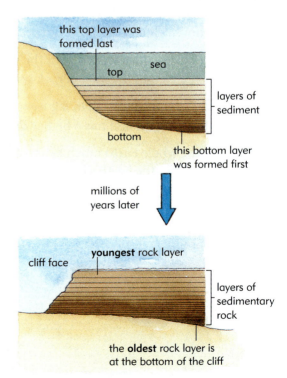

this top layer was formed last
sea
top
layers of sediment
bottom
this bottom layer was formed first

millions of years later

youngest rock layer
cliff face
layers of sedimentary rock

the **oldest** rock layer is at the bottom of the cliff

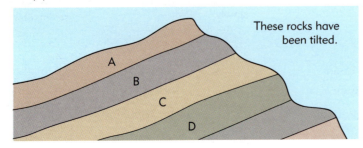

These rocks have been tilted.

A
B
C
D

Evidence for Earth movements

We often find layers of rock that have been tilted, folded and even broken or fractured. We call fractures in rocks <u>faults</u>. Folds and faults are evidence that the Earth's crust moves.
It takes very large forces to fold and fracture rocks. Earthquakes happen when rocks move at a fault.

All this is evidence that the Earth's crust is unstable.

11 Write down <u>three</u> different ways in which rocks can be moved.

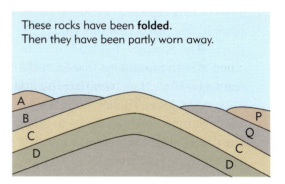

These rocks have fractured. They have broken and slipped. We say that they have been **faulted**.

Rocks that have moved

We can match the layers of rock that have been moved along a fault line or folded and partly worn away.

12 Find the rock on the right of the fault that is the same age as rock C.

13 Which rock in the folded rocks is the same as rock B?

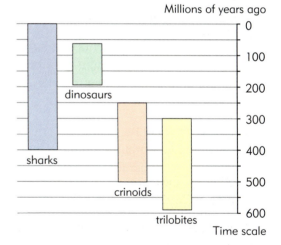

These rocks have been **folded**. Then they have been partly worn away.

Upside-down rocks

Sometimes layers of sedimentary rocks can get turned upside down by Earth movements. We know this has happened because of the ages of the fossils in the rocks.

How do we know when sediments are upside down?

In sedimentary rocks we often find **fossils**. These are the remains of plants and animals. Fossils tell us which kinds of plants and animals lived when the sediments were formed. Rocks that contain the same kinds of fossils are probably about the same age.

The bars show the ages of the fossils you might find.

14 (a) Which kinds of rock contain fossils?
(b) Why don't you find fossils in igneous rocks?

15 How old are the following rocks?
(a) Rocks that contain dinosaur bones.
(b) Rocks that contain trilobites but not crinoids.
(c) Rocks that contain both sharks and crinoids.

16 Look at the diagram. A friend says that these are sedimentary rocks so the oldest rocks must be at the bottom. Do you agree? Give a reason for your answer.

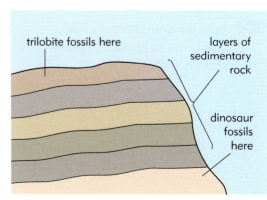

Movements that make mountains

Movements of the Earth's crust can push up rocks for thousands of metres. We find rocks at the tops of mountains that formed from sediments in the sea. It takes a long time and huge forces for this to happen.

1 Look at the diagram. How do we know that rocks can be pushed up thousands of metres?

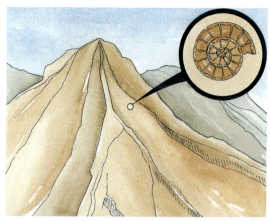

This fossil came from an animal that lived in the sea millions of years ago.

Some mountains are old, others are new

Different ranges of mountains were pushed up at different times during the Earth's history. As old mountains wore down in one place, new mountains were pushed up in others. The Scottish Highlands used to be part of a huge range of high mountains that formed 450 to 550 million years ago. Since then, **weathering** and **erosion** have worn them down. Higher mountains such as the Alps are much younger. They formed 7 to 25 million years ago. The Andes and the Himalayas are even higher and they are still being pushed up.

2 Younger mountain ranges are usually higher than older mountain ranges. Explain why.

We call the Earth movements that build mountains tectonic activity. Where there is tectonic activity, there are volcanoes. So, we find igneous rocks as well as sedimentary rocks in mountains.

> **Remember**
>
> We call rocks that are altered by heat and pressure metamorphic rocks.

The movements in the Earth's crust during mountain-building produce high temperatures and pressures. So we find **metamorphic** rocks in places where mountains are forming. We also find metamorphic rocks where mountains formed in the past.

3 Why does mountain-building also result in the formation of metamorphic rocks?

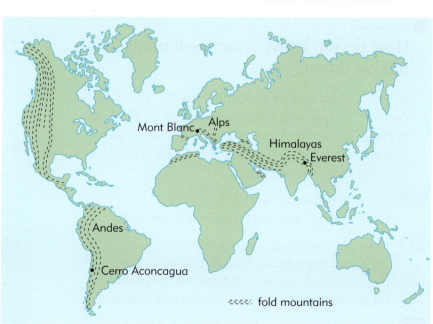

Mont Blanc Alps Himalayas Everest Andes Cerro Aconcagua ⟨⟨⟨⟨ fold mountains

Earth movements cause earthquakes

Earthquakes happen when there is movement at faults. The most unstable parts of the Earth's crust are where the biggest earthquakes happen. Earthquakes sometimes kill thousands of people so finding out when one will happen is important. If scientists can predict when and where an earthquake will happen, people can move to a safer place.

Movement of rocks along the San Andreas fault caused the earthquake that destroyed this bridge in California.

The problem of predicting earthquakes

Scientists have set up **seismic stations** all round the world. These stations automatically record earthquakes. The scientists use the records to find out exactly where each Earth movement happened. They also look for patterns in the records to try to predict when and where earthquakes will happen. Another earthquake is expected in California soon. Unfortunately, scientists can't predict exactly where or when.

4 A lot of time and money is spent measuring earthquakes and trying to predict when and where they will happen.
Why do you think this is?

Predicting when a volcano will erupt

Predicting **volcanic eruptions** is also hard. Scientists also keep a close watch on volcanoes. They measure temperatures, pressures and the gases given off. This can be difficult and dangerous work. They can sometimes measure an increase in pressure and say that there will be an eruption in the next few months. But there are so many factors involved that, often, they cannot be more accurate.

5 Scientists cannot predict earthquakes and volcanic eruptions accurately. Write down a list of reasons for this.

Why are some parts of the Earth's crust more unstable than others?

The Earth's crust and the upper part of the mantle are called the **lithosphere**.

The lithosphere is not made of one big piece of rock. Cracks split it into very large pieces called **tectonic plates**. The map shows some of them. The plates move all the time. They do not move very fast, just a few centimetres (cm) each year. But these small movements add up to big movements over a long time.

The lithosphere is most unstable in the places where two plates meet. These are the places that have earthquakes and volcanic eruptions.

Did you know?

Tectonic plates move hardly any faster than your finger nails grow. They move at a slower rate than your hair grows.

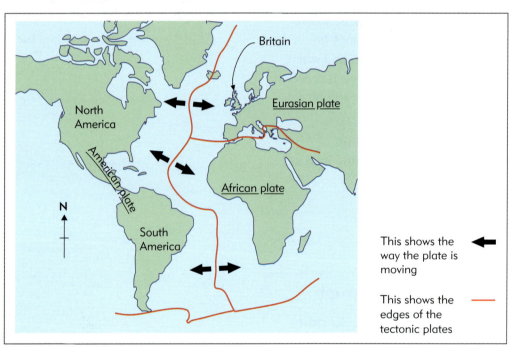

Look at the map.

6 Why do we have few earthquakes in Britain?

7 (a) Which way is the American plate moving?
  (b) Which way is the Eurasian plate moving?
  (c) What is happening to the distance between America and Europe?

8 A plate moves about 5 cm each year.
   How far will the plate move in 1000 years?

How do we know that plates move?

South America and Africa are on different plates. These plates have been moving away from each other for millions of years. Long ago, South America and Africa must have been together. Look at the diagrams.

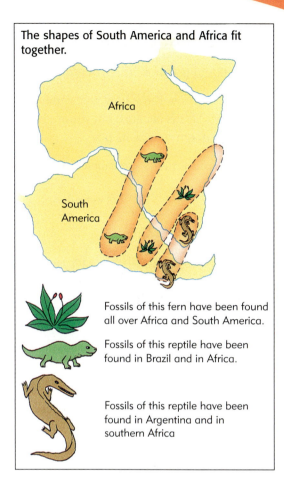

The shapes of South America and Africa fit together.

Africa

South America

The Earth today

The Earth millions of years ago

Fossils of this fern have been found all over Africa and South America.

Fossils of this reptile have been found in Brazil and in Africa.

Fossils of this reptile have been found in Argentina and in southern Africa

9 The shapes of South America and West Africa tell us that they were once together. Explain why.

10 The rocks in South America and Africa also tell us that they were once joined together. Explain why.

How moving plates can make new mountains

Some of the plates of the Earth's crust are moving away from each other. Other plates are pushing into each other. When two plates move towards each other they can force rocks upwards to give new mountains.

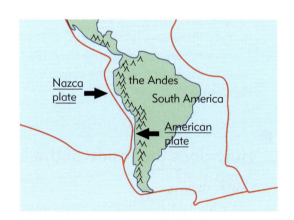

Nazca plate → the Andes   South America

American plate ←

11 Look at the map.
  (a) Which <u>two</u> plates are pushing into each other?
  (b) Which mountains have been formed by these moving plates?

What keeps the Earth's crust moving?

Earthquake vibrations pass only through solids. So we know that the Earth's mantle is solid. But it is so close to its melting point that it can flow like a liquid. It flows very slowly. When there are movements in the mantle, the tectonic plates above also move.

Remember

The Earth's lithosphere is cracked into pieces called tectonic plates. These plates move around slowly.

What makes liquids move?

If a liquid gets hot, it moves around.

The diagrams show how water moves around when you heat it. This is because hot water rises and cold water moves down to take its place.

These movements are called **convection currents**.

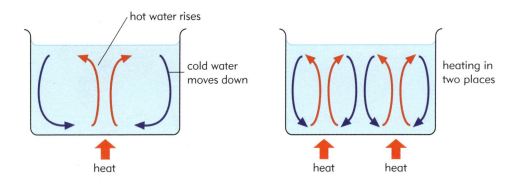

Convection currents inside the Earth

Heat produced inside the Earth causes slow convection currents in the mantle.

The diagram shows how convection currents in the mantle make the plates move. Plates A and B are moving together. Plates B and C are moving apart. The plates can move because they 'float' on top of the mantle. There are very slow convection currents in the mantle.

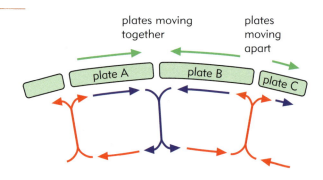

Movements of the mantle move the tectonic plates.

How does the inside of the Earth keep hot?

Something must be heating up the mantle or there wouldn't be any convection currents.

Radioactive substances inside the Earth produce the heat that is needed. The diagram shows how they do this.

12 What does a uranium atom have to do to release energy?

13 Radioactive substances will keep on heating up the inside of the Earth for a long time to come. Explain why.

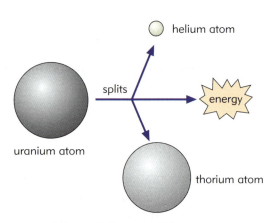

The Earth formed about 4.5 billion (4 500 000 000) years ago. Since then, about half of the uranium atoms have split up.

What happens when the tectonic plates move apart?

When tectonic plates push against into other, new mountains form.The diagram shows what happens when plates move apart.

14 (a) What type of new rock spreads out through the cracks between plates?
(b) Why is this type of rock formed?

15 Look at the map. Write down the name of a country where basalt is forming.

The edges of the plates that are moving apart are usually under the sea.

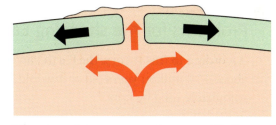

When two plates move apart, magma fills the gap. The magma quickly cools to form the igneous rock called basalt.

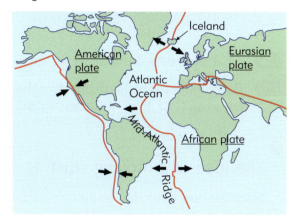

Changing ideas about the Earth

Until about 200 years ago, most people believed that the mountains, valleys and seas on the Earth had always been just like they are today. Many of these people thought the Earth was created only a few thousand years ago.

Then geologists started to study the rocks and to think about how they were formed. They realised that the Earth must be many millions of years old.

1 Why did they think that the Earth must be millions of years old?

2 What else did they then need to explain?

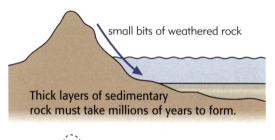

small bits of weathered rock

Thick layers of sedimentary rock must take millions of years to form.

Mountains would be completely worn away over millions of years. So geologists need to explain how new mountains are formed.

A cooling, shrinking Earth

The diagram shows one theory about how new mountains are formed.

3 Write down the following sentences in the correct order.

- The molten core carries on cooling, but more and more slowly. It shrinks as it cools.
- The Earth began as a ball of hot, molten rock.
- The shrinking core makes the crust wrinkle. The high places become mountains, the low places become seas.
- As the molten rock cooled, a solid crust formed.

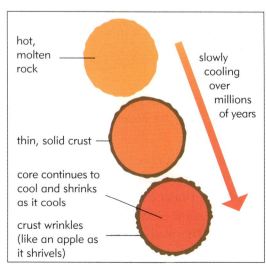

hot, molten rock

slowly cooling over millions of years

thin, solid crust

core continues to cool and shrinks as it cools

crust wrinkles (like an apple as it shrivels)

Problems for the shrinking Earth theory

According to this theory, the Earth can't be more than about 400 million years old or it would be cool and completely solid by now.

We now know that the Earth is a lot older than 400 million years. We know this because the Earth contains quite a lot of radioactive elements such as uranium. The atoms of these elements gradually decay (break up). Heat is released as they do so.

4 What effect does this have on the Earth's core?

5 The oldest rocks on Earth are more than 3.5 billion years old. How do scientists know this?

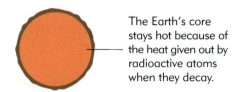

The Earth's core stays hot because of the heat given out by radioactive atoms when they decay.

Dating rocks

Scientists can measure

- the amounts of radioactive atoms in rocks
- the amounts of atoms produced when the radioactive atoms decay.

This tells them how old the rocks are.

The idea of a moving crust

Scientists now think that mountains are formed by the Earth's crust moving about. Alfred Wegener first suggested this idea during the early years of the twentieth century. But most scientists didn't agree for another 50 years. This idea was called the **theory of continental drift**.

6 Why was Wegener's theory of crustal movement called 'continental drift'?

7 What evidence did scientists have for continental drift?

8 Why did many scientists not agree?

During the 1950s, scientists started to explore the rocks at the bottom of the oceans. The diagrams show what they found and how they explained it.

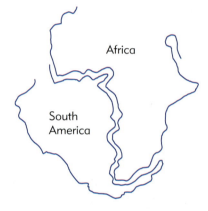

Some scientists suggested that South America and Africa must once have been together. Other scientists said that there was no way they could possibly have moved apart.

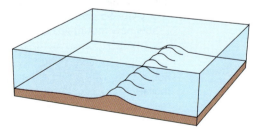

There are long mountain ridges underneath the ocean. They are made of young rocks.

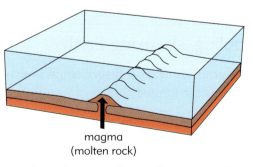

magma (molten rock)

Sections of crust on the sea floor are moving apart. New rock forms to fill the gap.

The new evidence convinced scientists that the Earth's crust is made of a small number of separate sections called **plates**. Under the oceans these plates are moving apart. But in some places these plates are moving towards each other. This pushes rock upwards to make new mountains.

Higher

More about tectonic plates

The Earth's lithosphere is cracked into a number of **tectonic plates**, which are slowly moving. Movement is caused by **convection currents** in the Earth's mantle. Tectonic plates vary in thickness from 50 to 200 km.

The boundaries between the plates are called **plate margins**. At these plate margins, three different things can happen:

- the plates can slide past each other;
- the plates can move towards each other;
- the plates can move apart.

9 Where on the map are the plates sliding past each other?

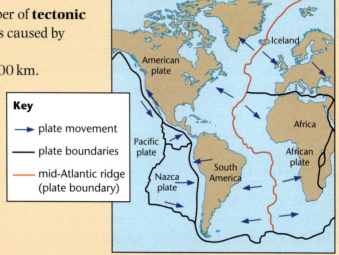

Key

→ plate movement

— plate boundaries

— mid-Atlantic ridge (plate boundary)

Plates sliding past each other

When two plates slide past each other, no crust is created or destroyed, but friction between the two plates leads to earthquakes.

San Francisco, in California, USA, sits on top of the San Andreas fault, where movement is about 5 cm per year. Small earthquakes happen often but Californians fear 'the big one'.

10 (a) What do we call the margin between two plates that are sliding past each other?

11 (b) What effect do the sliding plates have?

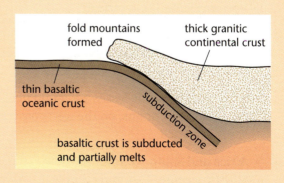

Plates moving towards each other

Oceanic crust is denser than continental crust. So when they are pushed against each other the oceanic crust is forced down or **subducted**. The huge forces crumple and push up the layers of rock in the continental crust to form a **fold mountain chain**. In places where the subducted crust melts, the magma rises and volcanoes form.

12 (a) Outline what happens when an oceanic and a continental plate move towards each other.

(b) Where, on the map at the top of the page, are mountains being made in this way?

Plates moving apart

Where two plates are moving apart, the oceanic crust cracks. Magma rises to fill the cracks. It cools and solidifies, forming new oceanic crust. New crust is added all the time as new cracks open. This is called **sea floor spreading**.

Sea floor spreading is happening in the middle of the Atlantic Ocean in a line between Europe and Africa and the Americas. We call it the **mid-Atlantic ridge**.

Occasionally the oceanic ridge rises above sea level to form volcanic islands. For example, the island of Surtsey off Iceland rose out of the Atlantic Ocean in 1963 from a large submarine volcanic eruption.

13 Explain the positions of the rocks on Iceland and describe what is happening at this plate boundary.

Scientists have found some interesting support for the theory of sea floor spreading. The Earth's magnetic field reverses from time to time. When magma solidifies, the iron-rich minerals in it line up with the magnetic field at the time.

Scientists looked at the magnetism of the rocks on either side of oceanic ridges. They found a pattern of magnetic stripes parallel to the ridges. This **magnetic reversal pattern** was the same on both sides of the ridges. In sea floor spreading, new rock moves away on both sides of a ridge, so the pattern was fairly symmetrical.

14 Describe the pattern of magnetic stripes found in rocks in certain places on the sea bed and explain how the stripes were formed.

15 Explain why marine fossils are found high in the Andes mountains, on the west coast of South America.

16 Africa and South America are now about 5000 km apart. Assuming that plates move apart at an average speed of 2.5 cm per year, how long ago did they form one land mass?

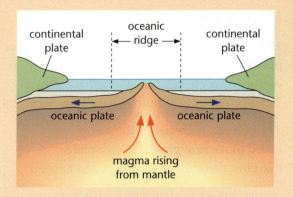

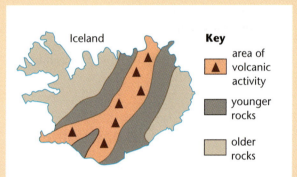

The island of Iceland sits astride an oceanic ridge called the mid-Atlantic ridge.

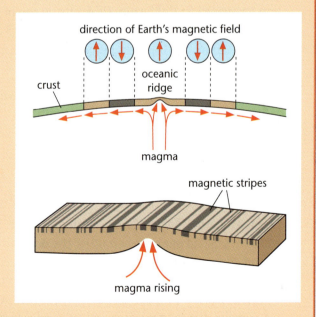

Structure and bonding

How substances can change their state

Water is often a liquid. But it can also be a solid or a gas. Many other substances can also be a solid, liquid or gas. What state they are in depends on the temperature.

1 What do we call water when it is in the:
   (a) solid state?
   (b) gas state?

2 What do we call the change when:
   (a) a solid changes into a liquid?
   (b) a liquid changes into a solid?
   (c) a liquid changes into a gas?
   (d) a gas changes into a liquid?

Scientists use the idea of **particles** to explain changes of state.

Why do solids melt?

If we heat up a solid we give the particles more <u>energy</u>. They can move from side to side more quickly. When they get enough energy the particles break away from their fixed positions. They are then free to move around each other.

3 Look at the diagrams.
   Describe what happens to the particles in a solid as it melts.

- Substances can be solids, liquids or gases. We call these the three states of matter.

- All substances are made of very small particles.

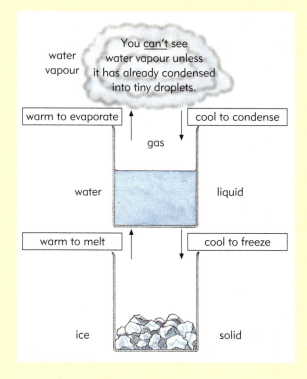

water vapour

You <u>can't</u> see water vapour unless it has already condensed into tiny droplets.

warm to evaporate   cool to condense

gas

water   liquid

warm to melt   cool to freeze

ice   solid

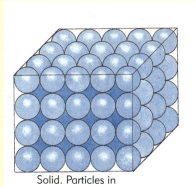

Solid. Particles in fixed positions.

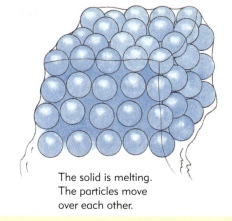

The solid is melting. The particles move over each other.

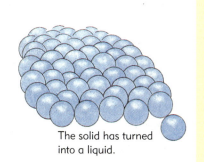

The solid has turned into a liquid.

The temperature at which a solid melts is called the <u>melting point</u> of that solid.

4 What is the melting point of:
(a) ice?
(b) gold?

Gold melts if it is hotter than 1064 °C.

Ice melts when it is warmer than 0 °C.

Why do liquids evaporate?

Liquids can evaporate even when they are quite cool. If you spill petrol on to a garage floor it 'dries up' quickly. The petrol <u>evaporates</u>.

We can make liquids evaporate more quickly if we heat them up.

5 Look at the diagram.
(a) What happens to the particles in a liquid if we heat it up?
(b) What happens to the particles in a liquid when they gain enough energy?
(c) What do the escaped particles form?

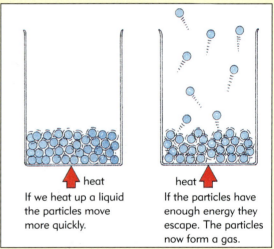

If we heat up a liquid the particles move more quickly.

If the particles have enough energy they escape. The particles now form a gas.

What happens when a liquid boils?

If we heat a liquid its temperature will rise. If we keep on heating it, the liquid will <u>boil</u>. When it is boiling the temperature of the liquid stays the same. This temperature is called the **boiling point**.

6 A kettle is switched on to make a cup of tea. The graph shows the temperature of the water inside the kettle.
(a) At what temperature did the water boil?
(b) How long did it take for the water to reach the boiling point?

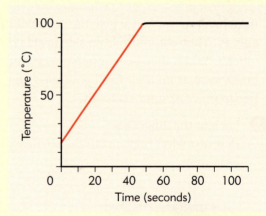

7 The diagram shows what happens when a liquid boils.
(a) How can you <u>see</u> when a liquid is boiling?
(b) Where have the bubbles of gas come from?

When a liquid is boiling, <u>all</u> the energy we supply is used to change the liquid into gas. So the temperature stays the same.

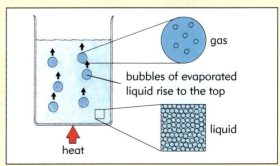

gas

bubbles of evaporated liquid rise to the top

liquid

A boiling liquid.

Elements, compounds and reactions

All chemical substances are made from tiny atoms. There are about 100 different kinds of atoms in nature. If a substance is made from just one kind of atom, we call it an element.

1 How many elements do you think there are? Give a reason for your answer.

Carbon is an element. It contains only carbon atoms.

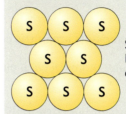
Sulphur is an element. It contains only sulphur atoms.

Using letters to stand for elements

We can save time and space by using our initials instead of writing our full name.

Kenneth Gill's briefcase.

These are Kenneth Gill's initials

In science we often use the initials of an element instead of the whole word. We call these letters the **symbols** of the elements.

The table shows some of these symbols.

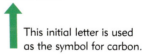
Carbon

This initial letter is used as the symbol for carbon.

2 What is the symbol for:
  (a) carbon?
  (b) sulphur?

3 (a) What are the symbols for calcium and for silicon?
  (b) Why do you think these elements need to have a second, smaller letter in their symbol?

Some of the symbols we use come from the old names of the elements.

4 (a) What is the name of the element that used to be called cuprum?
  (b) What is the old name for sodium?
  (c) What is the symbol for sodium?
  (d) What is the symbol for copper?

Element	Symbol we use
carbon	C
calcium	Ca
copper	Cu from *cuprum*, the old name
nitrogen	N
neon	Ne
sulphur	S
silicon	Si
sodium	Na from *natrium*, the old name

What are compounds?

When atoms of different elements join together we get substances called **compounds**.
Most substances are compounds.

The diagrams show some compounds. Each compound has its own **formula**. The formula of a compound tells us two things:

- It tells us which elements are in the compound.

- It tells us how many atoms of each element there are in the compound.

5 (a) What elements are in carbon dioxide?
(b) How many hydrogen atoms are shown in the formula for the compound we call water?

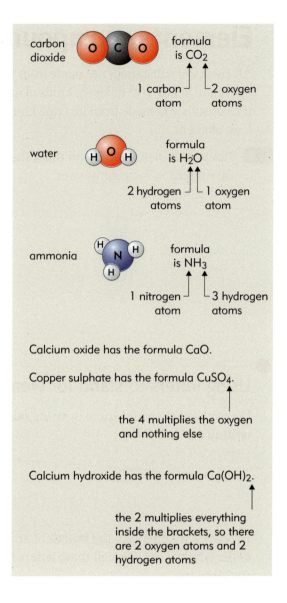

Calcium oxide has the formula CaO.

Copper sulphate has the formula $CuSO_4$.

the 4 multiplies the oxygen and nothing else

Calcium hydroxide has the formula $Ca(OH)_2$.

the 2 multiplies everything inside the brackets, so there are 2 oxygen atoms and 2 hydrogen atoms

How to describe chemical reactions

The diagram shows the chemical reactions in a barbeque. In a chemical reaction the **reactants** are the substances you use at the start. These turn into **products**, the substances left at the end.

In the barbecue reaction:

- the reactants are oxygen and carbon;
- the product is carbon dioxide.

6 Magnesium reacts with dilute hydrochloric acid to produce hydrogen gas and a solution of magnesium chloride.
(a) Name the reactants in this reaction.
(b) Name the products in this reaction.

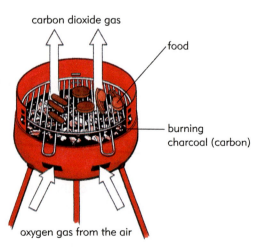

In a barbecue, carbon reacts with oxygen to produce carbon dioxide gas.

Writing word equations

In the barbecue reaction, carbon reacts with oxygen to produce carbon dioxide.

We can write this:

carbon + oxygen ⟶ carbon dioxide

We call this a **word equation**.

7 Zinc reacts with copper sulphate to produce zinc sulphate solution and copper metal. Write down a word equation for the reaction between zinc and copper sulphate solution.

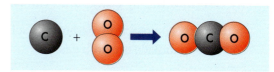

Carbon reacts with oxygen to produce carbon dioxide.

Understanding symbol equations

There is another way to write down what happens in a chemical reaction. We can replace the <u>name</u> of each reactant and product with a **formula**.

For the barbecue reaction the two kinds of equation look like this:

carbon   +   oxygen   ⟶   carbon dioxide

C + O_2 ⟶ CO_2

We call the second one a **symbol equation**.

The box shows the symbol equations for the reactions in questions 6 and 7.

8 Copy each of the symbol equations from the box. Write the name of each reactant and product underneath the right formula.

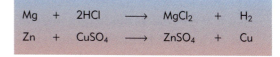

$$Mg + 2HCl \longrightarrow MgCl_2 + H_2$$
$$Zn + CuSO_4 \longrightarrow ZnSO_4 + Cu$$

The atoms of elements that are gases often go round in pairs.

9 Write down the names of <u>three</u> gases, besides oxygen, with atoms that go round in pairs.

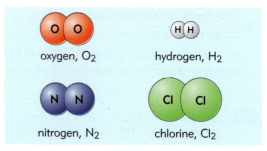

oxygen, O_2 hydrogen, H_2

nitrogen, N_2 chlorine, Cl_2

Oxygen is a gas. Oxygen atoms go round in pairs. The atoms of some other gases also go round in pairs.

Adding state symbols

Reactants and products can be solids, liquids or gases, or can be dissolved in water.
We can show this by using state symbols.

In the barbecue reaction, carbon is a solid, and oxygen and carbon dioxide are both gases.

We can now write the equation like this:

$$C(s) \ + \ O_2(g) \longrightarrow CO_2(g)$$

(s)	means	solid
(l)	means	liquid
(g)	means	gas
(aq)	means	aqueous, this means solutions of substances dissolved in water e.g. HCl(aq)

What state symbols mean.

10 Add state symbols to the symbol equations for the reactions in questions 6 and 7. Remember that solutions need the state symbol (aq).

Where did the idea of atoms come from?

The most important theory in chemistry is the **atomic theory**. An Englishman called John Dalton put forward this theory about 200 years ago. His idea was that there is only a small number of different types of atoms. These atoms can join together in different ways to produce millions of new substances.

Earlier ideas about atoms

The idea of everything being made of small particles is actually very old. It was first thought of 2000 years ago by the ancient Greeks (see Box). Much later, in the seventeenth century, Robert Boyle used the idea to explain why you can squeeze air and other gases into a smaller space. Dalton had read about Boyle's work and realised that he could use the idea of particles to explain discoveries that chemists had made during the eighteenth century. One of these discoveries was that there is a small number of simple substances, or **elements**, from which all other substances are made.

Where the word 'atom' comes from

The Greek word *atomos* means 'can't be cut'.

1 (a) When and where was the idea of atoms first used?
  (b) What does the word 'atom' mean?
  (c) Where did Dalton meet the idea of particles?

The idea of elements

The idea of everything being made from a small number of elements also dates back to the ancient Greeks. But their idea of what these elements were – air, earth, fire and water – was completely wrong. But alchemists believed in this idea for nearly 2000 years. This is why they kept on trying to do impossible things such as changing lead into gold.

2 Why did alchemists think that they should be able to change lead into gold?

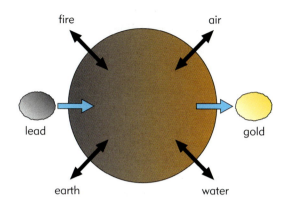

The alchemist's quest

By getting the right mix of the four elements you could turn lead into gold.

How chemists discovered elements

During the eighteenth century, chemists realised that there were some substances that they couldn't split up into anything simpler. So they decided that these substances must be the simplest possible substances. In other words these substances were <u>elements</u>.

The diagram shows the kinds of observations that chemists made to find out which substances were elements.

3 (a) Is the white powder an element?
　　　Give a reason for your answer.
　(b) Is the yellow powder an element?
　　　Give a reason for your answer.
　(c) Is the shiny bead of metal an element?
　　　Give a reason for your answer.

The following word equations describe the reactions shown in the diagram:

lead carbonate → lead oxide + carbon dioxide

lead oxide + carbon → lead + carbon dioxide

4 (a) What is the white powder?
　(b) What is the yellow powder?
　(c) What is the shiny metal?

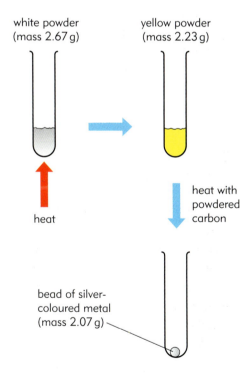

white powder
(mass 2.67 g)

yellow powder
(mass 2.23 g)

heat

heat with
powdered
carbon

bead of silver-
coloured metal
(mass 2.07 g)

No known chemical reaction using the silver-cloured metal will produce a substance with a smaller mass.

So, by the end of the eighteenth century, chemists had the idea of an element based on their experiments. An element was a substance that could not be split up into simpler substances (see Box). All other substances were compounds made from two, or more, different elements. By using chemical reactions, you could split up compounds into the elements they were made from.

5 By the end of the eighteenth century, chemists knew that it was impossible to turn lead into gold. Explain how they knew this.

Some of the substances that chemists in the late eighteenth century knew were elements

oxygen	sulphur	copper
hydrogen	charcoal	gold
	iron	
	lead	
	silver	
	mercury	

Dalton's atomic theory

Dalton explained <u>why</u> there are different elements by using these ideas:

- elements are made up of small, indestructible particles which can be called **atoms**;

- the atoms of a particular element are **identical**, for example in size and weight (mass);

- the atoms of different elements are **different**;

- **compounds** are formed by atoms of different elements joining together.

6 How would Dalton explain why it is impossible to change lead into gold?

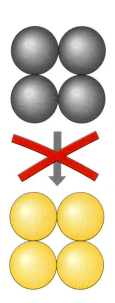

lead is made of indestructible lead atoms

changing lead into gold is impossible

gold is made of indestructible gold atoms

Dalton's ideas also explained another important discovery that chemists had made during the eighteenth century. They had found that when a chemical reaction happens inside a closed container there is never any change in mass.

7 Use Dalton's atomic theory to explain:
(a) why there is no change in mass when mercury reacts with oxygen inside a closed container;
(b) why there is a change in mass when mercury reacts with oxygen in the open.

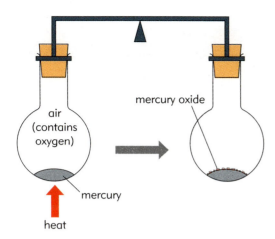

A chemical reaction in a sealed container. There is no change in mass.

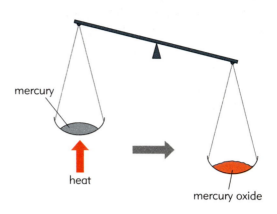

When the reaction occurs in the open there is an increase in mass.

What makes Dalton's theory scientific?

The problem with the ancient Greeks' idea of atoms was not that it was wrong but that they had no way of testing it. It was <u>just</u> an idea. There was no way of finding out whether it was right or wrong. So it wasn't really a scientific theory.

But Dalton's idea of atoms was not just an idea. It was a <u>scientific theory</u> which explained many observations and measurements that scientists had made. As scientists made lots more observations and measurements, the theory explained all of them too. So scientists became more and more sure that the theory was correct.

8 Explain why:
    (a) Dalton's atomic theory is a genuine <u>scientific</u> theory;
    (b) most other chemists at the time quickly accepted Dalton's theory;
    (c) Dalton's theory is now very firmly established.

How the news about Dalton's theory spread

Dalton earned his living as a schoolteacher and private tutor. He had to use his spare time to study science and to do scientific experiments. He spoke in public about his scientific ideas at the Manchester Literary and Philosophical Society. These talks were then published in print so many other people, including other scientists, soon knew about his new atomic theory.

9 How did other chemists very quickly find out about Dalton's atomic theory?

Higher

Dalton's theory succeeds again

By the end of the eighteenth century, chemists had also discovered that each particular chemical compound always contains the same elements in exactly the same proportions by weight (mass).

Water, for example, is always made up of 8 grams of oxygen combined with 1 gram of hydrogen.

Dalton explained this by suggesting that the atoms of different elements in a particular compound always join together in the same proportions. In some compounds there might be equal numbers of atoms of each element. In other compounds there might be two atoms of one element for each atom of another element.

In water, for example, Dalton thought that there were probably equal numbers of oxygen and hydrogen atoms.

We now know that in water there are in fact two hydrogen atoms for every oxygen atom.

= hydrogen

= oxygen

These are Dalton's own symbols for the atoms.

10 How many times heavier is an oxygen atom than a hydrogen atom:
(a) for Dalton?   (b) for modern chemists?

Some families of elements

A table of elements

During the nineteenth century, chemists discovered how to compare the weights of different atoms. In other words, they found the **relative atomic mass** of each element.

Mendeleev then produced a **Periodic Table**:

- by listing the elements in order starting with the element that had the lightest atoms;
- by arranging the list of elements in rows as shown on this page.

Each column in the table contains similar elements. These columns are called **Groups**.
[You will find more about how chemists developed the Periodic Table on pages 261–263.]

1 Write down the names of the first three elements:
(a) in Group 1;   (b) in Group 0.

2 Potassium atoms have a <u>smaller</u> mass than argon atoms. But potassium was placed <u>after</u> argon in the Periodic Table. Explain why.

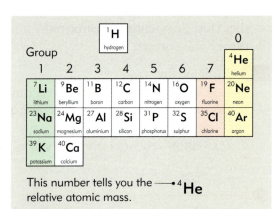

Periodic Table showing the first 20 elements.

Note

Potassium atoms actually have a smaller relative atomic mass than argon atoms. But we still put potassium after argon in the Periodic Table.
We do this because potassium is similar to the elements in Group 1 and argon is similar to the elements in Group 0.

Group 1 elements – the alkali metals

The elements in Group 1 of the Periodic Table are all metals. They are good conductors of heat and electricity. They can also be bent or hammered into shape without breaking. But they are softer than most other metals and have lower melting points and boiling points.

The diagram shows what happens when an alkali metal reacts with water.

3 How do we know that the colourless solution is alkaline?

4 Sodium gives an alkali with water. What is the name of this alkali?

All alkali metals react with water to give **alkaline** solutions. This is why we call these metals the **alkali metals**.

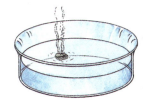

The sodium moves about on top of the water, making it fizz. A colourless solution of sodium hydroxide is produced.

dropper containing universal indicator

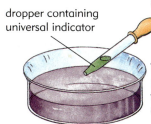

The indicator turns purple, which shows that the solution is alkaline.

Group 0 elements – the noble gases

The elements in Group 0 aren't very interesting. They are colourless gases so you cannot see them. You cannot use them to make new substances, because they are unreactive. This is why we call them **noble gases**.

Because they are so unreactive, or <u>inert</u>, noble gases have some important uses.

- Helium is lighter than air. It is non-flammable (does not burn). It is used in balloons and airships.

- Argon is used to fill the space inside the glass of a light bulb. It will not react with the metal filament of the lamp, even when it is hot.

The diagram shows another difference between noble gases and other elements that are gases at room temperature (20 °C).

5 (a) How do atoms of noble gases go around?
(b) How is this different from the atoms of other elements that are gases?

Atoms of noble gases go around by themselves. We say that they are **monatomic**. [mon = 1]

Atoms of other gases join in pairs to form molecules. We say they are **diatomic**. [di = 2]

Group 7 elements – the halogens

We have already learnt about two families of elements, the alkali metals in Group 1 of the Periodic Table of the elements, and the noble gases in Group 0.

In Group 7 there is another family of elements. We call this family of elements the **halogens**.

6 Write down the names of <u>four</u> elements in the halogen family.

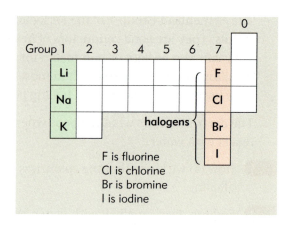

F is fluorine
Cl is chlorine
Br is bromine
I is iodine

What are the halogens like?

The halogens are **non-metals**. The pictures show what some of the halogens look like at room temperature.

7 At room temperature:
  (a) which halogen is a solid?
  (b) which <u>two</u> halogens are gases?
  (c) which halogen is a liquid that gives off a gas?

The halogens are all coloured when they are gases. Other elements that are gases have no colour, so we say they are **colourless**. An example is oxygen gas.

8 What is the colour of:
  (a) fluorine gas?
  (b) chlorine gas?
  (c) bromine gas?
  (d) iodine gas?

9 What is <u>one</u> difference between halogens and other elements that are gases?

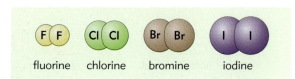

fluorine   chlorine   bromine   iodine

Halogen atoms are joined together in pairs. We call these pairs molecules.

10 Look at the diagram above.
How many halogen atoms join to make a halogen molecule?

Fluorine.

Chlorine.

Bromine.

heat

Iodine crystals produce iodine gas when you heat them.

Iodine crystals.

Halogens can react with metals

Halogens react with metals to form compounds called
halides. Halides are part of a family of compounds called
salts. The word 'halogen' means 'salt maker'.
The photographs show how you can make ordinary salt,
the kind you might use in cooking.

11 What is the chemical name for ordinary salt?

12 Which <u>two</u> elements are in ordinary salt?

13 Write down a word equation for the reaction to
make salt.

The other alkali metals react with halogens in the same
kind of way. The box shows how we name the salts.

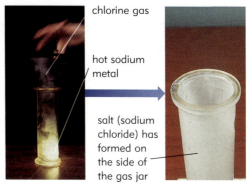

chlorine gas

hot sodium
metal

salt (sodium
chloride) has
formed on
the side of
the gas jar

$2Na(s) + Cl_2(g) \longrightarrow 2NaCl(s)$

> **The names we give to salts from the halogens**
> - fluorine gives salts called fluorides
> - chlorine gives salts called chlorides
> - bromine gives salts called bromides
> - iodine gives salts called iodides

14 What do we call the salt made from:
(a) potassium and iodine?
(b) lithium and bromine?

Halogens can react with other non-metals

Halogen atoms can join up with atoms of other
non-metals such as hydrogen and carbon. The diagram
shows some of the new compounds that can be made.

15 Write down the name and the formula of a halogen
compound that we use to make a plastic.

16 (a) Draw a molecule that contains two
different halogens.
(b) What was this compound once used for?
(c) Why don't we use this compound any more?

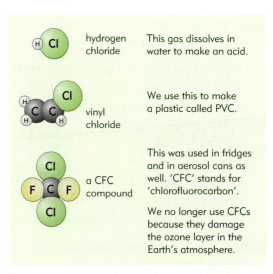

hydrogen
chloride — This gas dissolves in
water to make an acid.

vinyl
chloride — We use this to make
a plastic called PVC.

a CFC
compound — This was used in fridges
and in aerosol cans as
well. 'CFC' stands for
'chlorofluorocarbon'.

We no longer use CFCs
because they damage
the ozone layer in the
Earth's atmosphere.

Some compounds of the halogens with other
non-metals.

Differences between elements in the same Group

Elements in the same Group of the Periodic Table are similar to each other, but they are not exactly the same.

Differences between alkali metals

Alkali metals all react with water to produce the same kinds of products. Even so, there are some differences in how they each react.

Look at the pictures of three alkali metals reacting with water.

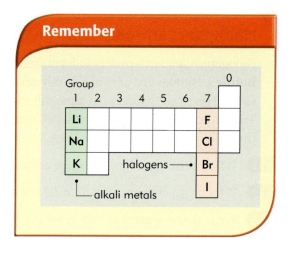

Lithium floats and bubbles.
Hydrogen gas forms.
The water becomes alkaline.

Sodium floats and gets so hot it melts.
Hydrogen gas forms rapidly.
The water becomes alkaline.

Potassium floats. It reacts violently.
The hydrogen gas burns.
The water becomes alkaline.

17 Write down <u>two</u> ways in which all these reactions are the same.

18 Write down <u>one</u> way in which they are different.

19 Put the three metals in order to show how reactive they are. Start with the one that reacts most quickly.

We say that there is a **pattern**, or **trend**, in the way the alkali metals react. The further down Group 1 you go, the more reactive the metals are.

The table shows the melting points and boiling points of some elements.

Use the figures in the table to answer these questions. (Hint: there are only three alkali metals shown in the table. The rest of the elements are not alkali metals.)

20 Which alkali metal has the:
(a) highest melting point?
(b) lowest melting point?
(c) highest boiling point?
(d) lowest boiling point?

21 What is the trend in melting points and boiling points as you go down Group 1?

Element	Melting point in °C	Boiling point in °C
bromine	−7	59
calcium	840	1484
chlorine	−101	−35
copper	1084	2570
iodine	114	184
lithium	180	1340
magnesium	650	1110
potassium	63	760
sodium	98	880
zinc	420	907

Trends in the halogens

The halogens are the elements in Group 7 of the Periodic Table. The halogens also show trends as you go down the Group.

22 What happens to the melting points and boiling points as you go down Group 7?

Element	Melting point in °C	Boiling point in °C
chlorine	−101	−35
bromine	−7	59
iodine	114	184

Some halogens are more reactive than others

One halogen can sometimes push a different halogen out of its compound. Here is an example.

chlorine + potassium ⟶ iodine + potassium
             iodide   chloride

We say that in this reaction chlorine **displaces** iodine. Chlorine pushes iodine out of its compound, potassium iodide. This happens because chlorine is more reactive than iodine.

As you go down Group 7 the halogens become more reactive.

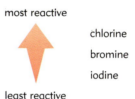

most reactive

chlorine

bromine

iodine

least reactive

What we learn from displacement reactions.

The Periodic Table of the elements

How the Periodic Table was discovered

The story of the Periodic Table tells us a lot about how scientists find things out.

By 1850, scientists knew:

- that everything is made from the atoms of a small number of elements;
- how heavy the atoms of different elements are compared to each other (we call this their **relative atomic mass**);
- that there are families of elements that have similar properties (see Box).

1 Write down the names of <u>two</u> families of elements that scientists knew about in 1850.

2 Scientists put the elements calcium and magnesium into the same family. Explain why.

> Scientists in 1850 knew about:
>
> - the family of elements that we call **alkali metals** [including lithium, sodium and potassium];
>
> - the family of elements that we call **halogens** [including chlorine, bromine and iodine].

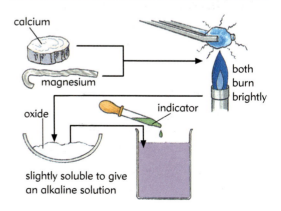

calcium

magnesium

oxide

indicator

both burn brightly

slightly soluble to give an alkaline solution

The first Periodic Table

In 1864, an English scientist called John Newlands wrote down the elements in order, starting with the lightest atoms. He only wrote down the elements he knew about at the time.

He noticed that if you count along seven from any element you reach another similar element.
So he wrote down the elements in rows of seven.

This was the first Periodic Table.

3   (a) Write down <u>three</u> differences between Newlands' Periodic Table and the Periodic Table that we use today (for the first twenty elements only).
     (b) Why do you think Newlands' table did not include the noble gases?

Improving the idea

Newlands' way of making a Periodic Table worked fine for the lighter elements. But it didn't work for heavier elements. At the time, most other scientists regarded Newlands' table as nothing more than a 'curiosity'.

A Russian scientist called Dmitri Mendeleev found a way to include all the elements he knew about. He did this:

● by putting any elements that didn't fit the table into a 'dustbin' column (he put many of the elements we call **transition metals** into this column);

● by putting each element into the Group where it fitted best, even when this meant leaving some blank spaces in his table.

Mendeleev didn't worry about the blank spaces. He just said that there must be some elements which hadn't been discovered yet.

4   Look at the box.
     (a) Where was there an element missing in Mendeleev's Periodic Table?
     (b) What did Mendeleev expect this element to be like?
     (c) Was Mendeleev right about this element?

Mendeleev's Periodic Table helped scientists to discover many new elements. So most scientists then realised that the Periodic Table was a very useful tool.

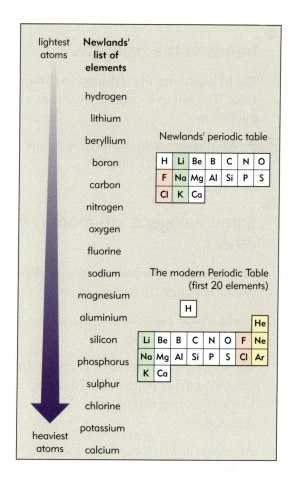

lightest atoms / Newlands' list of elements

hydrogen
lithium
beryllium
boron
carbon
nitrogen
oxygen
fluorine
sodium
magnesium
aluminium
silicon
phosphorus
sulphur
chlorine
potassium
calcium

heaviest atoms

Newlands' periodic table

H	Li	Be	B	C	N	O
F	Na	Mg	Al	Si	P	S
Cl	K	Ca				

The modern Periodic Table (first 20 elements)

H							
							He
Li	Be	B	C	N	O	F	Ne
Na	Mg	Al	Si	P	S	Cl	Ar
K	Ca						

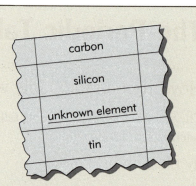

carbon
silicon
unknown element
tin

What Mendeleev said the unknown element would be like (in 1869):

• a grey metal
• its oxide would be white
• its chloride:
     would boil at less than 100 °C
     each cm^3 would have mass 1.9 g

The element germanium (discovered 27 years later):

• a grey metal
• a white oxide
• its chloride:
     boils at 86.5 °C
     each cm^3 has mass 1.8 g

A problem with a noble gas

When some new elements called **noble gases** were discovered, it was very easy to add a new Group to the Periodic Table. But the mass of argon atoms meant that it wasn't in quite the right place.

5 Where <u>should</u> argon have gone in the Periodic Table? Give a reason for your answer.

Many years later, scientists discovered a better way of arranging the elements in order. Argon then came before potassium, where it fits best in the Periodic Table.

							H		0

Group
1	2	3	4	5	6	7	He
Li	Be	B	C	N	O	F	Ne
Na	Mg	Al	Si	P	S	Cl	Ar
K	Ca						

relative atomic masses:
argon (Ar) = 40
potassium (K) = 39

What are atoms made of?

> **Remember**
>
> Everything is made from just over 90 different kinds of atoms.
>
> Elements are substances where all the atoms are of one kind.

The diagram shows what is inside a helium atom. In the centre of the atom is the **nucleus**.
Electrons move in the space around the nucleus.

6 (a) What <u>two</u> sorts of particles do you find in the nucleus of an atom?
   (b) What is the same about these two particles?
   (c) What is different?

7 The complete helium atom has no electrical charge overall. Why is this?

The number of protons is always the same as the number of electrons in an atom. This means that the positive and negative charges balance in an atom.

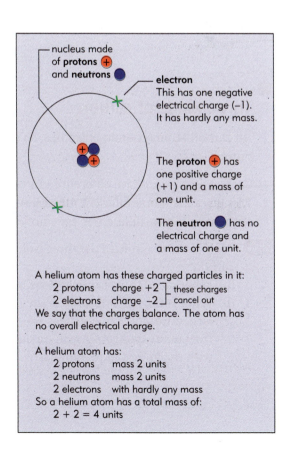

nucleus made of **protons** ⊕ and **neutrons** ●

electron
This has one negative electrical charge (−1). It has hardly any mass.

The **proton** ⊕ has one positive charge (+1) and a mass of one unit.

The **neutron** ● has no electrical charge and a mass of one unit.

A helium atom has these charged particles in it:
2 protons   charge +2 ⎤ these charges
2 electrons   charge −2 ⎦ cancel out
We say that the charges balance. The atom has no overall electrical charge.

A helium atom has:
2 protons   mass 2 units
2 neutrons   mass 2 units
2 electrons   with hardly any mass
So a helium atom has a total mass of:
2 + 2 = 4 units

The symbols that show what atoms contain

This diagram tells you everything you need to know about a helium atom.

The **proton number** for a helium atom is 2 so a helium atom must have 2 protons. This means it must also have 2 electrons. The helium atom has a **mass number** of 4. This means it must contain 2 neutrons in its nucleus.

The proton number of an atom tells us what element the atom is. So an atom with 2 protons must be a helium atom.

The diagram shows a hydrogen atom and a lithium atom.

8 Write down the following symbols. Add the mass number and the proton number for each one.
(a) H   (b) Li

9 Look at the symbol for a sodium atom.
(a) What is its proton number?
(b) How many protons does it contain?
(c) How many electrons does it have?
(d) What is its mass number?
(e) How many electrons does it contain?

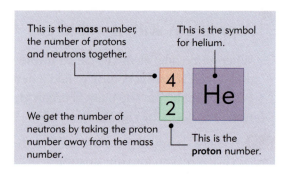

This is the **mass** number, the number of protons and neutrons together.

This is the symbol for helium.

We get the number of neutrons by taking the proton number away from the mass number.

This is the **proton** number.

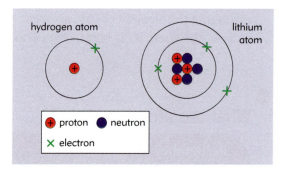

hydrogen atom   lithium atom

+ proton   ● neutron
× electron

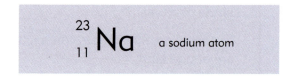

$^{23}_{11}$Na   a sodium atom

Three kinds of carbon

All carbon atoms contain 6 protons, so they have a proton number of 6.

Carbon atoms can have different numbers of neutrons. This gives the atoms different masses. Atoms of the same element that have different masses are called **isotopes**.

10 Look at the diagram of the three isotopes of the element carbon.
(a) What do all three isotopes have in common?
(b) What is the mass number of the isotope with the largest number of neutrons?
(c) How many neutrons are in an atom of the isotope with the largest mass number?

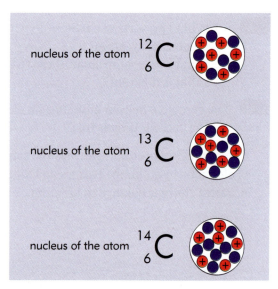

nucleus of the atom $^{12}_{6}$C

nucleus of the atom $^{13}_{6}$C

nucleus of the atom $^{14}_{6}$C

Three isotopes of the element carbon.

What is different about the modern Periodic Table?

> **Remember**
>
> this is the mass number → ^{40}
> this is the proton number → $_{18}$ Ar $^{39}_{19}K$
>
> So argon has a higher mass number than potassium but it has a lower proton number.

We now put the elements into the Periodic Table in the order of their proton numbers. The proton number also tells you the number of electrons in each atom.
The number of electrons is what gives an element its properties.

11 Why is it better to list the elements in order of their proton numbers than in order of their mass numbers?

Looking for patterns in a list of elements

The diagram below shows the first 20 elements in the order of their proton numbers.

$_1H$ $_2He$ $_3Li$ $_4Be$ $_5B$ $_6C$ $_7N$ $_8O$ $_9F$ $_{10}Ne$ $_{11}Na$ $_{12}Mg$ $_{13}Al$ $_{14}Si$ $_{15}P$ $_{16}S$ $_{17}Cl$ $_{18}Ar$ $_{19}K$ $_{20}Ca$

◯ halogen   ◯ alkali metal   ◯ noble gas

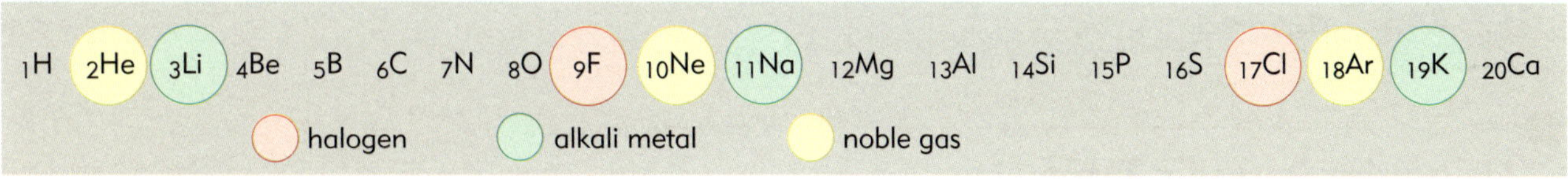

12 Look carefully at the list of elements.
(a) What kind of element comes straight after each noble gas?
(b) What kind of element usually comes just before each noble gas?
(c) How would the list be different if the elements were listed in order of their relative atomic masses (mass numbers)?

Making the list of elements into a Periodic Table

To make the list of elements into the Periodic Table:

- you place hydrogen and helium as shown in the diagram;
- you start a new row of the table every time you reach an element that is an alkali metal.

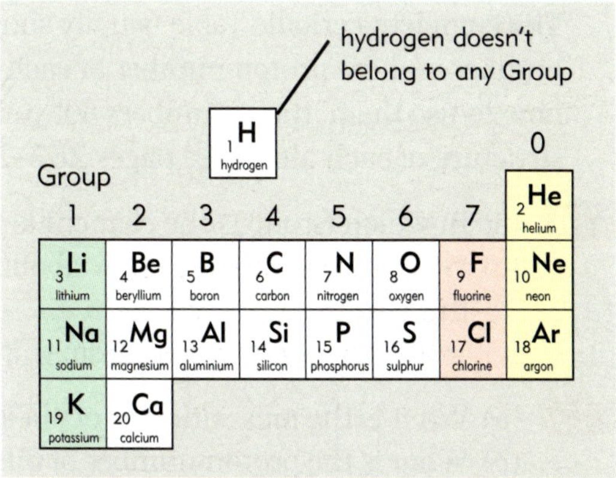

hydrogen doesn't belong to any Group

Group							0
							$_2$He helium
$_3$Li lithium	$_4$Be beryllium	$_5$B boron	$_6$C carbon	$_7$N nitrogen	$_8$O oxygen	$_9$F fluorine	$_{10}$Ne neon
$_{11}$Na sodium	$_{12}$Mg magnesium	$_{13}$Al aluminium	$_{14}$Si silicon	$_{15}$P phosphorus	$_{16}$S sulphur	$_{17}$Cl chlorine	$_{18}$Ar argon
$_{19}$K potassium	$_{20}$Ca calcium						

$_1$H hydrogen

| | 1 | 2 | 3 | 4 | 5 | 6 | 7 |

The first 20 elements in the modern Periodic Table.

Completing the Periodic Table

The complete Periodic Table shows all the elements that we know about. This makes it look more complicated.

13 There are lots of elements that are not placed in Groups 0 to 7. What do we call these elements?

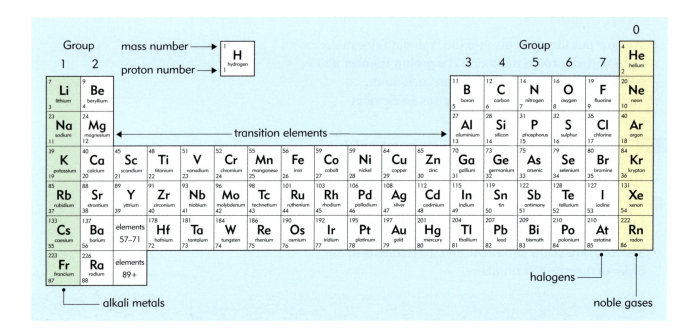

Using the Periodic Table

The Periodic Table is very useful. You can use it to make good guesses about elements you have never seen. This is because there are **patterns** we can understand in the table. For example, elements in the same Group are very much alike.

The complete Periodic Table usually shows us the mass number and the proton number of each atom. If you know how to use them, these numbers tell you a lot about the structure of each atom (see pages 267–268).

14 (a) In which Group is the element krypton (Kr)?
    (b) What do you already know about the elements in this Group?
    (c) What can you work out from this about krypton?

15 (a) What is the mass number of phosphorus?
    (b) What is the proton number of nitrogen?

Why are there families of elements?

What elements are like and the way they react depends on the electrons in their atoms. The proton number of an atom tells you how many protons there are in the nucleus. It also tells you how many electrons there are around the nucleus, because the number of protons in an atom is the same as the number of electrons.

Lithium	Sodium	Potassium
$_3$Li	$_{11}$Na	$_{19}$K

Some alkali metals.

16 How many electrons are there in:
(a) a lithium atom?
(b) a sodium atom?
(c) a potassium atom?

These alkali metals have different numbers of electrons, but the metals still react in a similar way. This is because the electrons are arranged in a similar way.

How are electrons arranged in an atom?

The electrons around the nucleus of an atom are in certain **energy levels**. The diagram shows the first three energy levels for electrons.

17 How many electrons can fit into:
(a) the first (lowest) energy level?
(b) the second energy level?
(c) the third energy level?

How electrons fill up the energy levels

The first energy level is the lowest. The electrons start to fill up this level first. When the first energy level is full, electrons start to fill up the second level.

The diagram shows where the electrons are in the first three elements.

18 Draw the same kind of diagram for:
(a) a carbon atom $_6$C;
(b) an oxygen atom $_8$O.

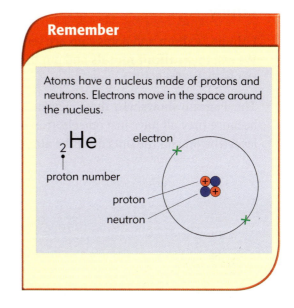

Remember

Atoms have a nucleus made of protons and neutrons. Electrons move in the space around the nucleus.

$_2$He — proton number

electron
proton
neutron

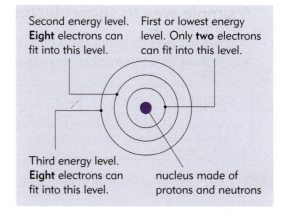

Second energy level. **Eight** electrons can fit into this level.

First or lowest energy level. Only **two** electrons can fit into this level.

Third energy level. **Eight** electrons can fit into this level.

nucleus made of protons and neutrons

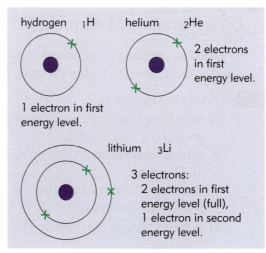

hydrogen $_1$H

helium $_2$He

2 electrons in first energy level.

1 electron in first energy level.

lithium $_3$Li

3 electrons:
2 electrons in first energy level (full),
1 electron in second energy level.

Why the alkali metals are in the same family

Lithium, sodium and potassium are very similar elements. We call them **alkali metals** and put them in Group 1 of the Periodic Table.

The diagram shows why these elements are similar. It is because they all have just one electron in their top energy level (the one on the outside of the atom).

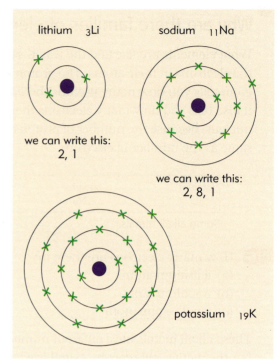

The arrangement of electrons in the alkali metals of Group 1.

A simple way to show electrons

Drawing electron diagrams takes time. Here is a quicker way to show how electrons are arranged in atoms.

> sodium $_{11}$Na is 2, 8, 1
>
> two electrons in lowest energy level which is full   eight electrons in second level which is full   one electron in top energy level

19 Write down the electron arrangement for potassium.

Other families of elements

Elements in the same Group always have the same number of electrons in their top energy levels.

20 Look at the diagrams.
(a) How many electrons are there in the top energy level of the alkali metals?
(b) Which group in the Periodic Table contains the alkali metals?
(c) How many electrons are there in the top energy level of the halogens?
(d) Which group in the Periodic Table contains the halogens?

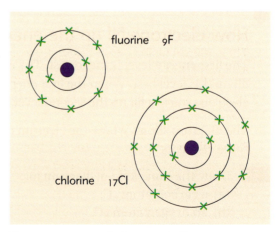

The arrangement of electrons in Group 7 (halogen) atoms.

Why elements react to form compounds

Atoms of different elements react together to form **compounds**. For example, sodium reacts with chlorine to produce the compound sodium chloride.

Elements react because of the electrons in their atoms. The diagram shows the arrangement of electrons in a sodium atom and in a chlorine atom.

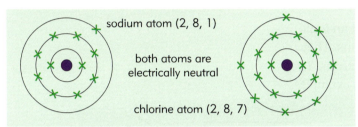

sodium atom (2, 8, 1)

both atoms are electrically neutral

chlorine atom (2, 8, 7)

1   How many electrons are there in the top energy level of:
(a) a sodium atom?
(b) a chlorine atom?

Atoms like to have each energy level either completely full or completely empty, just like they are in the noble gases. The atoms are then more **stable**. This is why sodium reacts with chlorine.

What happens when sodium reacts with chlorine?

A sodium atom has just 1 electron in the top energy level. The easiest way for it to become stable is to lose this single electron. The next energy level is now the top one, and is completely full.

A chlorine atom has 7 electrons in its top energy level. The easiest way for it to become stable is to find 1 more electron. This makes the top energy level completely full.

The diagram shows what happens when sodium reacts with chlorine.

The sodium atom gives up the electron in its top energy level to the chlorine atom. Both atoms now have an electrical charge.

2   What is the difference between a sodium ion and a sodium atom?

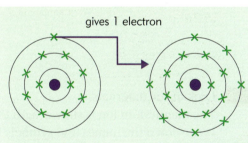

gives 1 electron

sodium atom (2, 8, 1)   chlorine atom (2, 8, 7)

so we get:

sodium ion (2, 8)   chloride ion (2, 8, 8)
        Na$^+$ Cl$^-$

The sodium atom now has one electron missing, so it has a positive electrical charge of +1. It is now a sodium ion, which we write Na$^+$.

The chlorine atom now has one extra electron, so it has a negative electrical charge of –1. It is now a chloride ion, which we write Cl$^-$.

3 Copy the diagram of a lithium atom and a fluorine atom. Then add an arrow to show how the electron moves when they react together.

Substances made from ions are called **ionic substances**.

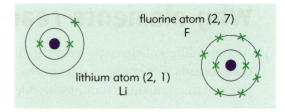

Some more ionic substances

When a metal reacts with a non-metal we get an ionic substance. The metal atoms give away electrons. They form **positive ions**. The non-metals take electrons. They form **negative ions**. The diagrams show two examples.

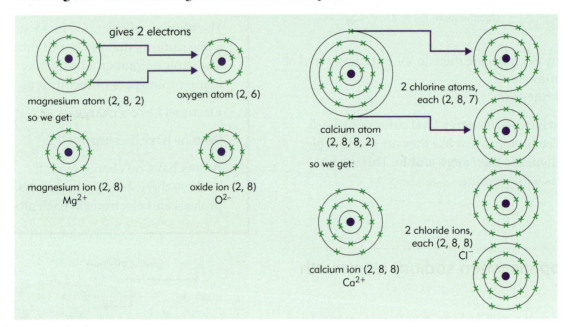

4 Look at the diagrams.
The magnesium ion is shown with the symbol Mg^{2+}.
Write down the names and symbols for all the other ions shown in the diagrams.

5 Draw a diagram to show how sodium oxide is formed.

The formula of an ionic substance

Sodium chloride has 1 chloride ion for each sodium ion.
We write its formula as NaCl. In calcium chloride there are 2 chloride ions for each calcium ion.
We write its formula as $CaCl_2$.

6 Write down the formula for magnesium oxide.

7 Sodium oxide has 2 Na^+ ions for every O^{2-} ion.
Write down the formula of sodium oxide.

Explaining trends in reactivity within Groups

Remember

Elements in the same Group have similar properties.

In Group 1, metals lower in the Group are more reactive.

In Group 7, non-metals lower in the Group are less reactive.

Group								
1	2	3	4	5	6	7	0	
Li						F		
Na						Cl		
K						Br		

alkali metals   halogens

Atoms give, take or share electrons so that they have full outer shells of electrons. They are then stable, like atoms of noble gases (Group 0).

The way that elements react depends largely on the number of electrons in their highest energy level (outer shell).

Elements in Group 1

8 List the ways in which elements in Group 1 are similar.

9 Write down a word equation for the reaction of lithium with water.

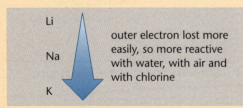

Li

Na

K

outer electron lost more easily, so more reactive with water, with air and with chlorine

Reactivity increases further down the Group because potassium loses its outer electron more easily than sodium does, and sodium more easily than lithium.

10 Describe how the increasing size of the atoms as you go down Group 1 affects their reactivity.

Electron arrangement in atoms of Group 1 elements

lithium Li	2, 1
sodium Na	2, 8, 1
potassium K	2, 8, 8, 1

Group 1 elements all have <u>one</u> electron in the highest energy level (outer shell).

All the metals in Group 1 <u>lose</u> the outer electron easily to form a positive ion, so they are all very reactive and react in the same way. For example:

sodium + water → sodium hydroxide + hydrogen

potassium + water → potassium hydroxide + hydrogen

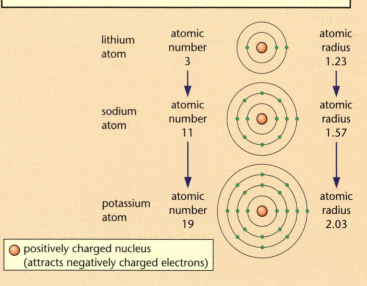

lithium atom — atomic number 3 — atomic radius 1.23

sodium atom — atomic number 11 — atomic radius 1.57

potassium atom — atomic number 19 — atomic radius 2.03

○ positively charged nucleus (attracts negatively charged electrons)

Elements in Group 7

The elements in Group 7 all react with metals to form compounds called **halides**.

11 List the ways in which elements in Group 7 are similar.

12 Write down a word equation for the reaction between fluorine and sodium.

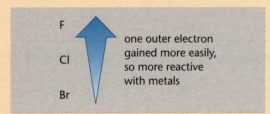

F
Cl
Br
one outer electron gained more easily, so more reactive with metals

Reactivity increases further <u>up</u> the Group because fluorine gains an electron more easily than chlorine does, and chlorine more easily than bromine.

13 Describe how the increasing size of the atoms in Group 7 affects their reactivity.

Halogens can also react with hydrogen. They do this by sharing electrons and forming **covalent** compounds.

14 How do the reactivities of the halogens with hydrogen compare with their reactivities with metals?

15 Beryllium, magnesium and calcium are three metal elements in Group 2 of the Periodic Table. Their electronic structures are:

Be (2, 2)   Mg (2, 8, 2)   Ca (2, 8, 8, 2)

(a) Would you expect these elements to have similar properties? Explain your answer.
(b) Arrange the three elements in order of reactivity and give reasons for your order.

Electron arrangement in atoms of Group 7 elements

fluorine F	2, <u>7</u>
chlorine Cl	2, 8, <u>7</u>
bromine Br	2, 8, 18, <u>7</u>

Group <u>7</u> elements all have <u>seven</u> electrons in the highest energy level (outer shell).

All the halogens in Group 7 react with metals by <u>gaining</u> an electron to form a negative ion.
For example:

chlorine + iron → iron chloride
bromine + iron → iron bromide

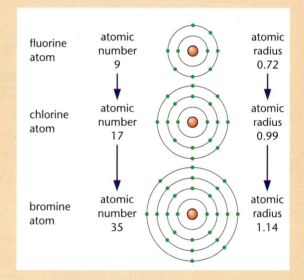

fluorine atom	atomic number 9		atomic radius 0.72
chlorine atom	atomic number 17		atomic radius 0.99
bromine atom	atomic number 35		atomic radius 1.14

How halogens react with hydrogen

Fluorine: reacts explosively in dark conditions.
Chlorine: reacts in dim light, but explosively in sunlight.
Bromine: no reaction in sunlight, but reacts if heated to 200°C.

How can atoms of non-metals join together?

A non-metal such as chlorine can react with a metal such as sodium. This produces an ionic compound called sodium chloride. Chlorine can also react with another non-metal, such as hydrogen.

What happens when chlorine and hydrogen react?

When two non-metals such as chlorine and hydrogen react, they do it by **sharing** electrons. The diagram shows what happens to the shared electrons.

When a hydrogen atom and a chlorine atom react, they share one pair of electrons. Each atom is then more stable. The hydrogen atom has a total of two electrons in its first energy level. This level is now full.
The chlorine atom has a total of eight electrons in its third energy level. This level is also full.

16   What is the name of the particle produced when an atom of hydrogen reacts with an atom of chlorine?

17   Write down the formula of hydrogen chloride.

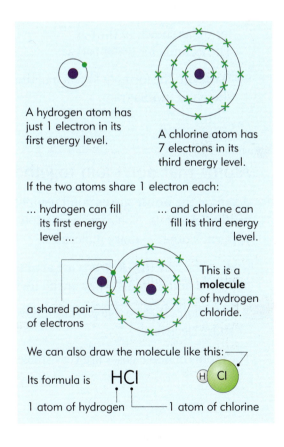

A hydrogen atom has just 1 electron in its first energy level.

A chlorine atom has 7 electrons in its third energy level.

If the two atoms share 1 electron each:

... hydrogen can fill its first energy level ...

... and chlorine can fill its third energy level.

This is a **molecule** of hydrogen chloride.

a shared pair of electrons

We can also draw the molecule like this:

Its formula is HCl

1 atom of hydrogen   1 atom of chlorine

Molecules of other substances

Molecular substances are substances that are made of **molecules**, like hydrogen chloride. Atoms of different non-metals can join together to make molecules. Atoms of the <u>same</u> non-metal element can also share electrons to make molecules.

The diagram shows some molecules of each type.

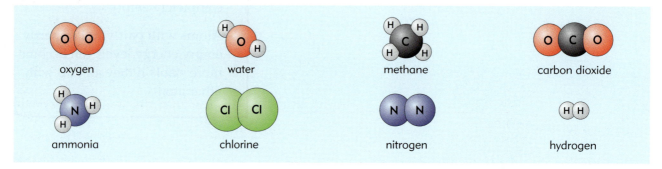

oxygen   water   methane   carbon dioxide

ammonia   chlorine   nitrogen   hydrogen

18 Compare the diagrams of the molecules to the formulas shown in the box.
The formula for water is H_2O.

Write out the formula for each of the molecules shown in the diagram.

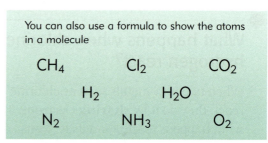

You can also use a formula to show the atoms in a molecule

CH_4 Cl_2 CO_2

H_2 H_2O

N_2 NH_3 O_2

Atoms that don't join together

Atoms of the **noble gases** don't usually join with atoms of other elements. Noble gas atoms don't even join up with each other to make molecules.

Look at the way the electrons are arranged in helium, neon and argon atoms. They are all noble gases.

19 Why don't these noble gas atoms give, take or share electrons?

The atoms of noble gases are very stable.

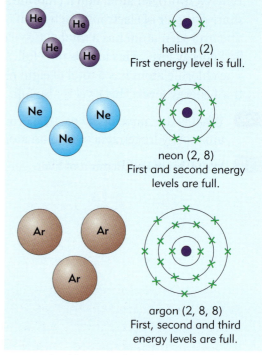

helium (2)
First energy level is full.

neon (2, 8)
First and second energy levels are full.

argon (2, 8, 8)
First, second and third energy levels are full.

Atoms of noble gases don't join up to form molecules.

Why different types of substances have different properties

Differences between ionic and molecular substances

The table shows some of the properties of ionic and molecular substances. The diagrams show a test we can do.

		Melting point in °C	Boiling point in °C
ionic substances	sodium chloride	801	1413
	calcium chloride	782	1600
	magnesium oxide	2852	3600
molecular substances	methane	−182	−161
	ammonia	−77	−34
	water	0	100

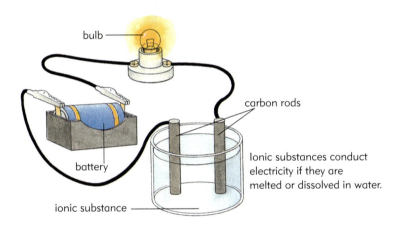

bulb

carbon rods

Ionic substances conduct electricity if they are melted or dissolved in water.

battery

ionic substance

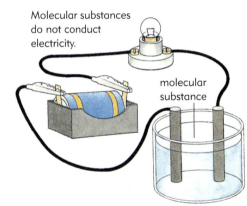

Molecular substances do not conduct electricity.

molecular substance

1 Write down <u>three</u> differences between ionic and molecular substances. Use the information in the table and the diagrams to help you.

Why do ionic compounds have high melting points and high boiling points?

Sodium chloride is an ionic compound. There are strong forces of attraction between Na$^+$ and Cl$^-$ ions. Each Na$^+$ ion is surrounded by six Cl$^-$ ions in a **giant structure**. So ionic structures have high melting points and high boiling points.

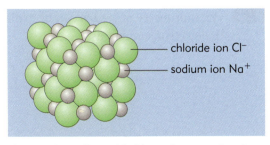

chloride ion Cl$^-$
sodium ion Na$^+$

The ions in sodium chloride make a regular giant ionic lattice.

2 Explain the following properties of sodium chloride.
    (a) Crystals of sodium chloride have a regular shape.
    (b) Sodium chloride has a high melting point (801 °C) and a high boiling point (1413 °C).

Molecular substances are made of separate molecules. These are easily pulled apart so molecular structures have low melting points and low boiling points.

Sodium chloride crystals.

Why do ionic substances conduct electricity?

Ionic substances are made up of **ions**. These are particles with electrical charges. An ionic substance will only conduct electricity if its charged particles can move about.

Look at the diagram.

3 Why must we melt or dissolve an ionic substance before it will conduct electricity?

4 Why can't molecular substances conduct electricity?

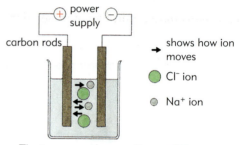

The ions can only move if we **melt** the substance or **dissolve** it in water. Molecules do not have electrical charges.

Explaining the properties of metals

Metals allow electric currents and heat (thermal energy) to pass through them easily. We say that they are good **conductors** of electricity and heat. Metals can also be hammered or bent into shape without breaking.

The diagram shows why metals have these properties.

Electrons in the metal that are free to move:

- can carry an electric current through the metal;
- can carry heat (thermal energy) through the metal.

5 What property of metals can be explained in terms of the electrons in metals?

In a piece of metal, electrons from the highest energy level of each atom are free to move anywhere in the metal. These electrons bind all of the metal atoms together into a single giant structure.

The metal atoms can slide over each other but are still held together by the free electrons.

mobile electrons

metal atoms

Higher

Explaining the properties of molecular substances

In molecular substances atoms join together by sharing electrons from the highest energy level (outer shell). These shared pairs of electrons are called **covalent bonds**.

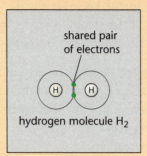

hydrogen molecule H$_2$

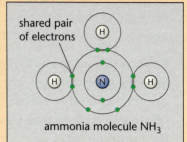

ammonia molecule NH$_3$

The diagrams show two molecules which are held together in this way.

Molecular substances have low melting points and low boiling points. They do not conduct electricity.

The diagrams explain why a molecular substance has these properties.

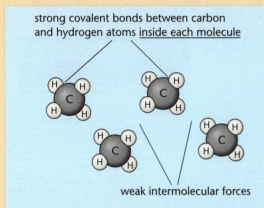

strong covalent bonds between carbon and hydrogen atoms <u>inside each molecule</u>

weak intermolecular forces

Because there are only weak forces <u>between the molecules</u>, melting and boiling points are low. The molecules do not carry an overall electric charge, so molecular substances do not conduct electricity.

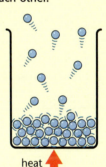

This molecular solid is melting. The molecules do not need much energy to be able to move over each other.

This molecular liquid is boiling.
The molecules do not need much energy to be able to escape from the surface of the liquid into the air.

heat

6 Why do molecular substances:
(a) have a low melting point?
(b) have a low boiling point?

7 Explain why molecular substances do not conduct electricity.

Graphite and diamond – giant structures

Graphite and diamond are two very different forms of the element carbon. Even though carbon atoms form covalent bonds with each other, these two forms have a <u>giant</u> structure rather than a molecular structure.

8 Diamond has a melting point of 3550 °C. It is used in cutting and drilling tools. Explain its properties by reference to its structure.

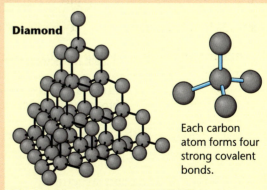

Diamond

Each carbon atom forms four strong covalent bonds.

Diamond has a rigid giant covalent structure of atoms (a lattice of atoms).
All the bonds are strong bonds, so this is a strong 3D structure with no weak links.

Higher

Graphite is unusual as a non-metal, because it will conduct an electric current. It is used for electrodes in electric furnaces and for 'brushes' in electric motors.

Each carbon atom has <u>four</u> outer-shell electrons, but in graphite layers only <u>three</u> of these electrons are used for bonding with other carbon atoms. So the other electron is free to move. These **free electrons** in the graphite structure can carry an electric current, which is a flow of electrons.

In graphite, bonds between carbon atoms in the layers are strong covalent bonds. But the bonds between layers of carbon atoms are weak, so the layers slide over each other. This makes the surface flaky and soft. Graphite has a high melting point.

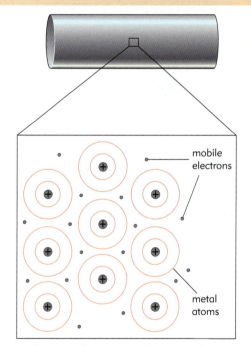

Graphite

weak bond between layers

strong covalent bond within layers

In graphite layers, each carbon atom forms three covalent bonds.

9 Graphite is used as a lubricant and as the 'lead' in soft pencils. Explain why graphite flakes easily.

10 Explain why graphite is an electrical conductor.

Comparing giant structures

There are three types of giant structure, each formed by a particular type of substance:

- giant molecular;
- metals;
- ionic compounds.

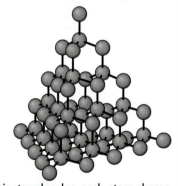

In giant molecules, each atom shares electrons with several other atoms. So there is a network of covalent bonds throughout the whole structure.

mobile electrons

metal atoms

In metals, atoms lose outer electrons to become ions. The positively charged ions are held together by a 'sea' of negatively charged free electrons.

11 Most substances with a giant structure are solids with a high, or fairly high, melting point. Explain why.

12 Which type of solid giant structures:
(a) always conduct electricity?
(b) never conduct electricity?

13 Which solid giant molecular substance conducts electricity, and why?

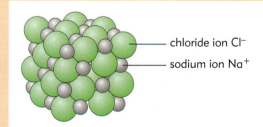

chloride ion Cl⁻

sodium ion Na⁺

In ionic compounds, each ion is surrounded by, and strongly attracted to, oppositely charged ions.

Salt – a very useful substance

Ordinary salt is a very important chemical.
We use salt for lots of things, and we can make other
useful substances from salt.

On food.

On icy roads.

To preserve food.

1 Write down <u>three</u> different ways
to use salt.

2 Write down <u>three</u> materials we
can make from salt.

To make soap.

For margarine and plastics.

Where do we get the salt we need?

There is a lot of salt dissolved in the sea. You can get salt
by letting sea water evaporate. In some places there are
large amounts of salt underground. Salt is a cheap raw
material because it is easy to collect.

3 How can you get salt from sea water?

4 (a) Where did the underground salt come from?
(b) How do we usually get salt from under the ground?

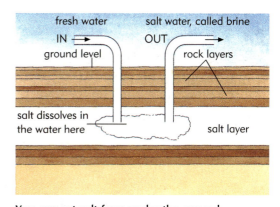

You can get salt from under the ground.

What are the elements in ordinary salt?

The chemical name for ordinary salt is sodium chloride.

5 What are the <u>two</u> elements in salt?

6 The two elements by themselves are very different
from salt. Write down <u>three</u> differences between salt
and its elements.

sodium
A very reactive alkali
metal which we keep
under oil.

sodium chloride
A harmless unreactive
substance. We put salt
on our food. We can
swim in salt water.

chlorine
A poisonous halogen gas.

What kind of substance is salt?

The diagram shows the arrangement of the particles in sodium chloride.

7 Sodium chloride is an ionic substance. What <u>two</u> particles is it made up of?

When we dissolve salt in water, the sodium ions and chloride ions can move about and conduct electricity.

Sodium chloride is an ionic compound.
Na^+ is a sodium ion,
Cl^- is a chloride ion.
The ions are arranged in a regular pattern.

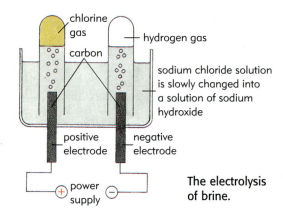

— chloride ion Cl^-
— sodium ion Na^+

How can we turn salt into other substances?

The diagram shows how we can make other chemicals from salt. We dissolve salt in water to give a solution called **brine**. Next we pass an electric current through the brine. We call this **electrolysis**.

The electrolysis of brine produces three useful substances:

- chlorine gas;
- hydrogen gas;
- a solution of the alkali sodium hydroxide.

8 When an electric current is passed through brine two different gases are produced.
(a) Which gas is produced at which electrode?
(b) How can you test for the gases produced?

chlorine gas

hydrogen gas

carbon

sodium chloride solution is slowly changed into a solution of sodium hydroxide

positive electrode

negative electrode

power supply

The electrolysis of brine.

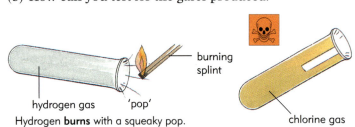

burning splint

hydrogen gas   'pop'
Hydrogen **burns** with a squeaky pop.

chlorine gas

Chlorine takes the colour out of damp indicator or litmus paper. It **bleaches** the indicator paper.

How to test the gases.

How do we use chlorine?

Chlorine is one of the three useful materials we produce from salt water. Chlorine is a poisonous gas.
This is useful when we want to kill harmful bacteria, but it can also be dangerous.

9 Write down <u>two</u> places where we can kill bacteria with the help of chlorine.

Very small amounts of chlorine in water kill dangerous bacteria.

Disinfectants are made from chlorine. They can kill bacteria.

We can use chlorine to make a plastic called **PVC**.

10 (a) Write down <u>two</u> ways in which we can use PVC.
(b) What do the letters PVC stand for?

You can also make bleach from chlorine.

11 (a) What is bleach used for?
(b) Why is it a bad idea to use bleach with brightly coloured clothes?

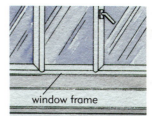

floor tiles

PVC plastic contains chlorine. PVC is short for poly(vinyl chloride).

Bleach can be made from chlorine. Bleach removes stains from cloth and makes colours fade.

How do we use hydrogen?

Hydrogen is also made from salt water by electrolysis. Hydrogen is the lightest gas of all, and many years ago airships were filled with hydrogen.

12 Airships filled with hydrogen were very dangerous. Explain why.

We can use hydrogen to make ammonia.

13 (a) Which element do we react with hydrogen to make ammonia?
(b) What useful material is made from ammonia?

Margarine is made using hydrogen.

Hydrogen gas is flammable. It catches fire very easily.

Ammonia is made from hydrogen and nitrogen. Ammonia is turned into fertiliser to grow more crops.

+ hydrogen

vegetable oil

margarine

14 What do you react the hydrogen with to make margarine?

How do we use sodium hydroxide?

The third material made from salt water is sodium hydroxide. Sodium hydroxide is used in oven cleaner. It is **corrosive**.

15 (a) Write down <u>one</u> use of sodium hydroxide in the home.
(b) Why should we use safety glasses and gloves when handling sodium hydroxide?

16 Sodium hydroxide is used to make some other useful products. Write down <u>three</u> examples.

Sodium hydroxide attacks and destroys skin and eyes.

vegetable oil

soap

sodium hydroxide

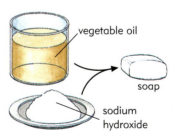

Sodium hydroxide:
● helps turn woodpulp into paper;
● is used in making ceramics (pottery).

Making hydrochloric acid

When hydrogen chloride gas dissolves in water, we get a solution of **hydrochloric acid**. The equation shows how we can make hydrogen chloride gas. This is a dangerous reaction, which we do not carry out in school.

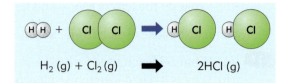

$$H_2 \text{ (g)} + Cl_2 \text{ (g)} \longrightarrow 2HCl \text{ (g)}$$

17 Write down a word equation for this reaction.

18 How can you show that a solution of hydrogen chloride is an acid?

Solutions of hydrogen fluoride, hydrogen bromide and hydrogen iodide are also acidic. Because these compounds are all made from hydrogen and the halogen elements, they are called hydrogen **halides**.

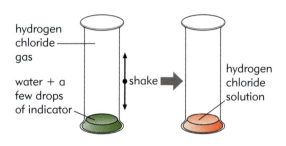

The chemicals we use to make photographs

We need special chemicals that react to light to make photographs. These chemicals are placed in layers on a film (or paper) that can be used in a camera. The simplest kind of film produces black and white photographs.

Colour photographs use the same basic chemical reactions, but use lots of other reactions too.

Remember

Halogens react with metals to form compounds called halides.

Chemicals that react to light

One of the chemicals we can use to make photographs is silver chloride.
The diagram shows how we can make silver chloride.
It also shows what happens when light shines on silver chloride.

19 Write down a word equation to show how silver chloride is made.

20 (a) What colour is freshly prepared silver chloride?
(b) How does this colour change in the light?

We say that the light **reduces** the silver chloride to silver metal. The same kind of change happens when the other silver halides react to light.

21 Write down the names of <u>two</u> other silver halides.

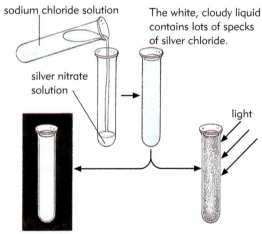

sodium chloride solution

The white, cloudy liquid contains lots of specks of silver chloride.

silver nitrate solution

light

In the dark, the white specks of silver chloride stay white.

In the light, the white specks of silver chloride turn into black specks of silver metal.

What is photographic film?

The diagram shows photographic film is made from specks of silver halide on a transparent plastic sheet.

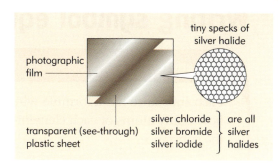

How does photographic film work?

The diagrams show what happens if you take a photograph of a black cross.

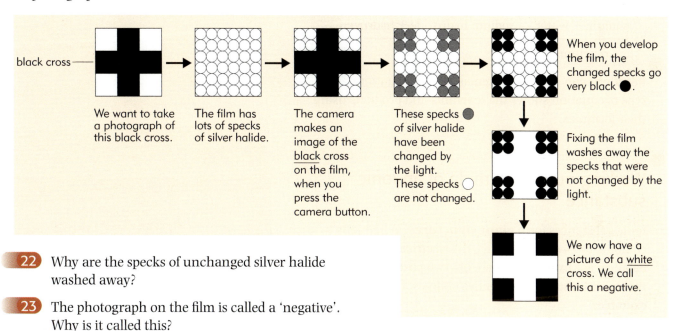

black cross

We want to take a photograph of this black cross.

The film has lots of specks of silver halide.

The camera makes an image of the <u>black cross</u> on the film, when you press the camera button.

These specks ● of silver halide have been changed by the light. These specks ○ are not changed.

When you develop the film, the changed specks go very black ●.

Fixing the film washes away the specks that were not changed by the light.

We now have a picture of a <u>white</u> cross. We call this a negative.

22 Why are the specks of unchanged silver halide washed away?

23 The photograph on the film is called a 'negative'. Why is it called this?

Photographic paper works in the same way.

What else can change silver halides?

X-rays and the radiation from **radioactive materials** will also reduce silver halides.

24 How can X-rays make a photograph of bones inside your body?

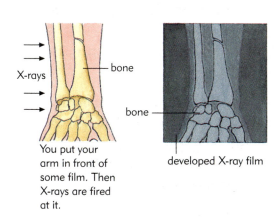

X-rays — bone

You put your arm in front of some film. Then X-rays are fired at it.

bone — developed X-ray film

Writing symbol equations for reactions

> **Remember**
>
> Each substance has a formula which shows how many atoms of each element there are in the substance.
>
> For example, the formula for water is H_2O: for every 1 atom of oxygen there are 2 atoms of hydrogen.

> You are expected to remember the formulas of these common molecular substances:
>
> | | | | |
> |---|---|---|---|
> | oxygen | O_2 | methane | CH_4 |
> | hydrogen | H_2 | ammonia | NH_3 |
> | nitrogen | N_2 | carbon dioxide | CO_2 |
> | chlorine | Cl_2 | carbon monoxide | CO |
> | water | H_2O | hydrogen chloride | HCl |
>
> You also need to be able to work out the formulas of ionic substances.

A word equation shows the reactants and products in a chemical reaction. For example:

$$\underbrace{\text{sodium + water}}_{\text{reactants}} \rightarrow \underbrace{\text{sodium hydroxide + hydrogen}}_{\text{products}}$$

Reactions can also be written as <u>symbol equations</u>. This type of equation shows the reactants and products using their formulas.

Working out the formula of an ionic substance

Ionic substances form **giant structures**. When ions combine to form compounds, the electrical charges must balance. For example, if there are two positive charges, there must also be two negative charges. Look at the examples.

1 Write down the formula for each of these:
  (a) potassium bromide;   (d) calcium hydroxide;
  (b) magnesium sulphide; (e) sodium hydroxide;
  (c) aluminium chloride;   (f) aluminium oxide.

Na^+	balances	Cl^-	to give the formula	$NaCl$
Ca^{2+}	balances	$\begin{cases} Cl^- \\ Cl^- \end{cases}$	to give the formula	$CaCl_2$
Mg^{2+}	balances	O^{2-}	to give the formula	MgO
Mg^{2+}	balances	$\begin{cases} OH^- \\ OH^- \end{cases}$	to give the formula	$Mg(OH)_2$

Some common ions

sodium	Na^+	chloride	Cl^-
potassium	K^+	bromide	Br^-
calcium	Ca^{2+}	hydroxide	OH^-
magnesium	Mg^{2+}	oxide	O^{2-}
aluminium	Al^{3+}	sulphide	S^{2-}

Why symbol equations need to be balanced

Atoms don't just disappear during chemical reactions. So the number of each type of atom must be exactly the same in the products as in the reactants.
In other words, symbol equations must be <u>balanced</u>.

2 Copy the symbol equation below. Then show that it is balanced.

$$Mg + 2HCl \rightarrow MgCl_2 + H_2$$

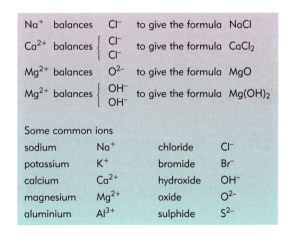

hydrogen + oxygen   water
$2H_2 + O_2$ $2H_2O$

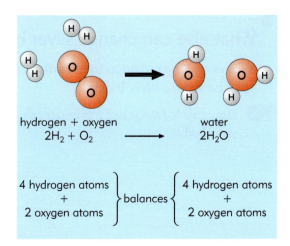

4 hydrogen atoms		4 hydrogen atoms
+	balances	+
2 oxygen atoms		2 oxygen atoms

How to write a balanced symbol equation

Step 1. Write down the word equation for the reaction.

Step 2. Write down the formulas for the reactants and products.

Step 3. Check to see if the equation is balanced. Count the atoms on both sides of the equation.

[You do not need to write this down.]

If the equation is not balanced, you need to go on to Step 4.

Step 4. Balance the equation. This is done by writing a number in front of one or more of the formulas. This number increases the numbers of all of the atoms in the formula.

Check that the equation is now balanced.

Example: The reaction between sodium metal and water

sodium + water → sodium hydroxide + hydrogen

$$Na + H_2O \rightarrow NaOH + H_2$$

Reactants		Products
1	sodium atoms	1
2	hydrogen atoms	3
1	oxygen atoms	1

The equation is <u>not</u> balanced because the number of hydrogen atoms is not the same on both sides.

We can balance the hydrogen atoms by doubling up the water + sodium hydroxide.

$$2Na + 2H_2O \rightarrow 2NaOH + H_2$$

2NaOH means 2 Na atoms, 2 O atoms and 2 H atoms.

This means 4 H atoms and 2 O atoms. So the O atoms also balance (two on each side).

This 2 is then needed so that there are 2 Na atoms on each side.

Check

Reactants		Products
2	sodium atoms	2
4	hydrogen atoms	4
2	oxygen atoms	2

The equation now balances. There is the same number of each type of atom on both sides.

---

3 (a) Write down the word equation and the unbalanced symbol equation for the following reaction:

calcium + water → calcium hydroxide + hydrogen

$$Ca + H_2O \rightarrow Ca(OH)_2 + H_2$$

(b) Balance the symbol equation.

(c) Add state symbols to your equation (see Box).

4 Write balanced symbol equations for these reactions. [Show all the steps.]

(a) potassium + chlorine → potassium chloride

(b) copper oxide + hydrogen → copper + water

Adding state symbols

When you have balanced an equation, you should then add state symbols. For example:

$$2Na(s) + 2H_2O(l) \rightarrow 2NaOH(aq) + H_2(g)$$

Remember: (s) = solid, (l) = liquid, (g) = gas, (aq) = in solution in water

---

Writing chemical equations for electrolysis reactions

When ionic compounds are dissolved in water or melted, they conduct electricity. Chemical changes occur when an electric current is passed through them. This is <u>electrolysis</u>.

At the negative electrode, positively charged ions gain electrons from the electrode to produce neutral atoms or molecules.

At the positive electrode, negatively charged ions lose electrons to the electrode to produce neutral atoms or molecules.

The diagram shows the electrolysis of copper chloride solution. The information below the diagram shows how the change at each electrode can be represented by a **half-equation**.

You may be given some information about a half-equation, then asked to balance it. The example in the box shows you how to do this.

5 Copy and then balance these half-equations for the electrolysis of sodium bromide.
(a) $Br^- - e^- \rightarrow Br_2$ [bromine molecule]
(b) $Na+ \rightarrow Na$ [sodium atom]

6 Complete and balance these half-equations which occur during the extraction of aluminium from aluminium oxide.
(a) $Al^{3+} + e^- \rightarrow Al$
(b) $O^{2-} - \rightarrow O_2$

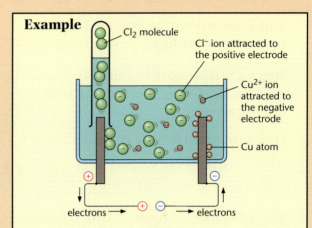

Example
Cl$_2$ molecule
Cl$^-$ ion attracted to the positive electrode
Cu^{2+} ion attracted to the negative electrode
Cu atom
electrons → ← electrons

In the electrolysis of copper chloride solution:

- copper atoms are released at the negative electrode;

- chlorine molecules are released at the positive electrode.

At the negative electrode copper ions gain electrons to become copper atoms:

$$Cu^{2+} + e^- \rightarrow Cu$$

When balanced, this half-equation is:

$$Cu^{2+} + 2e^- \rightarrow Cu$$
copper ion   two electrons from the electrode   copper atom (neutral)

At the positive electrode chloride ions lose electrons and form chlorine molecules:

$$Cl^- - 2e^- \rightarrow Cl_2$$

When balanced, this half-equation is:

$$2Cl^- - 2e^- \rightarrow Cl_2$$
two chloride ions   two electrons to the electrode   one chlorine molecule (neutral)

8 Patterns of chemical change

How can we make a reaction go faster?

Some chemical reactions are very fast, others are slow.
The reactions go at different speeds or **rates**.

When we cook food there are chemical reactions going on.
How fast the food cooks depends on how hot we make it.

boiling water at 100 °C

potato pieces

The pieces of potato take about 20 minutes to cook.

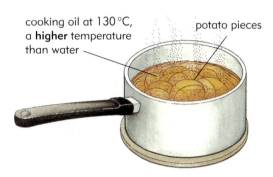

cooking oil at 130 °C, a **higher** temperature than water

potato pieces

The potatoes take less than 10 minutes to cook.

1 Look at the pictures.
  (a) Which is faster, cooking in boiling water or in cooking oil?
  (b) Why do you think this is?

How much difference does temperature make?

Look at the colour change reaction in the picture.
The table shows how long it takes for the mixture to change colour.

Temperature (°C)	20	30	40	50
Time taken to go blue (seconds)	400	200	100	50

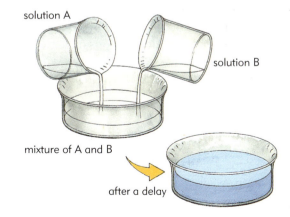

solution A

solution B

mixture of A and B

after a delay

2 What is the effect of increasing the temperature on the time taken for the reaction?

3 How long do you think the reaction will take at:
  (a) 60 °C?
  (b) 10 °C?

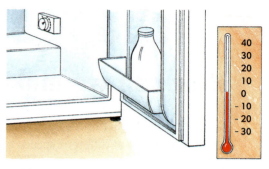

Using temperature to control reactions

If you increase the temperature by 10°C, chemical reactions go about twice as fast. To slow down a chemical reaction you must reduce the temperature.

4 Where can you put milk to slow down the chemical reactions that make it go bad?

5 About how long will it take the milk to go sour in the fridge?

6 (a) How many times faster do the potatoes cook in the pressure cooker?
  (b) What does this tell you about the temperature of the water inside the pressure cooker?

Inside a fridge, the milk takes many days to go sour.

water boiling at 100 °C

pressure cooker

The potatoes take about 24 minutes to cook.

The potatoes take about 6 minutes to cook.

Outside, the milk goes sour in two days.

Making solutions react faster

> **Remember**
>
> A reaction that takes a short time has a high speed or rate.

Some substances will dissolve in water to make a **solution**. You can use solutions for many chemical reactions. The speed of these chemical reactions depends on how strong the solutions are.

7 What is the chemical solution in a car battery?

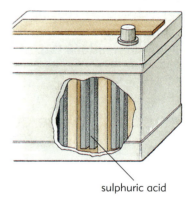

sulphuric acid

The chemical reactions in a car battery need sulphuric acid of just the right strength.

'Strong' and 'weak' solutions

Your friend likes her tea to taste sweet, but not too sweet.

8 What did the tea taste like from the cup where the sugar solution was too strong?

9 A mug of tea is 1.5 times bigger than one of the cups shown above. How many spoonfuls of sugar should your friend put into a mug of tea? Give a reason for your answer.

one spoonful   two spoonfuls   three spoonfuls

sugar

not sweet enough   just right   too sweet

sugar solution is too weak

sugar solution is too strong

Dilute to taste

We call a 'strong' solution a **concentrated** solution.
To make a solution 'weaker', we **dilute** it with water.

10 Look at the diagrams.
Why do you need to dilute the orange drink
with water?

orange drink

50 cm³

concentrated solution

dilute with water to
make it good to drink

400 cm³

water

How does concentration affect the speed of a chemical reaction?

Look at the pictures of the reaction between a chemical we
call 'thio' and an acid. The table shows the results.

Solution	Time for cross to disappear
A	8 minutes
B	4 minutes
C	2.5 minutes

11 (a) How many spatulas of the crystals are in the most
concentrated solution?
(b) Which concentration of thio solution reacts in the
shortest time?

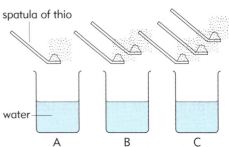

spatula of thio

water

A   B   C

Some students make 3 different strengths of
thio solution.

Making gases react faster

Some gases will react together to make new substances.
For example, you can make ammonia gas by reacting
together a mixture of nitrogen and hydrogen gases.

You can squeeze gases into a smaller space. This is like
making a more concentrated solution. The gases will then
react together faster. A **high** pressure gas is like a very
concentrated solution.

12 A chemical factory makes ammonia gas.
They already make the hydrogen and nitrogen
as hot as they can. What else should they do to
make the reaction go faster?

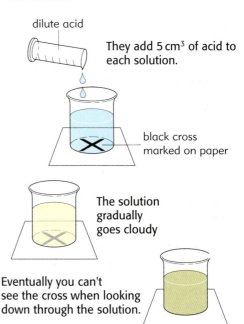

dilute acid

They add 5 cm³ of acid to
each solution.

black cross
marked on paper

The solution
gradually
goes cloudy

Eventually you can't
see the cross when looking
down through the solution.

Making solids react faster

The photographs show a chemical reaction between a solid and a solution.

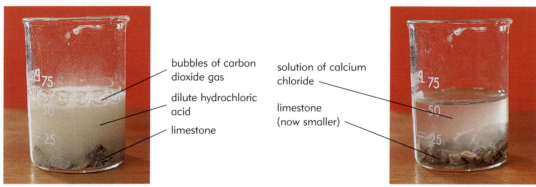

bubbles of carbon dioxide gas

dilute hydrochloric acid

limestone

During the reaction.

solution of calcium chloride

limestone (now smaller)

When all the acid has been used up, the reaction stops.

13 Write down:
(a) the name of the solid in the reaction;
(b) the name of the solution used;
(c) the name of the gas produced.

14 Write down the word equation for the reaction between limestone and dilute hydrochloric acid.

Making the reaction faster

One way to make the reaction faster is by using more concentrated acid. But how fast the limestone reacts also depends on how big the pieces of limestone are.

15 Look at the photographs.
What is the effect of the size of the pieces of limestone on how quickly the reaction takes place?

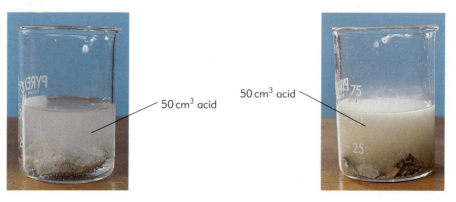

50 cm³ acid

50 cm³ acid

With one large piece of limestone, the gas bubbles continue for 10 minutes.

With smaller pieces, the gas bubbles continue for 1 minute. The bubbling is faster.

50 cm³ acid

With very small pieces, the gas bubbles continue for a few seconds. The bubbling is very fast.

Do you suck sweets or crush them?

Think about eating a hard sweet. If you suck the sweet in one piece it lasts quite a long time. If you crush the sweet into little pieces it doesn't last so long.

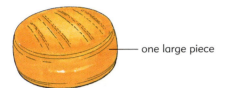

one large piece

Sucking your sweet. Your saliva can only get at the outside surface of the sweet.

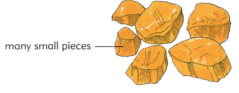

many small pieces

Crushing your sweet. Your saliva can get at more of the sweet at once.

16 Why does the crushed sweet dissolve faster? Explain your answer as fully as you can.

Why small bits react faster

The same amount of limestone in smaller bits reacts faster. The acid can get at smaller bits better. This is because they have more **surface area**.

17 Look at the large 'cube of limestone'.
(a) How many little squares are there on one face of the large cube?
(b) How many faces are there on the cube?
(c) What is the total number of little squares on the surface of the large cube? This is the surface area of the cube.

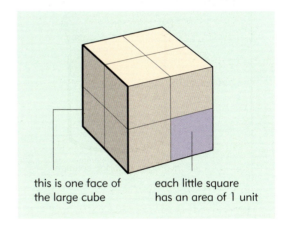

this is one face of the large cube

each little square has an area of 1 unit

18 Now look at the large cube broken up into smaller cubes.
(a) What is the surface area of each small cube?
(b) What is the total surface area of all the small cubes added together?
(c) How many times more surface area do the small cubes have than the large cube?

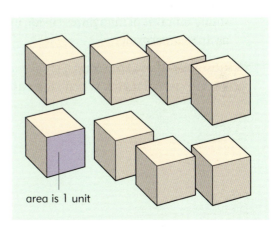

area is 1 unit

Substances that speed up reactions

People use hydrogen peroxide to bleach hair. It does this by releasing oxygen. This reaction turns the hair a very pale blonde colour.

19 Write a word equation for the reaction where hydrogen peroxide splits up.

In the bottle, the hydrogen peroxide very slowly splits up into oxygen gas and water.

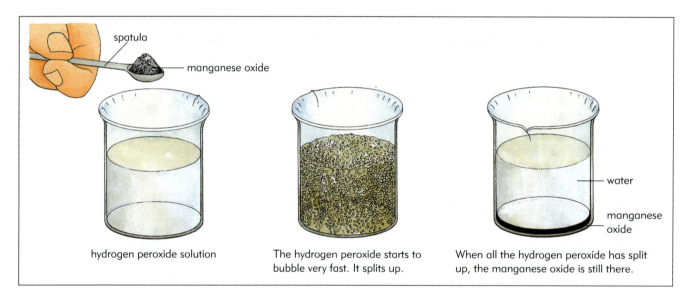

hydrogen peroxide solution

The hydrogen peroxide starts to bubble very fast. It splits up.

When all the hydrogen peroxide has split up, the manganese oxide is still there.

20 Look at the diagrams. What happens if you put a tiny amount of manganese oxide into some hydrogen peroxide?

A substance which speeds up a chemical reaction in this way has a special name. We call it a **catalyst**.

Why don't you need much of the catalyst?

The manganese oxide is not used up in the chemical reaction. It is still there at the end. It can be used over and over again to split up more hydrogen peroxide.

21 How could you collect the catalyst so that you could use it again?

22 How does this experiment show that a catalyst is not used up in the reaction?

You can show that the catalyst is not one of the ordinary chemicals that react, by writing your equation like this:

$$\text{hydrogen peroxide} \xrightarrow{\text{manganese oxide}} \text{oxygen} + \text{water}$$

We write the name of the catalyst above the arrow.

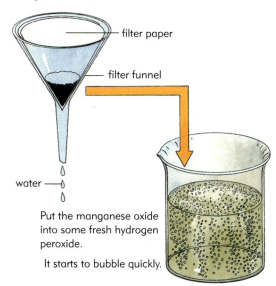

You can use the same manganese oxide over and over again. First filter the water and manganese oxide.

filter paper

filter funnel

water

Put the manganese oxide into some fresh hydrogen peroxide.

It starts to bubble quickly.

What can we make using catalysts?

We can make lots of useful materials using catalysts. These materials cost less to make when you use a catalyst. Usually each chemical reaction needs its own special catalyst.

23 What substance do we make using a catalyst called vanadium oxide?

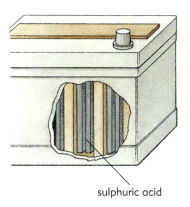

sulphuric acid

Car batteries contain sulphuric acid. We make this acid using a catalyst called vanadium oxide.

Why do cars have catalytic converters?

Look at the diagrams at the bottom of the page.

24 Why do we fit cars with catalytic converters?

25 You often have to fill up a car's fuel tank.
You don't have to add more catalyst to the converter.
Why is this?

exhaust gases: unburnt fuel and nitrogen oxides, with carbon dioxide, water and nitrogen gas

exhaust gases: carbon dioxide, water and nitrogen gas

catalytic converter

The catalytic converter changes harmful gases into safer gases. The catalyst is not used up in the reactions.

Investigating the speed of reactions

Looking and timing

All you need to measure the speed of many chemical reactions is a clock. You can then watch the reaction carefully to see how it changes.

You need to look out for different things in different reactions.

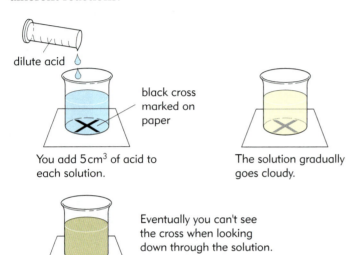

dilute acid

black cross marked on paper

You add 5 cm³ of acid to each solution.

The solution gradually goes cloudy.

Eventually you can't see the cross when looking down through the solution.

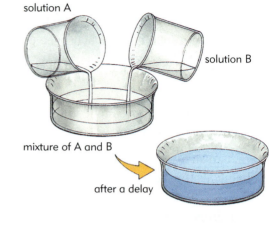

solution A

solution B

mixture of A and B

after a delay

Remember

When limestone reacts with dilute hydrochloric acid, the limestone pieces get smaller during the reaction (see page 290). When all the acid has been used up the reaction stops.

1 Write down <u>three</u> different things you might look for when you are timing a chemical reaction.

How much gas is produced?

Some chemical reactions produce a gas.

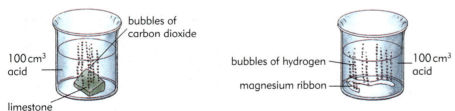

bubbles of carbon dioxide

100 cm³ acid

limestone

bubbles of hydrogen

magnesium ribbon

100 cm³ acid

2 Write down the name of the gas produced when:
   (a) limestone reacts with acid;
   (b) magnesium reacts with acid.

You can collect the gas and measure how much there is. Then you can use your results to draw a graph.

3 How can you collect and measure a gas produced during a reaction?

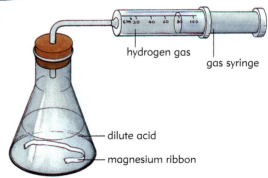

hydrogen gas

gas syringe

dilute acid

magnesium ribbon

Look at the graph. It shows the results of the experiment of magnesium reacting with acid.
A gas syringe was used to collect the gas.

4 What is happening to the reaction:
   (a) during the first two minutes?
   (b) between the second and third minutes?
   (c) after the fourth minute?

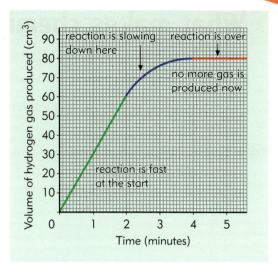

How does the mass change?

You can also measure the rate of reaction by **weighing**.
If a gas escapes into the air during a reaction, the mass of what is left goes down.

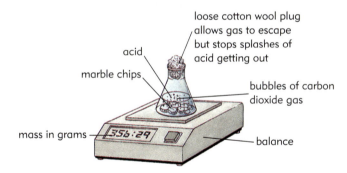

The graph shows some students' results for this experiment.

5 Look at the graph.
   (a) Which reaction takes longer to finish?
   (b) Which reaction has the faster rate?
   (c) How much carbon dioxide gas is produced in each reaction?

6 Why is there a cotton wool plug in the neck of the flask?

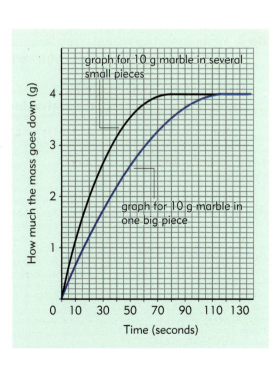

What makes chemical reactions happen?

Chemical reactions can only happen when the particles of different substances **collide** with each other.

The diagrams show what happens when carbon burns in oxygen.

1 How many carbon atoms does a molecule of oxygen join with to make a molecule of carbon dioxide?

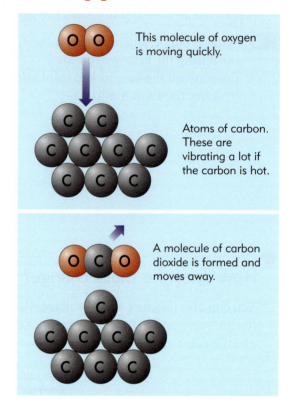

This molecule of oxygen is moving quickly.

Atoms of carbon. These are vibrating a lot if the carbon is hot.

A molecule of carbon dioxide is formed and moves away.

Why do reactions speed up when you increase the temperature?

The higher the temperature, the faster the oxygen molecules move.

2 Write down <u>two</u> reasons why faster-moving oxygen molecules react more easily with carbon.

The smallest amount of energy particles must have for a reaction to occur is called the **activation energy**.

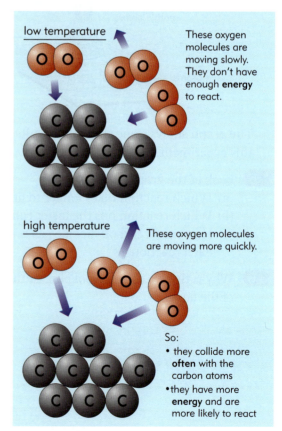

low temperature

These oxygen molecules are moving slowly. They don't have enough **energy** to react.

high temperature

These oxygen molecules are moving more quickly.

So:
• they collide more **often** with the carbon atoms
• they have more **energy** and are more likely to react

Why does breaking up a solid make it react faster?

A lump of iron doesn't react very quickly with oxygen, even if it is very hot. But the tiny specks of iron in a sparkler burn quite easily.

3 Why do tiny specks of iron react more easily than a big lump of iron?

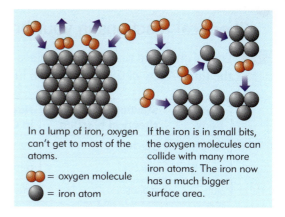

In a lump of iron, oxygen can't get to most of the atoms.

= oxygen molecule

= iron atom

If the iron is in small bits, the oxygen molecules can collide with many more iron atoms. The iron now has a much bigger surface area.

Why do strong solutions react faster?

Magnesium metal reacts with acid.

The reaction is faster if the acid is made more concentrated.

4 Explain why the reaction is faster in more concentrated acid.

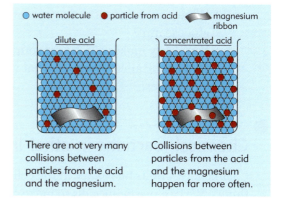

○ water molecule ● particle from acid   magnesium ribbon

dilute acid

concentrated acid

There are not very many collisions between particles from the acid and the magnesium.

Collisions between particles from the acid and the magnesium happen far more often.

Another way to make gases react faster

Gases react faster if they are hot.

The diagram shows another way to make gases react faster.

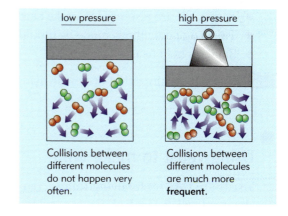

low pressure

high pressure

Collisions between different molecules do not happen very often.

Collisions between different molecules are much more **frequent**.

Living things can do our chemistry for us

Lots of people think that chemistry happens only in laboratories. They imagine that chemicals react in strange bits of glass – a bit like in a horror film!
But chemistry happens wherever we change the substances we start with into new substances.

So, there are lots of places where chemical reactions happen. One of these places is inside your body.

Chemical reactions in your body

1 Lots of chemical reactions take place in your body all the time. A few of them are shown in the table.

Part of body	Starting material in chemical reaction	New material made in chemical reaction
mouth	starch	maltose (a sugar)
muscles	glucose	carbon dioxide + water
gut	maltose	glucose

(a) Which part of your body turns starch into maltose?
(b) What new substances do your muscles change glucose into?
(c) Your body makes glucose in your gut.
What substance do you eat that your gut makes into glucose?

Chemical reactions in other living cells

Chemical reactions take place in the cells of all living things. We can use living cells to make chemicals for us.

- Yeast cells are used to make bread.
- Yeast cells are also used to make beer.
- The drug penicillin is made by moulds.
- Bacteria help us to make yogurt.

2 Name <u>two</u> things that are made using yeast cells.

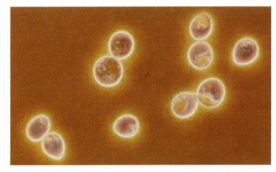

These are yeast cells. When they have plenty of food (sugar) they can grow and divide quickly. They turn the sugar into alcohol and the gas carbon dioxide.

Using yeast to make wine

Yeasts are very useful living things. Yeast cells make the chemical called **alcohol** in wine and beer.

We call this reaction **fermentation**.

3 Look at the photograph of the yeast cells under the microscope.

Write a word equation for the fermentation reaction.

4 Look at the picture of wine being made.

How can you tell that the grape juice is fermenting?

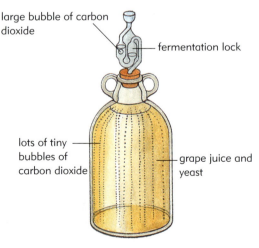

large bubble of carbon dioxide

fermentation lock

lots of tiny bubbles of carbon dioxide

grape juice and yeast

Making wine. The sugar for the reaction comes from the grape juice.

Using yeast to make bread

The carbon dioxide that the yeast makes is useful too. This gas helps to make bread rise. When you slice through bread you can see lots of tiny holes.

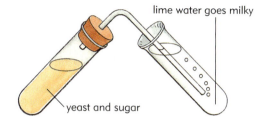

lime water goes milky

yeast and sugar

A simple test for carbon dioxide.

5 What do you think causes these holes?

6 Look at the diagram showing a simple test for carbon dioxide.

How can you tell that yeast produces carbon dioxide?

Using bacteria to make yogurt

Bacteria are living cells. Some bacteria can be harmful to us. Other bacteria are very helpful.

The diagram shows how you can make yogurt.

7 What do yogurt bacteria produce when they use up the sugar in milk?

8 Where do the bacteria come from to make the next batch of yogurt?

Living cells make new substances fastest if they are at just the right temperature. If the cells are too cold then the reactions only happen slowly. If cells are too hot, they can be damaged.

9 What is the best temperature for making yogurt?

Why are living things such good chemists?

Living things contain **enzymes**, which help them to make new materials. Enzymes do this by speeding up chemical reactions. Because of enzymes, reactions in cells are quite fast, even though the cells are not very hot.

There are thousands of different enzymes, just as there are thousands of chemical reactions that happen in living things.

10 What do we call substances (like enzymes) that speed up chemical reactions?

yogurt bacteria

milk

Milk contains a **sugar** called lactose.

Yogurt bacteria use up the sugar and make **lactic acid**.

keep in a warm place (about 40 °C) for 12 hours

there are now many more bacteria, so we can use a little of this yogurt to start the next batch

yogurt

The milk turns into yogurt.

Remember

A catalyst can speed up a chemical reaction.

Why mustn't enzymes be made too hot?

Enzymes are made from **protein**. They are big molecules and they have special shapes that help them to work.

There are proteins in the white of an egg. If you put an egg in hot water the protein changes, and you can't change it back again.

11 Why do you think heat damages or destroys enzymes?

12 Look at the pictures about making bread.
 (a) What makes the dough rise?
 (b) Why must the dough be left in a warm place for this to happen?
 (c) What happens to the live yeast in the hot oven?

> **Remember**
>
> The enzymes in yeast release carbon dioxide when they split up sugar.

Make the dough.

(flour + water + sugar + yeast + salt)

Put dough in the tin.

Leave in a warm place. (25 to 30 °C)

bubbles of carbon dioxide make the dough rise

Bake in a hot oven. (230 °C)

What makes food go bad?

Food will not stay fresh for ever, sooner or later it starts to break down or go bad.

This is because the living cells of moulds, yeasts and bacteria start to feed on it. They use enzymes to break down the food. Then they take the food into their cells. They use it to grow and to reproduce.

So in a warm room, living cells like moulds and bacteria grow and multiply quite fast.

13 Why can moulds and bacteria break down food quickly in a warm room?

14 The temperature in our bodies is 37 °C.
 This temperature is ideal for any bacteria that get inside us. Explain why.

> **Remember**
>
> Living cells use enzymes to speed up their chemical reactions.
>
> Most enzymes work best at 35 to 40 °C.

Keeping food fresh

To stop food from going bad we often keep it in a fridge or a freezer. The pictures explain why this works.

3 °C

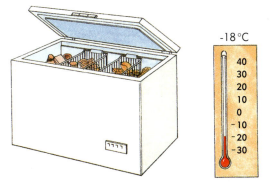

-18 °C

In the fridge, living cells can only break down food slowly. Their enzymes do not work well at this low temperature. Food stays fresh for several days.

In the freezer, living cells cannot break down food at all. Their enzymes do not work at this very low temperature. Food stays fresh for weeks or even months.

15 Why does food stay fresh for a long time in a freezer?

Nice and warm

We need to keep living things warm when we use their enzymes to make things for us. So when we use yeast to make wine, we must keep it at the right temperature.

16 Joe set up three identical fermenting bottles to make strawberry wine. He put one in the kitchen, one in the bedroom and one in the garage.

Five days later he looked at the bottles.
The table shows what he saw.

Where Joe put the fermenting bottles	Number of bubbles per minute on day 5
bedroom (15°C)	15
kitchen (25°C)	30
garage (10°C)	10

(a) In which bottle was the yeast working most quickly?
(b) Why do you think the yeast in the other two bottles was working more slowly?
(c) Which gas were the bubbles made from?
(d) After 10 days, no more bubbles were made by the bottle in the kitchen. Why not?

you can count the bubbles of carbon dioxide gas which escape from here

fermenting bottle

fruit juice, sugar and yeast

Making strawberry wine.

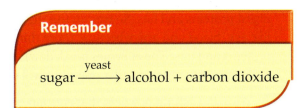

Remember

$$\text{sugar} \xrightarrow{\text{yeast}} \text{alcohol} + \text{carbon dioxide}$$

The right pH

The pH affects how well enzymes work. Different enzymes work best in different conditions.

17 The diagram shows the best pH conditions for some enzymes in your digestive system. Write down the name of an enzyme that works best in:
(a) acid conditions;
(b) alkaline conditions.

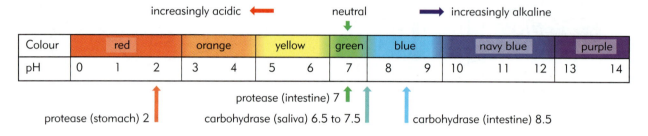

increasingly acidic ←   neutral   → increasingly alkaline

Colour	red		orange		yellow		green	blue		navy blue		purple	
pH	0	1	2	3	4	5	6	7	8	9	10	11	12	13	14

protease (stomach) 2

protease (intestine) 7

carbohydrase (saliva) 6.5 to 7.5

carbohydrase (intestine) 8.5

More uses of enzymes

Remember

- Enzymes in the cells of bacteria and fungi do some of our chemistry for us.

- Temperatures above 45 °C damage most enzymes.

- Examples of enzymes: Proteases digest proteins.
  Lipases digest fats (lipids).

Where do the enzymes come from?

Most of the enzymes that we use come from microorganisms. For some jobs we extract the enzymes from the microorganisms. For other jobs, we use the whole microorganism.

Enzymes that we use at home

Lots of washing powders contain enzymes that come from bacteria. They are the enzymes that the bacteria use to digest proteins and fats. So we can use them to digest stains like egg, blood and gravy because those things contain proteins and fats. The enzymes break down the large molecules into much smaller ones that will rinse out.

1 We call washing powders that contain enzymes 'biological' washing powders. Why is this?

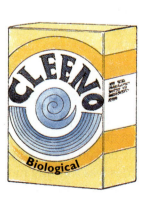

Biological powders work at lower temperatures than ordinary powders. So less energy is used to heat water.

2 Look at the table.
  (a) At which temperature is 'Cleeno' best at removing blood stains? Why is this?
  (b) 'Cleeno' can't remove blood stains at all at 55 °C. Why not?

Temperature of wash in °C	What happened to blood stain
15	
25	
35	
45	
55	

Biological powders are good at removing certain stains. But they can cause problems. Your skin is made of protein. So you shouldn't use biological powders for washing clothes by hand. Also, some people get rashes because they are allergic to the enzymes.

3 The enzymes in biological washing powders work best at fairly low temperatures.
Write down <u>one</u> benefit and <u>one</u> problem of using biological washing powders.

Enzymes that we use in industry

In the chemical industry many processes need high temperatures and pressures. So they need expensive equipment as well as expensive energy. Sometimes chemists choose enzymes to carry out a process because this is much cheaper. Enzyme reactions happen at normal temperatures and pressures.

4 Many industrial processes are expensive.
Write down <u>two</u> reasons for this.

Extracting copper

Some ores have only a small percentage of copper in them. We call them **low-grade** ores. Often we don't use low-grade ores because it costs too much to get the copper out. We say that it is **uneconomic**.

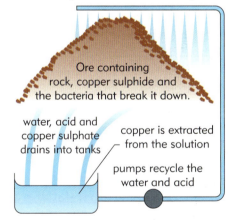

A sprinkler system sprays water and sulphuric acid onto the heap.

Ore containing rock, copper sulphide and the bacteria that break it down.

water, acid and copper sulphate drains into tanks

copper is extracted from the solution

pumps recycle the water and acid

Now scientists have found a new, cheap way of getting the copper out of low-grade ores. They use the enzymes in bacteria. The bacteria change insoluble copper sulphide to soluble copper sulphate.

copper sulphide	+	oxygen	+	water	$\xrightarrow[\text{bacteria}]{\text{enzymes in}}$	copper sulphate	+	sulphuric acid	+	energy

Scientists also use this method to clean up waste tips at copper mines.

5 Look at the diagram and the word equation.
What change do the bacteria make to the copper compound in the ore?

6 How do bacteria that use copper sulphide help the environment?

Enzymes in the food industry

Our food comes from plants and animals. A lot of the food we buy now is changed. We say that it is **processed**. We call the people who work out how to do this processing **food technologists**. Often food technologists change cheap raw materials into expensive products. We say that they **add value** to the raw materials.

Making sugar syrups

One example of added value is the sugar syrup in sweets, cakes and many other foods. Food technologists make these syrups by digesting cheap starch. They used to digest the starch using acid. Then they found that it was cheaper to use enzymes.

7 Look at the label. Write down the names of <u>two</u> sugar syrups.

8 (a) This word equation shows one way of digesting starch to make glucose syrup.

$$\text{starch} \xrightarrow{\text{acid}} \text{glucose}$$

Write a word equation to show a second way of digesting starch to make glucose syrup.
(b) Which of these ways is used most? Explain why.
(c) Write down <u>two</u> plants that we get the cheap starch from.

INGREDIENTS
Pear Slices, Water, Syrup, Glucose Syrup, Corn Syrup, Citric Acid.

NUTRITION INFORMATION

Average values per 100g of product	
Energy:	281kJ
	66kcal
Protein:	0.5g
Carbohydrate:	16.0g
of which sugars:	16.0g
Fat:	Trace
of which saturates:	Trace
Fibre:	1.5g
Sodium:	Trace

A label from tinned pears.

Which enzymes digest starch?

Each enzyme only does one job. So the enzymes that food technologists use to digest starch are not the same as the ones they use to digest proteins.

Starch is a **carbohydrate**. **Carbohydrases** are the only enzymes that digest carbohydrates. Food technologists get these enzymes from microorganisms. The fungi and bacteria that they use have to be the kind that don't cause disease.

9 Write down the name of the group of enzymes that we can use to change starch to glucose.

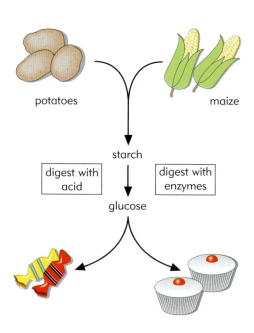

Making an even sweeter syrup

Food technologists often use fructose syrup in slimming foods. It is more expensive than glucose, but it is sweeter too. This means that they use only small amounts. So the foods contain less energy.

They make fructose syrup using enzymes too.

$$\text{glucose} \xrightarrow{\text{isomerase}} \text{fructose}$$

10 Write down the name of an enzyme that we use to change glucose to fructose.

11 We use fructose to sweeten foods for slimmers and diabetics. Explain why.

INGREDIENTS
Fructose Syrup, Cocoa Mass, Maltitol, Sorbitol, Cocoa Butter, Glycerine, Flavouring, Emulsifier (Soya Lecithin)

NUTRITIONAL INFORMATION
Typical Values Grams per 100g

Energy Value:	1432kJ (344kcal)
Protein:	2.3g
Carbohydrate:	66.8g
of which Sugars:	*1.1g*
Fructose:	15.6g
of which Polyols:	50.1g
of which Starch:	1.0g
Total Fat:	16.3g
of which Saturates:	9.9g
Fibre:	4.4g
Sodium:	0g

People with diabetes have to limit the amount of sugars and other carbohydrates that they eat. This is a label from a diabetic food.

Enzymes for tenderising meat

We can also use enzymes to make meat tender. Meat tenderisers contain **proteases** that digest some of the protein. This makes the meat softer.

South American Indians have used an enzyme called **papain** to tenderise meat for centuries. The enzyme is in papaya fruits and in the papaya leaves that they wrap their meat in before they cook it.

12 Is papain a carbohydrase, an isomerase or a protease? Explain your answer.

We use proteases in the food industry. Proteases make cheap cuts of meat tender and cooking times shorter. The proteins in beans are not very easy to digest. So we use proteases to pre-digest some bean products. Babies don't digest some protein foods very easily. So baby foods are mashed up and sometimes partly digested too.

13 Why is the protein in baby foods often pre-digested?

Ingredients

Rice, Vegetables (4%) in variable proportion (Red Pepper, Green Pepper, Mushrooms), Mushroom Extract, Flavourings, Hydrogenated Vegetable Oil, Salt, Hydrolysed Vegetable Protein, Vegetable Bouillon, Parsley, Garlic.

CONTAINS MILK, WHEAT AND SOYA.

You can find out if a food contains pre-digested proteins by looking at the label. It will say 'extract of protein' or 'hydrolysed protein'. Both these mean digested protein. This savoury rice mix contains hydrolysed protein.

Making the best use of enzymes

In good conditions, an enzyme molecule does the same job over and over again. But an enzyme is easily damaged. High temperatures or the wrong pH change the shape of an enzyme. We say that it is **denatured**. A denatured enzyme no longer works and it can't be repaired.

14 Make a copy of the graph. On your copy, mark:
  (a) the part that shows the rate of reaction increasing;
  (b) the optimum or best temperature for this reaction.

15 In an experiment, a sample of the same enzyme was kept at 55 °C. Then it was cooled to 35 °C and mixed with some starch. It didn't digest the starch. Explain this as fully as you can.

Carbohydrases from different microorganisms work best in slightly different conditions. Some enzymes are harder to damage than others. We say that they are more **stable**. Enzymes are expensive. So in industry, scientists choose the most stable enzymes. Then they keep them working by carefully controlling the conditions.

16 Write down <u>two</u> conditions that scientists control in processes that use enzymes.

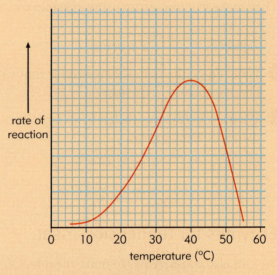

How temperature affects the rate of starch digestion by a carbohrase.

Two kinds of industrial processes

Look at the flow chart about a **batch process**. In this kind of process, we make one batch of product. When that is finished, we start again.

17 What is a batch process? Describe, as fully as you can, how beer is made by a batch process.

Making beer by a batch process

Each batch is fermented in a large vat. The raw materials are malt, sugar, hops and water. The enzymes that are used for the process are inside yeast cells.

Batch processes

We put the raw materials and enzymes in a container.

⬇

The reaction happens.

⬇

We separate the enzymes from the product.

⬇

We sterilise the container ready for the next batch.

Sometimes we can keep a process going for months. So we call it a **continuous process**. We do this by trapping the enzyme, then trickling a solution of the raw material over it. The reaction happens and the product trickles out. The enzyme stays trapped in the container.

18 Look at the diagram. Write down <u>three</u> ways that a continuous process is better than a batch process.

How do we trap the enzymes?

We trap enzymes in materials that don't affect the enzyme or the reaction. We say that these materials are **inert**. The enzymes can't move out of the materials. So we call them **immobilised** enzymes.

19 (a) What is an immobilised enzyme?
(b) Why do we use immobilised enzymes?

20 Look at the pictures. Write down <u>two</u> materials that we use to immobilise enzymes.

21 Look at the data for the reaction

$$\text{maltose} \xrightarrow{\text{carbohydrase}} \text{glucose}$$

	Concentration of glucose after 3 hours (%)
immobilised enzyme	6
non-immobilised enzyme	8

(a) <u>From this data only</u>, which way of using the enzyme seems to be better?

(b) After 3 months, the yield for the immobilised enzyme was twice as much as the yield for the non-immobilised enzyme. Explain this as fully as you can.

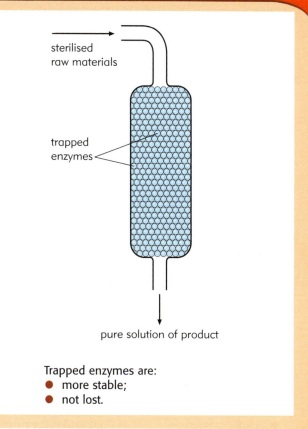

sterilised raw materials

trapped enzymes

pure solution of product

Trapped enzymes are:
- more stable;
- not lost.

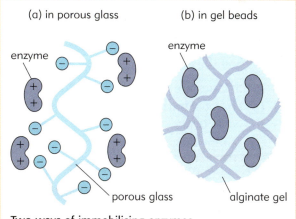

(a) in porous glass

(b) in gel beads

enzyme

enzyme

porous glass

alginate gel

Two ways of immobilising enzymes.

Chemical reactions and energy

On a barbecue you burn charcoal to cook the food.
Burning charcoal releases energy in the form of heat.
Burning charcoal is an example of a **chemical reaction**.

1 Write a word equation to describe what happens
when carbon burns.

Substances that we burn to release energy are called
fuels.

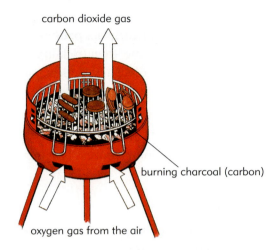

carbon dioxide gas

burning charcoal (carbon)

oxygen gas from the air

Are all fuels the same?

All fuels release energy as they burn. Different fuels give
different waste gases:

- If a fuel contains carbon it produces carbon dioxide
  when it burns.

- If a fuel contains hydrogen it produces water vapour
  when it burns.

- Many fossil fuels such as coal and oil contain a little
  sulphur. If a fuel contains some sulphur, it produces
  sulphur dioxide gas when it burns.

2 Look at the pictures.

What gases are produced when:
(a) charcoal burns?
(b) coal burns?
(c) butane burns?

3 Write a word equation to describe what happens when
camping gas burns.

chimney

waste gases are carbon
dioxide and water
vapour and some
sulphur dioxide

burning coal

What other reactions release energy?

Many chemical reactions happen in solutions.
These reactions may also release some energy.

Chemical reactions that release energy are called
exothermic reactions.

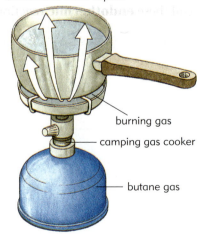

waste gases are carbon
dioxide and water vapour

burning gas

camping gas cooker

butane gas

Look at the diagram.

4 How do you know that this reaction releases energy?

5 Write a word equation to describe what happens when zinc is added to nitric acid.

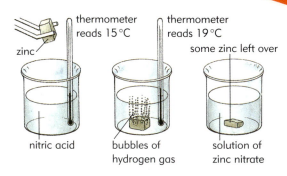

Other types of energy released by chemical reactions

Some chemical reactions release different kinds of energy. Look at the pictures.

6 What other kinds of energy do these chemical reactions release?

A candle.   A stone quarry.

Do chemical reactions always release energy?

> ### Remember
>
> When fuels burn or when metals react with acids, heat energy is released. We say that these reactions are exothermic.

Many chemical reactions release energy.

7 Write down <u>two</u> examples of chemical reactions that release energy.

8 What is the name for a reaction that releases energy?

Other reactions will happen only if we supply energy. We call these **endothermic** reactions.

9 Write down <u>one</u> everyday example of an endothermic reaction.

You have to supply energy to cook the egg.

Using heat to make reactions happen

We can make some chemical reactions happen by supplying energy in the form of **heat**. This is why a Bunsen burner is so useful; it supplies heat energy.

10 Look at the diagrams.

Write word equations for the two endothermic reactions shown in the diagrams.

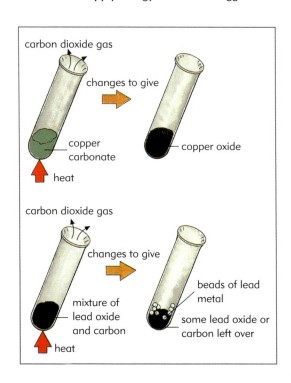

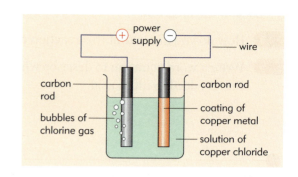

Using electricity to make reactions happen

We can make some chemical reactions happen by supplying energy in the form of **electricity**.

We can use electricity to obtain copper metal from a solution of copper chloride.

11 Write a word equation for the reaction that happens when electricity is passed through a solution of copper chloride.

Extracting metals from their natural ores

You need to supply energy to extract metals from ores.

12 What is the name for the type of reaction we use to extract metals from their ores?

13 How do you supply the energy you need to extract aluminium from aluminium ore?

14 How do you supply the energy you need to extract iron from iron ore in a blast furnace?

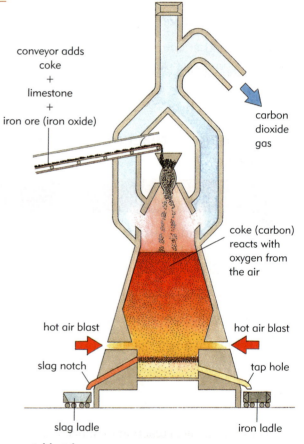

A blast furnace.

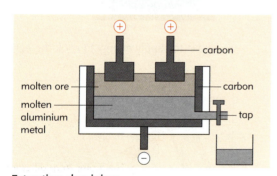

Extracting aluminium.

Reactions that use light energy

When you take a photograph, **light energy** changes the chemicals in the film. Films contain silver halides. The word equation is:

silver halide + energy ⟶ silver metal + halogen

15 What does light energy do to a silver halide when you take a photograph?

What makes reactions exothermic or endothermic?

Chemical reactions that <u>give out</u> energy into the surroundings are called **exothermic** reactions. **Endothermic** reactions will only take place if energy is <u>taken in</u> from the surroundings.

Energy is transferred by chemical reactions because of the forces of attraction that hold atoms together. These forces of attraction between atoms are called **bonds**.

The chemical bonds between atoms in a molecule can be represented by lines:

- a **single bond** (one shared pair of electrons) is shown by –
- a **double bond** (two shared pairs of electrons) is =
- a **triple bond** (three shared pairs of electrons) is ≡

16 Copy the following drawings of water, carbon dioxide and ammonia molecules. Below each one, draw a diagram to show the bonds between the atoms.

water
(single bonds)

carbon dioxide
(double bonds)

ammonia
(single bonds)

The diagrams in the box show what happens when methane and oxygen react.

$$\text{methane} + \text{oxygen} \rightarrow \text{carbon dioxide} + \text{water}$$
$$CH_4 + 2O_2 \rightarrow CO_2 + 2H_2O$$

For this chemical reaction to happen, the existing bonds in the methane and oxygen molecules must first be broken. New bonds can then be formed between different atoms to produce new substances, carbon dioxide and water.

17 Draw a similar set of diagrams to show the decomposition of ammonia into nitrogen and hydrogen.

$$\text{ammonia} \rightarrow \text{nitrogen} + \text{hydrogen}$$
$$2NH_3 \rightarrow N_2 + 3H_2$$

Add captions to your diagrams to explain what is happening.

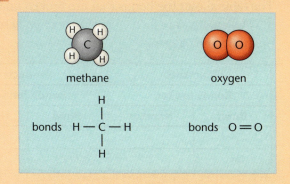

methane

oxygen

bonds   H — C — H

bonds   O = O

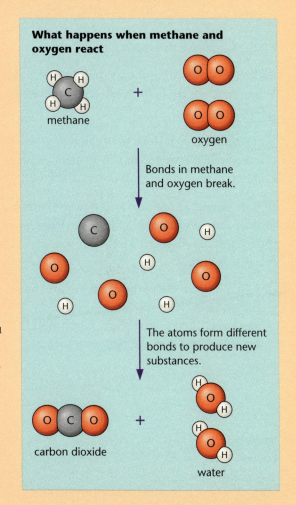

What happens when methane and oxygen react

methane

oxygen

+

Bonds in methane and oxygen break.

The atoms form different bonds to produce new substances.

carbon dioxide

+

water

Higher

The table shows the bonds that are broken and the new bonds that are formed in the reaction between methane and oxygen.

Reactants		Products	
Bonds broken		Bonds formed	
Bond	Number	Bond	Number
C–H	4	C=O	2
O=O	2	O–H	4

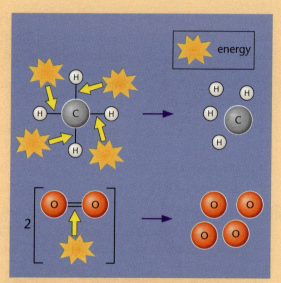

18 Copy the following.

$$H_2N-NH_2 + H_2N-NH_2 \rightarrow N\equiv N + H-H + H-H + H-H$$

Now make a table showing the bonds broken and the bonds formed during this reaction.

To break bonds, energy must be supplied (taken in). When new bonds are formed, energy is released (given out).

If more energy is needed to break bonds than is released when new bonds are formed, the overall reaction takes in energy. It is endothermic.

If more energy is released when new bonds are formed than was needed to break existing bonds, the overall reaction gives out energy. It is exothermic.

19 Explain, in terms of the breaking and forming of bonds:
  (a) what makes a reaction exothermic;
  (b) what makes a reaction endothermic.

Energy is needed to break bonds in reactant molecules.

Energy is released when bonds of product molecules are formed.

Higher

Calculating energy changes in reactions

In a chemical reaction, energy is needed to break bonds. Energy is released when new bonds form. The energy change in a reaction is the difference between the two. We call this difference ΔH ('delta aitch').

ΔH = energy needed to break bonds − energy released when new bonds form

In an exothermic reaction, more energy is released when new bonds form than is needed to break existing bonds, so ΔH is negative (−).

Bond	Energy needed (or released) when bond is broken (or formed) (kJ per formula mass)
C–C	347
C=C	612
C–H	413
C=O	805
N≡N	945
O=O	498
H–O	464
H–H	436
H–N	391

Example

$$\Delta H = -55 \text{ kJ}$$

The minus sign tells you the reaction is exothermic.   This is how much energy is transferred for a formula mass of bonds. (kJ = kilojoule)

The **formula mass** is the **relative molecular mass** in grams.
For example, for methane, CH_4, it is

[12 + (4 × 1)] = 16 grams

For carbon dioxide, CO_2, it is

[12 + (2 × 16)] = 44 grams

In an endothermic reaction, more energy is needed to break bonds than is released when new bonds form, so ΔH is positive (+).

The overall energy change in a chemical reaction (the **net energy transfer**) can be calculated by using **bond energy values**. These values are the amounts of energy needed to break bonds, or the amounts released when new bonds form.

20 (a) Which of the bonds in the table is the strongest (that is, needs the most energy for it to break)?
(b) How does the bond energy of a C=C bond compare with that of a C–C bond?

Examples 1 and 2 show how to calculate the energy needed (or released) to break (or form) bonds.

21 Calculate the energy released when a formula mass of water (18 g) is formed.

Example 1 (breaking bonds in methane)

$$H - \underset{\underset{H}{|}}{\overset{\overset{H}{|}}{C}} - H = 4 \text{ [C–H] bonds broken for 1 formula mass of } CH_4$$

Energy needed to break bonds
= 4 × 413
= 1652 kJ per formula mass

Example 2 (forming bonds in carbon dioxide)

O=C=O = 2 [C=O] bonds formed for 1 formula mass of CO_2

Energy released when bonds formed
= 2 × 805
= 1610 kJ per formula mass

Higher

Energy transfers in a complete reaction

Follow these steps if you need to find the net energy transfer in a reaction. The box on the right gives a worked example.

Step 1. Write down the balanced symbol equation for the reaction.
(You may be given this.)

Step 2. Write down the reaction using structural formulas to show the bonds in the reactants and the products.

Step 3. Show the bonds broken and formed and the number of each.

Step 4. Look up the bond energy for each type of bond.

Work out the energy needed to break the bonds of the reactants and to form the bonds of the products.

Step 5. Work out the net energy transfer in the reaction (ΔH).

Step 6. Write down your conclusion.

Example 3

Work out the net energy transfer in the reaction:

methane + oxygen → carbon dioxide + water

Is the reaction endothermic or exothermic?

$$CH_4 + 2O_2 \rightarrow CO_2 + 2H_2O$$

Reactants

$$H-\underset{\underset{H}{|}}{\overset{\overset{H}{|}}{C}}-H \quad + 2[O=O]$$

Products

$$O=C=O + 2[H-O-H]$$

Bonds broken		Bonds formed	
Type	Number	Type	Number
C–H	4	C=O	2
O=O	2	O–H	4

Energy needed to break the bonds (reactants):
C–H $= 413 \times 4 = 1652$
O=O $= 498 \times 2 = 996$
Total $= \underline{2648}$ kJ

Energy released when the products form:
C=O $= 805 \times 2 = 1610$
O–H $= 464 \times 4 = 1856$
Total $= \underline{3466}$ kJ

$$\Delta H = \begin{matrix} \text{energy needed} \\ \text{to break bonds} \end{matrix} - \begin{matrix} \text{energy released when} \\ \text{new bonds form} \end{matrix}$$

$$= 2648 - 3466 = -818 \text{ kJ}$$

Because ΔH is negative, the reaction is exothermic.

22 Calculate the net energy transfer when water is decomposed into hydrogen and oxygen:

$$2H_2O \longrightarrow 2H_2 + O_2$$

(Work it out for <u>one</u> formula mass of water.)
Use your results to find ΔH for this reaction and say whether it is exothermic or endothermic.

23 Repeat question 22 for the decomposition of ammonia:

$$2NH_3 \longrightarrow N_2 + 3H_2$$

Using diagrams to show energy transfers

The energy changes that take place in a chemical reaction can be represented on an **energy level diagram**.

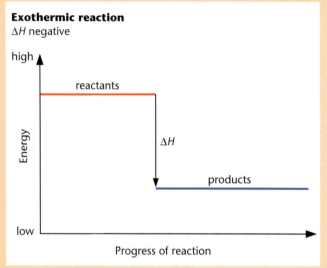

Exothermic reaction
ΔH negative

In an exothermic reaction, the reaction mixture loses thermal energy to the surroundings. This means that the energy level of the products of the reaction is less than the energy level of the reactants.

Endothermic reaction
ΔH positive

In an endothermic reaction, energy is taken in from the surroundings. So the energy level of the products is greater than the energy level of the reactants.

When showing a particular reaction, we write the names of the reactants and products on the energy level diagram. For example, the diagram on the right is for the reaction between methane and oxygen.

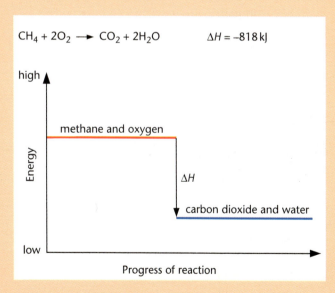

$$CH_4 + 2O_2 \longrightarrow CO_2 + 2H_2O \qquad \Delta H = -818\,kJ$$

24 Draw and label an energy level diagram for the following reaction:

sodium hydroxide + hydrochloric acid $\rightarrow$ sodium chloride + water

$$NaOH \quad + \quad HCl \quad \rightarrow \quad NaCl \quad + H_2O \qquad \Delta H = -55\,kJ$$

Higher

When you turn on the methane gas to a Bunsen burner, it does not begin to burn, no matter how long you wait. It is only when you put a flame to the gas that it lights.

For methane to react with oxygen, the existing chemical bonds between the atoms must first be broken. Putting a flame to the gas provides the energy that is needed to break some of these bonds and get the reaction started.

The energy needed to start a reaction is called the **activation energy**. It can be shown on an energy level diagram for the reaction.

25 (a) Why must we usually supply energy to get an exothermic reaction started?
   (b) What do we call the energy we have to supply?
   (c) An exothermic reaction will carry on by itself once it has been started. Explain why.

Some reactions will happen more quickly if a **catalyst** is used. Catalysts are substances that increase the rate of a chemical reaction but are not changed by the reaction.

Transition metals are often used as catalysts. For example, iron pellets are used as a catalyst in the manufacture of ammonia.

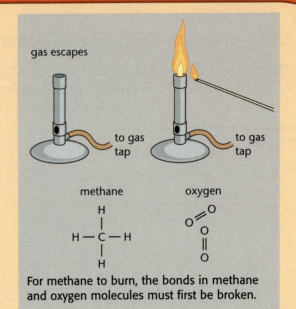

For methane to burn, the bonds in methane and oxygen molecules must first be broken.

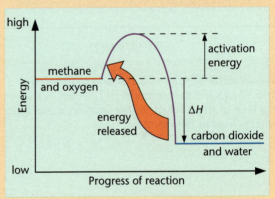

The energy released when new bonds form breaks more of the old bonds and keeps the reaction going.

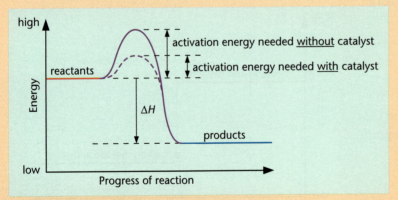

The thermal energy change (ΔH) for the reaction is unchanged, whether a catalyst is used or not.

26 Draw an energy level diagram for this reaction:

$$2NH_3 \longrightarrow N_2 + 3H_2 \quad \Delta H = +93\,kJ$$
(for the quantities shown in the balanced equation)

27 Hydrogen and oxygen will combine explosively if ignited by a spark, to form water (vapour).

$$2H_2 + O_2 \rightarrow 2H_2O \quad \Delta H = -486\,kJ$$

The activation energy for the reaction is 1370 kJ. Draw and fully label an energy level diagram for this reaction.

Reactions that go forwards and backwards

In <u>most</u> chemical reactions:

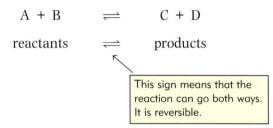

substances at the   new substances at
start of the reaction $\longrightarrow$ the end of the reaction

reactants $\longrightarrow$ products

In <u>some</u> reactions, the products can change back into the original reactants.

A + B $\rightleftharpoons$ C + D

reactants $\rightleftharpoons$ products

> This sign means that the reaction can go both ways. It is reversible.

This kind of reaction can go in both directions.
So we call it a **reversible** reaction.

What happens when we heat ammonium chloride?

1 Look at the diagram. When you heat ammonium chloride it decomposes to form two colourless gases. What are they?

2 Write a word equation for the reaction where ammonium chloride decomposes.

3 Look at the hazard warning label for the two gases. What do you need to do to heat ammonium chloride safely?

4 What does the symbol $\rightleftharpoons$ in the equation tell you?

Ammonia and hydrogen chloride gases are both irritant.

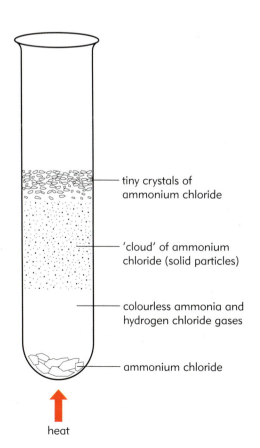

tiny crystals of ammonium chloride

'cloud' of ammonium chloride (solid particles)

colourless ammonia and hydrogen chloride gases

ammonium chloride

heat

What happens when we heat copper sulphate?

Look at the photographs of the two forms of copper sulphate.

These crystals of copper sulphate have water molecules in them as well as copper sulphate. We say that they are hydrated.

These crystals are anhydrous copper sulphate. This means that they have no water in them.

It is easy to change one into the other.
So this change is reversible.

5 Copy the equation.
The spaces are there for you to write in the colours.

hydrated   anhydrous
copper sulphate + ⟩energy⟨ ⇌ copper sulphate + water
[_____]   [_____]

6 Is making anhydrous copper sulphate this way an exothermic or an endothermic reaction?
Explain your answer.

The diagrams show what happens when you add water to anhydrous copper sulphate.

7 Describe the energy transfer when you add water to anhydrous copper sulphate.

So energy is transferred <u>to</u> the surroundings when you add water to anhydrous copper sulphate. The same amount of energy <u>from</u> the surroundings is needed to drive water out of the hydrated crystals.

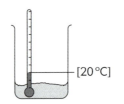

Crystals of anhydrous copper sulphate [20 °C].

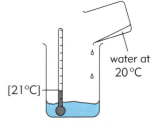

Crystals of hydrated copper sulphate [21 °C].

Remember

We can use anhydrous copper sulphate as a test for water.

Nitrogen chemistry

What use is nitrogen?

The biggest part of the air is made from a gas called **nitrogen**. Nitrogen is an unreactive gas. Our bodies don't use the nitrogen in the air, but it is really useful to us in other ways. Chemists can make it into many useful substances.

Name of gas	How much there is in air
nitrogen	78%
oxygen	21%
other gases	1%

1 Look at the table.
(a) How much of the air is nitrogen?
(b) Is this about $\frac{1}{2}$, $\frac{2}{3}$ or $\frac{4}{5}$ of the air?

2 Look at the pictures.

Write down <u>four</u> things that chemists can make with nitrogen.

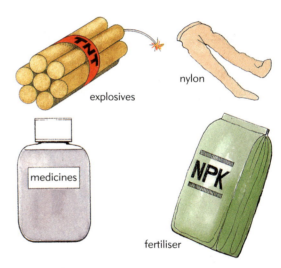

Plants need nitrogen

To grow healthy plants we must give them more than just water. Plants need nitrogen to help them grow well.

Plants can't use the nitrogen gas that we find in the air. The nitrogen must be joined with other elements in substances called **nitrates**. These nitrates dissolve in water. Plants can then take them in through their roots.

We need nitrogen to make all these things.

Why do farmers need fertilisers?

Fertilisers are important to farmers. Plants take nitrates out of the soil when they grow. Farmers often use the same fields year after year. It is important for the farmer to put nitrates back into the soil again.

Chemists can turn the nitrogen from the air into nitrates. This can happen in nature too. Farmers buy nitrates for fertiliser.

3 Look at the fertiliser labels. Write down:
(a) <u>one</u> way in which fertilisers A and B are the same;
(b) <u>two</u> ways in which the fertilisers are different.

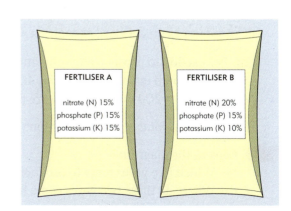

FERTILISER A

nitrate (N) 15%
phosphate (P) 15%
potassium (K) 15%

FERTILISER B

nitrate (N) 20%
phosphate (P) 15%
potassium (K) 10%

Big is best!

Farmers want to grow the best crops possible. The **yield** is the amount of crops that a farmer can grow. Farmers can increase the yield of their crops by using fertilisers.

4 Look at the chart.

What happens to the height of the crop as more fertiliser is used? Answer as carefully as you can.

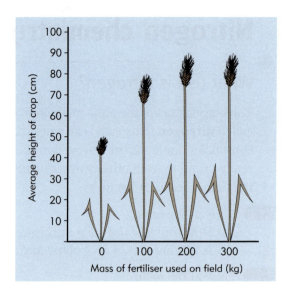

Catching nitrogen to feed plants

Plants need nitrogen to grow properly but they can't use nitrogen from the air. The nitrogen has to be changed into nitrate.

Chemists make nitrate fertiliser in several steps. In the first step they change nitrogen into a chemical called **ammonia**. They do this using the **Haber process**.

Making ammonia by the Haber process

In the Haber process,

nitrogen + hydrogen $\rightleftharpoons$ ammonia

5 What does the symbol $\rightleftharpoons$ in the equation tell you about the reaction?

The reaction is reversible. So not all the nitrogen and hydrogen change to ammonia. Chemists and chemical engineers had to work out how to get a reasonable yield of ammonia as quickly and cheaply as possible.

For the reaction to produce ammonia most economically, it must be at a high temperature (about 400 °C) and a high pressure (about 200 times the pressure of the atmosphere).

Hot iron is the catalyst for the reaction.

6 Look at the diagram.
(a) Which <u>two</u> gases react to produce ammonia?
(b) Where do these two gases come from?

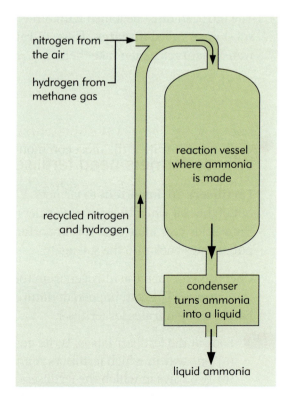

The Haber process for making ammonia.

7 Write down <u>three</u> things that help to produce ammonia faster.

Not all of the nitrogen and hydrogen react.

8 How is the ammonia separated from the unreacted nitrogen and hydrogen?

9 What then happens to the unreacted nitrogen and hydrogen?

Making nitric acid

Some ammonia from the Haber process is then changed into **nitric acid**. The diagram shows how this is done.

The ammonia is heated with oxygen to make a gas called **nitrogen monoxide**. This gas is then heated with water and oxygen to make nitric acid.

The first stage of the process is called an **oxidation** reaction.

10 What do we do to make the ammonia and oxygen react more quickly?

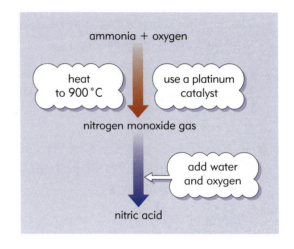

And finally, making ammonium nitrate

Farmers can't put corrosive nitric acid on the soil!

We can change the acid into **ammonium nitrate**. Ammonium nitrate is the most common fertiliser. We make it by adding the nitric acid to more ammonia. We call this reaction **neutralisation**. The ammonia neutralises the nitric acid.

11 Write a word equation for the reaction where ammonia is added to nitric acid to neutralise it.

Some reactions are reversible

In chemical reactions, the substances you start off with (the **reactants**) change into new substances (the **products**). But in some chemical reactions, the products can change back again into the original reactants. Reactions like this that can go in both directions are called **reversible** reactions.

The diagram shows a reversible reaction that you have seen many times before.

Adding acid to purple litmus changes it from purple to red. Adding alkali to purple litmus changes it from purple to blue.

> A forward reaction is: reactant(s) → product(s).
>
> A reverse reaction is: reactant(s) ← product(s).
>
> A reversible reaction can go both ways.
> It is shown like this: reactant(s) ⇌ product(s).

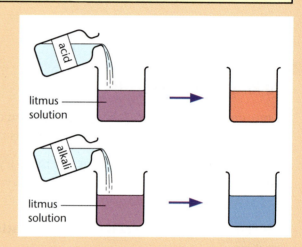

An important reversible reaction: the Haber process

Without nitrogen fertiliser, farmers couldn't produce so much food. The first step in making nitrogen fertiliser is to make ammonia from the nitrogen in the air, and hydrogen.

Almost all the ammonia used throughout the world is manufactured by the process developed by Fritz Haber at the beginning of the twentieth century. We shall now look at the Haber process in detail.

The reaction used in the Haber process is a reversible reaction. It can go both ways.

12 Look at the diagram.
   For the Haber process:
   (a) write down the equation for the forward reaction;
   (b) write down the equation for the reverse reaction;
   (c) write down an equation which shows both reactions at the same time.

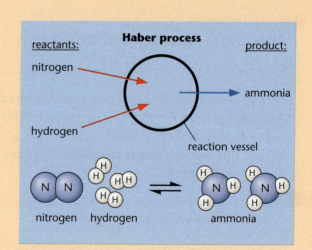

Equilibrium – a key idea

When the Haber process reaction begins, there will be many reacting molecules of nitrogen and hydrogen but few ammonia molecules. This means that the forward reaction will be fast, but the reverse reaction will be slow.

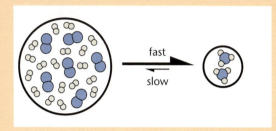

As the reaction continues, the number of nitrogen and hydrogen molecules will decrease and the number of ammonia molecules will increase. So the rate of the forward reaction will decrease and the rate of the reverse reaction will increase.

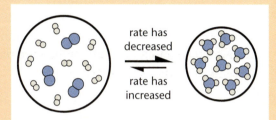

Eventually a point is reached where the rate of the forward reaction and the rate of the reverse reaction are **equal**. This point is called **equilibrium**.

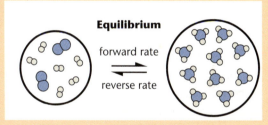

At equilibrium the rates of the forward and reverse reactions are the <u>same</u>.

For the reactants and products to reach equilibrium, they must be in a **closed** system.

13 What <u>three</u> substances will be present in the Haber process reaction vessel at equilibrium?

How much of the product there is in the mixture at equilibrium depends on the particular reaction and on the reaction <u>conditions</u> (that is, the temperature and pressure).

In the Haber process, we need to know what conditions will give a good **yield** of ammonia.

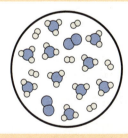

Equilibrium is reached only if both reactants and products are prevented from leaving the reaction vessel. We call this a <u>closed system</u>.

> Under normal temperature and pressure (25 °C and 1 atmosphere), the amount of ammonia at equilibrium is about 1 %.

Higher

Changing the position of equilibrium

The reaction used in the Haber process is reversible:

$$N_2 + 3H_2 \rightleftharpoons 2NH_3$$

Like other reversible reactions, this reaction reaches an equilibrium in a closed system.

At equilibrium, the forward and reverse reactions occur at the same rate.

In a reversible reaction such as the Haber process, it is possible to change the position of equilibrium by changing the conditions under which the reaction takes place.

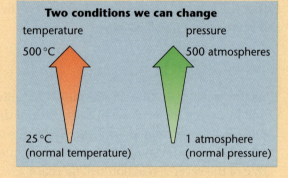

Two conditions we can change

temperature — 500 °C ... 25 °C (normal temperature)

pressure — 500 atmospheres ... 1 atmosphere (normal pressure)

14 Write down <u>two</u> ways in which we can change the conditions inside the reaction vessel used in the Haber process.

In the Haber process, ammonia is produced in the forward reaction:

$$N_2 + 3H_2 \rightarrow 2NH_3$$

To produce more ammonia we need to choose conditions that will favour the forward reaction.

Increasing the pressure

If we increase the pressure on an equilibrium mixture of nitrogen, hydrogen and ammonia, the rate of the forward reaction increases more than the rate of the reverse reaction. So a new equilibrium is reached which contains more ammonia than before.

15 (a) Which reaction in the Haber process, forward or reverse, is favoured by raising the pressure?

(b) Why is this reaction favoured?

If the pressure is increased, the reaction mixture changes so as to reduce it again. The pressure is reduced if there are fewer molecules in the reaction vessel. The forward reaction forms <u>two</u> ammonia molecules for every <u>four</u> reactant molecules (one nitrogen and three hydrogen). So <u>more</u> forward reaction <u>reduces</u> the pressure.

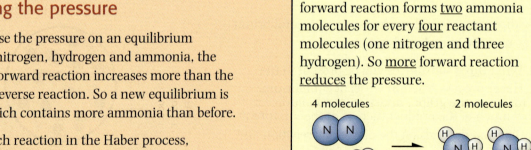

4 molecules ⇌ 2 molecules

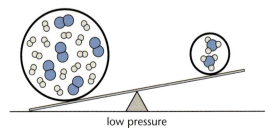

low pressure

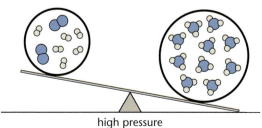

high pressure

Higher

Increasing the temperature

The energy changes in a reversible reaction are also reversible:

$$N_2 + 3H_2 \rightarrow 2NH_3 \qquad \Delta H = -93\,kJ$$

$$N_2 + 3H_2 \leftarrow 2NH_3 \qquad \Delta H = +93\,kJ$$

If we <u>increase</u> the temperature of an equilibrium mixture of nitrogen, hydrogen and ammonia, the equilibrium will shift so as to <u>reduce</u> the temperature. So the rate of the <u>endothermic</u> reaction is increased.

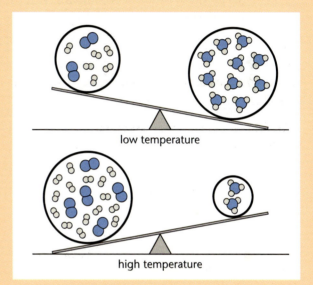

low temperature

high temperature

16 (a) Which reaction in the Haber process, forward or reverse, is favoured by increasing the temperature?
(b) Why is this reaction favoured?

Putting it all together

The graph shows the percentage of reacting gases converted into ammonia, at different temperatures and pressures.

17 (a) From the graph, under what conditions of temperature and pressure is the yield of ammonia greatest?
(b) Suggest a combination of temperature and pressure that would give an even higher yield.

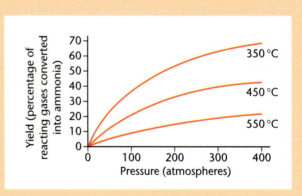

Higher

Economic and environmental considerations

All manufacturers try to make their products as economically as possible. Ammonia is usually manufactured at a temperature of about 400 °C and a pressure of about 200 atmospheres.

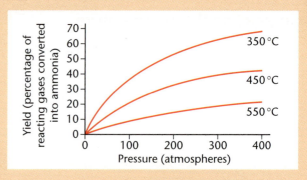

18 (a) How do these conditions compare with those for highest yield as shown on the graph?
(b) Estimate the yield under these conditions.

The reasons for using these conditions are mainly economic.

- The reaction vessel is made from reinforced steel and may be 20 metres high with a mass of up to 200 tonnes. The cost of making the reaction vessel is high, but it would be much higher if it was built to withstand higher pressures. It would need to be even thicker and stronger.

- Running costs would be greater at higher pressures, as the reacting gases would have to be pumped to a higher pressure.

- Operating at a higher pressure also increases safety risks.

For these reasons, an **optimum pressure** of 200 atmospheres is used. ('Optimum' means the most favourable when all factors are considered.)

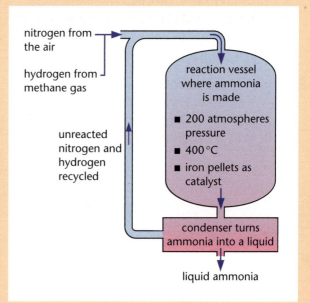

19 Explain why 200 atmospheres is the optimum pressure.

Although the yield of ammonia is higher when the temperature of reaction is lower, the rate of reaction is also lower. Slow production increases the costs of manufacture considerably.

For this reason, an **optimum temperature** of 400 °C is used.

20 Explain why 400 °C is the optimum temperature.

An iron **catalyst** is used to increase the rate of reaction. The catalyst increases the rates of both forward and reverse reactions equally in this reversible reaction.

Manufacturing ammonia economically

Ensure the <u>pressure</u> is:

- not too high, or the costs will be too high;
- not too low, or the yield will be low.

Ensure the <u>temperature</u> is:

- not too high, or the yield will be low;
- not too low, or the reaction will be too slow.

A catalyst ensures that equilibrium is reached more quickly, but it does not affect the equilibrium position.

21 Describe the impact of raising the pressure above 200 atmospheres or reducing the temperature below 400°C on the yield and running costs of ammonia production.

Looking at the whole process

Reactants are put in and products are removed continuously over a long period of time. So we call this a **continuous** process.

Energy released from the exothermic reaction, and from hot ammonia leaving the system, is used to heat up incoming reactants and produce steam to drive turbines.

The plant can be operated by a small number of staff.

The ammonia plant and the other plants which use ammonia for manufacturing of nitric acid and fertilisers are usually on the same site. This reduces transport costs and delivery times.

Research is carried out into more-efficient processes, so making energy savings.

22 The hydrogen used in the reaction is expensive to produce. How does the process ensure that none of it is wasted?

23 The table shows the percentage of yield of ammonia at different temperatures and pressures in the Haber process.
   (a) Under what conditions (shown in the table) would the maximum yield of ammonia be obtained?
   (b) At room temperature and pressure the yield of ammonia is only about 1%. Describe and explain how the yield is affected by the conditions used in the manufacturing process.
   (c) Why is the pressure used lower than the pressure that gives the maximum yield?

Environmental considerations

If manufacturers are not careful, some processes can damage the environment and be a danger to health. So, environmental considerations are important.

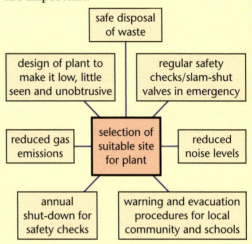

Sometimes reducing damage to the environment increases manufacturing costs.

Pressure (atm)	100 °C	300 °C	500 °C
25	91.7%	27.4%	2.9%
100	96.7%	52.5%	10.6%
400	99.4%	79.7%	31.9%

No chemicals, thank you

We make fertilisers such as ammonium nitrate in chemical factories. So we say that ammonium nitrate is an <u>artificial</u> fertiliser.

But we can grow healthy plants without using artificial fertilisers. Some people use natural fertilisers to put the nitrogen back into the soil.

Look at the pictures.

1   Write down <u>three</u> natural ways we can put nitrogen back into the soil.

Natural fertilisers do the same job as artificial fertilisers, because they contain similar chemicals. However, if they just relied on natural fertilisers, farmers wouldn't be able to grow enough food for us all. There just isn't enough natural fertiliser to grow all of the crops that we need.

A compost heap makes fertiliser from rotting waste.

Animal manure is rich in nitrogen and makes a good fertiliser.

Clover has bacteria in its roots. These can change nitrogen from the air into nitrates. Plants can take in these nitrates.

A problem with fertilisers

Many fertilisers contain **nitrates**. Nitrates are good for plants because they contain nitrogen that makes plants grow well. But if the nitrates don't stay in the soil then we have a problem. Rain can wash them into our rivers and ponds.

2   What type of nitrogen compound do many fertilisers contain?

3   What can happen if you swallow nitrates?

4   What can happen if nitrates get into rivers and lakes?

If you swallow nitrates, they can get into your blood. Your blood then cannot carry oxygen around your body properly.

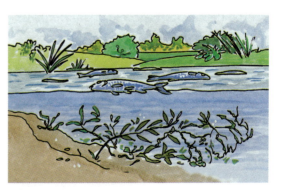

Nitrates in lakes and rivers can cause the plants and animals to die.

So are fertilisers good or bad?

There are two sides to every argument.
Look at the different things that people say about fertilisers.

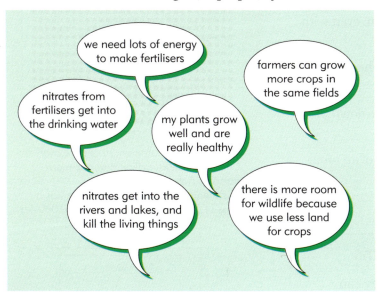

we need lots of energy to make fertilisers

farmers can grow more crops in the same fields

nitrates from fertilisers get into the drinking water

my plants grow well and are really healthy

nitrates get into the rivers and lakes, and kill the living things

there is more room for wildlife because we use less land for crops

5 List some advantages and disadvantages of using fertilisers.

6 Imagine that you are an environmental health officer. It is your job to make sure that the environment doesn't harm people's health. You have just read the two newspaper articles. What should you now do?

My best crop ever!

Farmer William Mitchell cannot believe his luck this year. At Mill Farm he has had a bumper wheat harvest in spite of the poor weather.
'I can only think it must be the new fertiliser that I used. These ears of wheat are enormous, they have to be seen to be believed.'

Poisoned water

Three children in Mill St. Newbarton have been admitted to hospital this week. They were all suffering from nitrate poisoning. They were all swimming in the lake by Mill Farm earlier this week.

How heavy are atoms and molecules?

Size of units

We choose units of measurement that make the numbers easy – not too big and not too small. For example, we measure the length of a piece of paper in centimetres. We measure the length of a room in metres and the length of a journey to the next town in kilometres.

1 Which units (centimetres, metres or kilometres) would you use to measure the length of:
(a) an air journey?
(b) a picture frame?
(c) a garden?

2 Look at the drawings.
Which mass would you measure in:
(a) grams?
(b) kilograms?
(c) tonnes?

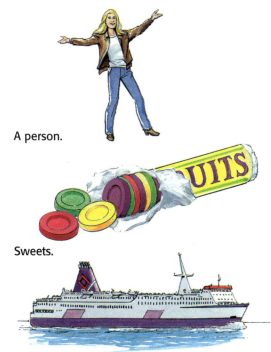

A person.

Sweets.

A large ship.

Can we weigh atoms?

Atoms are the very small particles that make up all of the elements. Atoms of different elements have different masses.

Atoms are so small that you can't weigh them, even with the best scientific balance.

3 What is the name of the element with the heaviest atoms?

Numbers as small as this are not easy to write down or use in calculations.

4 Why do we not usually measure the mass of an atom in grams?

Atoms of uranium are the heaviest atoms that we find in nature. Even so there is a huge number of atoms in just 1 gram of uranium. There are lots of dots in this box.

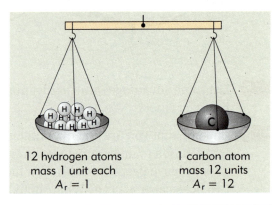

But in 1 gram of uranium, there are over 4 million, million, million times as many atoms as dots in the box.
This means that one uranium atom has a mass of 0.000 000 000 000 000 000 000 4 g.

Inventing a scale of mass for atoms

Chemists can't weigh separate atoms. But they can **compare** how heavy different atoms are.

For example, a carbon atom weighs 12 times as much as a hydrogen atom. So if we say that the lightest atom, hydrogen, has a mass of 1 unit, then a carbon atom has a mass of (12 × 1 units =) 12 units.

We call the mass of an atom in these units its **relative atomic mass**. We use the symbol A_r for short.

5 What is the symbol used for relative atomic mass?

6 What is the relative atomic mass of:
(a) hydrogen?
(b) carbon?
(c) helium?

7 Work out the relative atomic masses of the atoms shown in diagrams A–F.

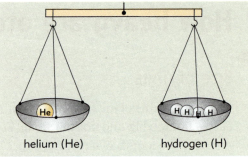

12 hydrogen atoms
mass 1 unit each
$A_r = 1$

1 carbon atom
mass 12 units
$A_r = 12$

helium (He)   hydrogen (H)

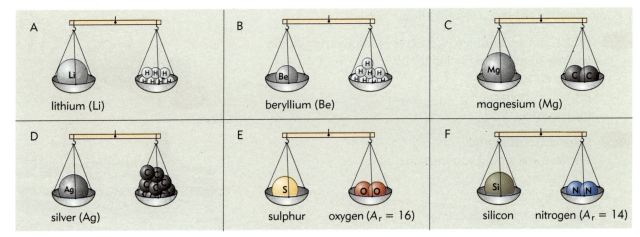

A   lithium (Li)

B   beryllium (Be)

C   magnesium (Mg)

D   silver (Ag)

E   sulphur   oxygen ($A_r = 16$)

F   silicon   nitrogen ($A_r = 14$)

Can we weigh molecules?

In molecules, atoms are joined together. Substances are called **compounds** if their molecules are made from atoms of different elements.

The first diagram shows a molecule of ammonia. This is a compound because it is made from atoms of different elements. The second diagram shows a molecule of the element nitrogen.

8 Which atoms make up a molecule of ammonia?

9 How many atoms are there in a molecule of nitrogen?

We use the relative atomic mass scale to compare the masses of different molecules.

We call the mass of the molecule its **relative molecular mass**, M_r.

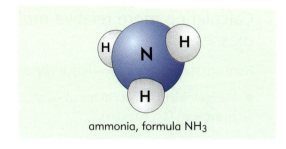

ammonia, formula NH_3

nitrogen, formula N_2

Calculating the mass of molecules

If we know the formula of a molecule then working out the relative molecular mass is easy.

We look up the relative atomic masses of the elements. Then we add the masses of all the atoms in the formula.

● Carbon dioxide has the formula CO_2.
   It contains one carbon atom and two oxygen atoms.

   Adding the relative atomic masses together, we get:

$$\begin{array}{cccc} C & O & O & CO_2 \\ \text{Relative molecular mass} = \ 12 & + \ 16 & + \ 16 & = \ 44 \end{array}$$

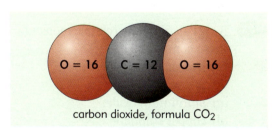

carbon dioxide, formula CO_2

● A molecule of oxygen, formula O_2, has got two oxygen atoms in the molecule.
   Each oxygen atom has a mass of 16.
   Therefore the two oxygen atoms have a total mass of 32.

$$\begin{array}{ccc} O & O & O_2 \\ \text{Relative molecular mass} = \ 16 & + \ 16 & = \ 32 \end{array}$$

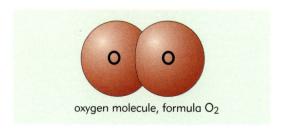

oxygen molecule, formula O_2

10 (a) Draw a molecule of ammonia.
(b) Write the relative atomic mass of each atom on your diagram.
(c) Now work out the relative molecular mass (M_r) for ammonia.

11 Calculate the relative molecular mass, M_r, for nitrogen in the same way.

Calculating more relative molecular masses

Here are some rules for reading a chemical formula.

- Each element has a chemical symbol (e.g. H = hydrogen, O = oxygen).

- A chemical symbol without a number stands for one atom of that element. So in H_2O (water) there is one atom of oxygen.

- The little number to the right of a symbol tells you how many atoms there are of that element only. So, in water there are two hydrogen atoms.

12 The formula for copper sulphate is $CuSO_4$.
(a) How many atoms of copper does it have?
(b) How many atoms of sulphur does it have?
(c) How many atoms of oxygen does it have?

- The number to the right of a bracket gives us the number of atoms of every element inside the brackets.

So, in $Ca(OH)_2$ there are two atoms of oxygen and two atoms of hydrogen.

13 Now calculate the relative molecular mass for each of these compounds:
(a) hydrogen chloride;
(b) water;
(c) aluminium oxide;
(d) copper sulphate;
(e) calcium hydroxide.

Element	Symbol	A_r
aluminium	Al	27
bromine	Br	80
calcium	Ca	40
carbon	C	12
chlorine	Cl	35.5
copper	Cu	64
helium	He	4
hydrogen	H	1
iron	Fe	56
krypton	Kr	84
magnesium	Mg	24
nitrogen	N	14
oxygen	O	16
sulphur	S	32

The relative atomic masses of some elements. You need some of these to answer question 13.

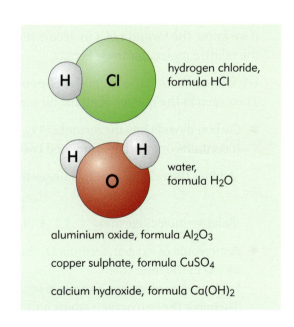

hydrogen chloride, formula HCl

water, formula H_2O

aluminium oxide, formula Al_2O_3

copper sulphate, formula $CuSO_4$

calcium hydroxide, formula $Ca(OH)_2$

Higher

Using chemical equations to calculate reacting masses

A balanced symbol equation is a useful shorthand way of describing what happens in a chemical reaction.

14 What does this equation tell you?

$$CH_4 + 2O_2 \rightarrow 2H_2O + CO_2$$

(CH_4 is the formula for methane.)

An equation does <u>not</u> tell us about the conditions necessary for the reaction, the rate of reaction, energy changes, and so on. However, we <u>can</u> use a balanced symbol equation to work out the masses of substances which react together and the masses of the products.

These are the steps to follow.

<u>Step 1.</u> Write down the balanced symbol equation.

<u>Step 2.</u> Decide what each formula tells you about the numbers of each kind of atom. That is, decide how many of each kind of atom are shown in each formula. (You may find it helpful to write this down.)

<u>Step 3.</u> Write in the relative atomic masses.

<u>Step 4.</u> Work out the reacting masses and product mass. (Note that the mass of reactant(s) equals the mass of product(s). This is because all the same atoms are still there.)

<u>Step 5.</u> Write in words what this means. (You can use <u>any</u> units. Normally, you should use the units given in the question.)

15 Follow steps 1–4 to work out the masses of reactants and products in these two chemical reactions.
(a) $C + O_2 \rightarrow CO_2$
(b) $CH_4 + 2O_2 \rightarrow 2H_2O + CO_2$

Set out your answers as in Example 1.

Element	Symbol	A_r
aluminium	Al	27
carbon	C	12
iron	Fe	56
magnesium	Mg	24
oxygen	O	16

Relative atomic masses (A_r) of some atoms.

Example 1

Magnesium reacts with oxygen to form magnesium oxide. Work out the reacting masses and the product mass.

$$2Mg + O_2 \quad \rightarrow \quad 2MgO$$

2 magnesium atoms → 2 magnesium atoms
+ 2 oxygen atoms + 2 oxygen atoms

$(2 \times 24) + (2 \times 16) \rightarrow [(2 \times 24) + (2 \times 16)]$
48   +   32   →   [48   +   32]
48   +   32   →   80
For the product, work out the inner ( ) brackets first.

$48 + 32 \rightarrow 80$

48 grams of magnesium react with 32 grams of oxygen to form 80 grams of magnesium oxide.

Higher

You may be asked to calculate the mass of a product from a given mass of reactant in a chemical reaction.

Use only the quantities of the substances about which you are asked.

Example 2 shows how you should set out your answer so that what you are doing is clear.

16 Calculate the mass of calcium oxide (CaO) that is produced from heating 10 g of limestone ($CaCO_3$).

$$CaCO_3 \rightarrow CaO + CO_2$$

Set out your answer as in Example 2.

Sometimes you will be asked to calculate the mass of one of the reactants.

Again, use only the quantities about which you are asked.

Example 3 shows how you should set out your answer so that what you are doing is clear.

17 How much copper oxide (CuO) is needed to produce 16 kg of copper in this reaction?

$$CuO + H_2 \rightarrow Cu + H_2O$$

Set out your answer in a similar way to Example 3. (But this time the calculation isn't exactly the same.)

Example 2

$$2Al + Fe_2O_3 \rightarrow Al_2O_3 + 2Fe$$

In this reaction, what mass of iron is produced from 8 grams of iron oxide (Fe_2O_3)?

$$Fe_2O_3 \rightarrow 2Fe$$

2 iron atoms + 3 oxygen atoms $\rightarrow$ 2 iron atoms

$$[(2 \times 56) + (3 \times 16)] \rightarrow (2 \times 56)$$
$$[112 + 48] \rightarrow 112$$
$$160 \rightarrow 112$$

So 160 g of iron oxide produces 112 g of iron.

So 1 g of iron oxide produces $\frac{112}{160}$ g of iron.

So 8 g of iron oxide produces $\frac{112}{160} \times 8 = 5.6$ g of iron.

Example 3

$$2Al + Fe_2O_3 \rightarrow Al_2O_3 + 2Fe$$

How much aluminium is needed to react completely with 8 kg of iron oxide in this reaction?

$$2Al + Fe_2O_3$$

2 aluminium atoms   2 iron atoms + 3 oxygen atoms

$$(2 \times 27) \qquad [(2 \times 56) + (3 \times 16)]$$
$$54 \qquad\qquad [112 + 48]$$
$$54 \qquad\qquad\qquad 160$$

So 160 kg of iron oxide reacts with 54 kg of aluminium.

So 1 kg of iron oxide reacts with $\frac{54}{160}$ kg of aluminium.

So 8 kg of iron oxide reacts with

$$\frac{54}{160} \times 8 = 2.7 \text{ kg of aluminium.}$$

More calculations using chemical equations

Experiments with gases show that at a given temperature and pressure, equal numbers of molecules of all gases occupy the same volume.

For example, the diagram shows water being split up into hydrogen and oxygen gases by passing an electric current through it.

18 (a) How many times more hydrogen molecules are produced than oxygen molecules?

(b) How many times more cm^3 of hydrogen are produced than oxygen?

(c) What do your answers to (a) and (b) tell you?

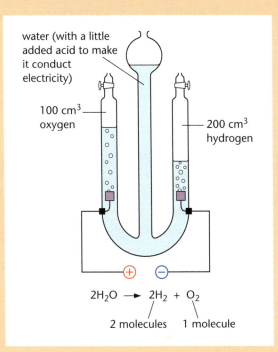

water (with a little added acid to make it conduct electricity)

100 cm^3 oxygen

200 cm^3 hydrogen

$$2H_2O \rightarrow 2H_2 + O_2$$

2 molecules   1 molecule

Calculating the volume of a gas in a reaction

The **relative formula mass** (M_r), in grams, of any gas has a volume of $24\,000\ cm^3$ (24 litres), at $25\,°C$ and 1 atmosphere.

> **Calculating the relative formula mass of gases**
>
> Hydrogen (H_2) $M_r = (2 \times 1) = 2$
> 2 grams of hydrogen has a volume of $24\,000\ cm^3$.
>
> Oxygen (O_2) $M_r = (2 \times 16) = 32$
> 32 grams of oxygen has a volume of $24\,000\ cm^3$.

You can use this information about volumes of gases to calculate the volumes of gases in a chemical reaction.

Example 4 shows how you can do this.

19 What volume of hydrogen is produced when 0.072 g of magnesium reacts with hydrochloric acid?

$$Mg + 2HCl \rightarrow MgCl_2 + H_2$$

Set out your answer as in Example 4.

> **Example 4**
>
> What volume of carbon dioxide is produced when 6 g of carbon burns in oxygen?
>
> $$C + O_2 \rightarrow CO_2$$
>
> A_r for carbon is 12
>
> so 12 g → one formula mass
> of carbon   of carbon dioxide
>
> that is 12 g → $24\,000\ cm^3$
>
> So 1 g → $(24\,000 \div 12)\ cm^3$
>
> So 6 g → $(24\,000 \div 12) \times 6$
>
> $= 12\,000\ cm^3$
>
> 6 g of carbon produces $12\,000\ cm^3$ of carbon dioxide.

Higher

Calculating the masses and volumes of substances produced during electrolysis

In the electrolysis of potassium bromide (KBr), reactions take place at the two electrodes.

You can calculate the amount of chemical change at one electrode if you are given information about the change at the other electrode.

For example, during an experiment to electrolyse molten potassium bromide, the mass of potassium released at the negative electrode was 1.56 g. You can calculate the volume of bromine given off at the positive electrode like this.

Step 1. Write down the two balanced half-equations (these will be provided for you in questions of this type).

Step 2. Write in words what happens at each electrode.

Step 3. Work out the unknown quantity using the information about the other amount.

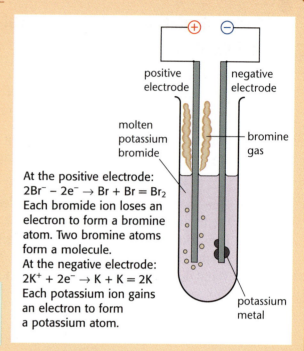

At the positive electrode:
$2Br^- - 2e^- \rightarrow Br + Br = Br_2$
Each bromide ion loses an electron to form a bromine atom. Two bromine atoms form a molecule.
At the negative electrode:
$2K^+ + 2e^- \rightarrow K + K = 2K$
Each potassium ion gains an electron to form a potassium atom.

Example 5

Positive electrode: $2Br^- - 2e^- \rightarrow Br_2$
Negative electrode: $2K^+ + 2e^- \rightarrow 2K$

One molecule (two atoms) of bromine is released at the positive electrode for every two potassium atoms that are deposited at the negative electrode.

M_r $Br_2(g)$ has a volume of $24\,000$ cm^3.
A_r K is 39, so 2K is 78.

So volume of bromine displaced while 1.56 g of potassium is formed is

$$\frac{1.56}{78} \times 24\,000 = 480 \text{ cm}^3$$

20 During the electrolysis of molten sodium chloride, 9.2 g of sodium are formed at the negative electrode. What volume of chlorine is formed at the positive electrode at normal temperature and pressure?

The half equations are:
At the positive electrode:
$2Cl^- - 2e^- \rightarrow Cl_2$

At the negative electrode:
$2Na^+ + 2e^- \rightarrow 2Na$

Set out your answer as in Example 5.

21 The equation shows the reaction between sodium metal and water.

$$2Na + 2H_2O \rightarrow 2NaOH + H_2$$

Calculate the mass and volume of hydrogen produced when 4.6 g of sodium reacts with water.

22 1.92 g of copper were deposited at the negative electrode during electrolysis of a solution of copper sulphate. What volume of oxygen at 25°C and 1 atmosphere was formed during the experiment?

The half-equations are as follows:
$4OH^- - 4e^- \rightarrow O_2 + 2H_2O$
$2Cu^{2+} + 4e^- \rightarrow 2Cu$

[A_r for copper = 63.5]

Elementary pie

Think about an apple pie you buy from the supermarket. There is usually a table of information on the packet. This tells us how much carbohydrate, fat and protein there is in each 100 g of the pie.

1 Write down how much of each type of food substance there is in 100 g of the pie. Write the list in order, starting with what there is most of.

Telling you how much of everything there is in each 100 g makes it easy to compare different foods.

2 How do the amounts of protein and fat in the apple pie compare with the amounts in the bread?

Another way of saying 8 g out of 100 g is to say 8 per cent. Per cent means 'out of one hundred'.

Apple pie Nutritional information Average values per 100g	
protein	3g
carbohydrate	54g
fat	11g

Bread Nutritional information Average values per 100g	
protein	8g
carbohydrate	31g
fat	2g

How much of an element is in a compound?

We can easily see how many units of mass of elements are in a compound.

For example, sulphur dioxide is SO_2.

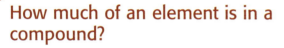

$$
\begin{aligned}
M_r & = \text{mass of S atom} + \text{mass of 2 O atoms} \\
\text{(relative} & = \quad 32 \quad + \quad 2 \times 16 \\
\text{molecular} & = \quad 32 \quad + \quad 32 \\
\text{mass)} & = \qquad\quad 64
\end{aligned}
$$

Sulphur gives 32 units of mass out of 64 for sulphur dioxide. Oxygen gives the other 32 units of mass.

This means that sulphur dioxide is 50 per cent sulphur and 50 per cent oxygen by mass.

3 Now work out the percentage by mass of carbon and hydrogen in methane one step at at time, like this:
(a) What is the mass of all the hydrogen atoms?
(b) What is the mass of the carbon atom?
(c) What is the relative molecular mass of methane?
(d) What is the percentage by mass of hydrogen in methane?
(e) What is the percentage by mass of carbon in methane?

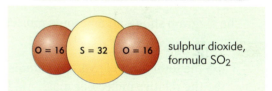
sulphur dioxide, formula SO_2

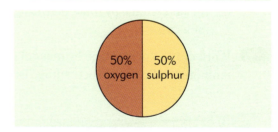

Half is the same as 50%.

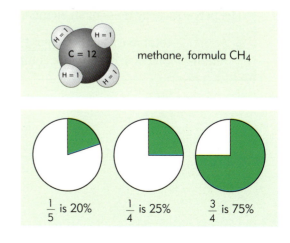
methane, formula CH_4

$\frac{1}{5}$ is 20% $\frac{1}{4}$ is 25% $\frac{3}{4}$ is 75%

How to calculate percentages

Percentages don't usually work out as easily as they do for sulphur dioxide and methane.

In water, for example, 2 parts out of 18 are hydrogen.

To calculate this as a percentage on a calculator:

- press the number 2
- then press ÷
- then press the numbers 1 then 8 (18)
- then press %

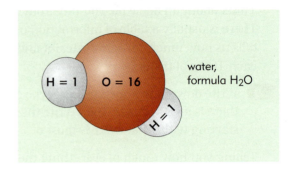

water, formula H_2O

4   What is 2 out of 18 as a percentage?

You can work out other 'awkward' percentages in a similar way.

The percentages by mass of elements in ammonia

The diagram shows an ammonia molecule.

5   Work out:
  (a) the total mass of hydrogen atoms in the molecule;
  (b) the relative molecular mass, M_r, for the molecule;
  (c) the percentage by mass of hydrogen in the molecule.

6   Work out the percentage by mass of nitrogen in ammonia. (Hint: what percentage isn't hydrogen?)

ammonia, formula NH_3

General percentage rule

The percentage by mass of an element in a compound

$$= \frac{\text{total mass of the element}}{\text{relative molecular mass of the compound}} \times 100$$

Higher

Working out the formulas of compounds

The formula of a compound tells you how many of each kind of atom there are in a compound.

7 What is the ratio of the atoms (or ions) in:
(a) an ammonia molecule, formula NH_3?
(b) a methane molecule, formula CH_4?
(c) the compound magnesium oxide, formula MgO?
(d) the compound aluminium oxide, formula Al_2O_3?

You can find the masses of the elements that combine by careful weighing in experiments. Using this information, you can find the **ratio** of atoms in a compound.

The example below shows how to do this.

Example

A chemist found that 0.12 g of magnesium combined with 0.8 g of bromine. What is the ratio of magnesium to bromine atoms in the compound magnesium bromide?

Step 1. Write down the ratio of the masses combining (from information in the question).

Step 2. Write down A_r for each element.

Step 3. Divide each mass by A_r to get the ratio of the atoms of each element.

Step 4. Work out the simplest whole-number ratio (in this case divide the larger number by the smaller).

The ratio of magnesium to bromine atoms is 1 : 2.
The ratio Mg : Br is 1 : 2.
The simplest formula for the compound is $MgBr_2$.
This is called the **empirical formula**.

The formula for carbon dioxide is

$$\underset{\text{1 carbon atom}}{\underbrace{\quad}} CO_2 \underset{\text{2 oxygen atoms}}{\underbrace{\quad}}$$

The ratio of carbon to oxygen atoms in a carbon dioxide molecule is 1 : 2.

The formula for sodium chloride is

$$\underset{\text{1 sodium atom}}{\underbrace{\quad}} NaCl \underset{\text{1 chlorine atom}}{\underbrace{\quad}}$$

Sodium chloride is an ionic compound.

— chloride ion Cl^-
— sodium ion Na^+

The ratio of sodium atoms (ions) to chlorine atoms (ions) is 1 : 1.

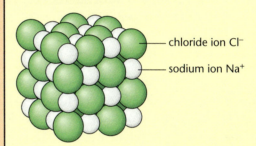

magnesium	:	bromine
0.12 g	:	0.8 g
$A_r = 24$	:	$A_r = 80$
$0.12 \div 24 = 0.005$	:	$0.8 \div 80 = 0.01$
1	:	2

Higher

8 1.28 grams of an oxide of sulphur contain 0.64 g of sulphur and 0.64 g of oxygen. Find the ratio of sulphur to oxygen atoms and work out the empirical formula for this compound.
[Set out your answer as in the example. A_r sulphur = 32; A_r oxygen = 16]

Finding a formula by experiment

The diagram shows an experiment to find the empirical formula of copper oxide. The results (weighings) taken are shown in the table.

9 (a) Copy the table of results and then complete it.

(b) Use the results to work out the empirical formula for copper oxide.
[Set out your answer as in the example. A_r copper = 63.5]

Note. You do not usually get exact whole-number ratios from the results of an experiment. So if, for example, you get a ratio of 2.1 : 1, you would assume that the correct answer is 2 : 1.

10 In an analysis of potassium oxide, a student found that 1.17 g of potassium combines with 0.24 g of oxygen. Find the ratio of potassium to oxygen atoms and the simplest (empirical) formula for potassium oxide. [A_r potassium = 39]

11 An experiment shows that 13.5 g of aluminium combine with 12.0 g of oxygen. Find the ratio of aluminium to oxygen atoms and work out the empirical formula for aluminium oxide.

12 A hydrocarbon consists of 75% carbon and 25% hydrogen by mass. Find the ratio of carbon to hydrogen atoms and the simplest (empirical) formula for the hydrocarbon. [A_r carbon = 12; A_r hydrogen = 1]

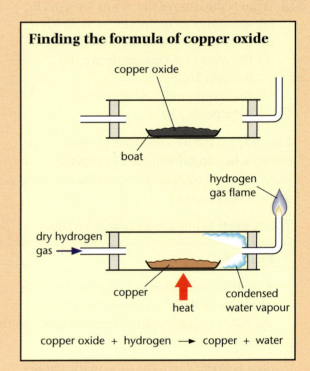

Finding the formula of copper oxide

copper oxide + hydrogen ⟶ copper + water

Results	
Mass of boat	= 15.43 grams
Mass of boat + copper oxide	= 23.38 grams
Mass of boat + copper	= 21.78 grams
Mass of copper oxide	=
Mass of copper	=
Mass of oxygen in the copper oxide	=

9 Energy

Transferring heat energy

How heat (thermal energy) is transferred

Heat (thermal energy) always moves from hot places to cold places. This is called **thermal energy transfer**. Sometimes you want to make it easy for thermal energy to go from one place to another. Sometimes you want to keep thermal energy in one place.

So you need to know how thermal energy travels.

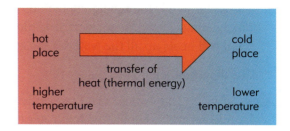

Conduction

If you put a solid between somewhere hot and somewhere cold, the thermal energy has to travel through the solid. This is called <u>conduction</u>. Thermal energy passes easily through some solids, for example metals. We call these **conductors**. Other solids conduct heat badly, and we call these **insulators**. Insulators include most non-metals, for example wood, plastic and glass.

1. Write down one example of:
   (a) a conductor   (b) an insulator.

2. Does thermal energy pass best through an insulator or a conductor?

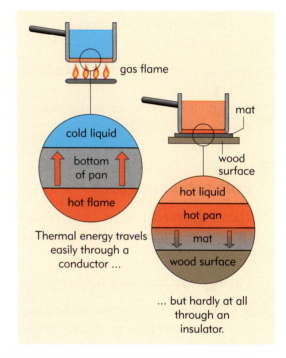

Thermal energy travels easily through a conductor ...

... but hardly at all through an insulator.

Higher

Gases are poor conductors of thermal energy. Even metals, when they are hot enough to be gases, are poor thermal conductors. This means that all gases are good thermal <u>insulators</u>. The diagrams explain why solid and liquid metals are good conductors and why gases are always poor conductors.

3. (a) Which substances are good thermal conductors?
   (b) What is special about the structure of metals that makes them good conductors?
   (c) Explain how the free electrons in metals enable them to transfer thermal energy by conduction.

Key
- e⁻ free electron
- + metal atom

The atoms in metals have some electrons that can move about anywhere in a piece of metal. Heating the metal gives these electrons more kinetic energy. They transfer energy to other electrons when they collide with them.

The particles in a gas move about and collide with each other. But there are a lot fewer particles in each cm³ than there are in a solid or a liquid. So gases are poor thermal conductors.

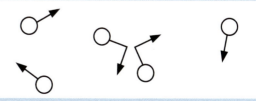

Convection in liquids and gases

The water in a kettle is a liquid. Liquids can flow.
The heating element in an electric kettle is at the bottom, but it still heats up all the water in the kettle.

The diagrams show how it does this.

4 Draw <u>one</u> large diagram of the kettle.
Add arrows to show how hot water rises and cold water falls.
Label them or colour them in.
Use red for hot and blue for cold.

Each time the water moves around the kettle it gets a little bit hotter.
Hot liquids move and carry thermal energy with them.
This is called <u>convection</u>.

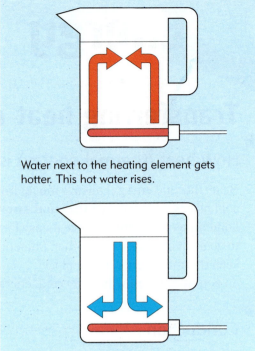

Water next to the heating element gets hotter. This hot water rises.

Colder water then falls down to take its place.

Convection in gases

The air in a room is a gas. Gases can also flow.
Heaters are usually near the floor, but the whole of the room gets heated.

The diagrams show how heaters do this.

5 Draw <u>one</u> large diagram of the room and heater.
Draw arrows to show the hot air rising and the cold air falling.
Label or colour the arrows.

Hot gases, like hot liquids, move around and transfer thermal energy by convection as they do so.

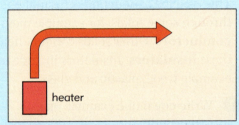

Air next to the heater becomes hotter. This hot air rises.

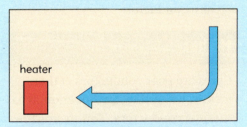

Colder air then falls down to take its place.

Higher

Explaining convection

Convection occurs because liquids and gases become less dense when they are warmer. To understand why this happens, you need to imagine a 'bubble' of liquid completely surrounded by the same liquid.

The 'bubble' of liquid doesn't move because the forces that act on it are <u>balanced</u>. The weight of the liquid pulling downwards is balanced by the force of the surrounding liquid pushing upwards.

If the 'bubble' of liquid is heated, energy is transferred to the particles. The particles move about faster; they have more **kinetic energy**. This means that they take up more space. So the 'bubble' of liquid expands. It still has the same mass so the weight of the 'bubble' is still the same. But the 'bubble' has a bigger volume, so the surrounding liquid pushes up on it more.

There is now an unbalanced upwards force, so the 'bubble' moves up. Colder liquid moves in to take its place. This movement is called a **convection current**.

6　Liquids and gases expand when they are heated.
　(a) Does the density of the liquid or gas get larger or smaller as it is heated?
　(b) Use your answer from part (a) to explain why warm liquids and gases rise and cool liquids and gases fall.

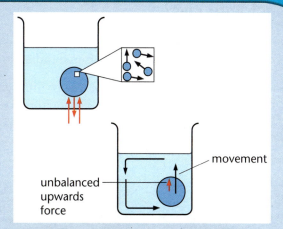

The same <u>mass</u> of particles is now in a larger <u>volume</u>. So the 'bubble' of liquid is now less <u>dense</u> than it was before.
Exactly the same thing happens when a gas is heated.

Radiation

When you stand near to a roaring bonfire, you can feel thermal energy from the bonfire falling on your face.

This happens because hot objects transfer thermal energy by sending out rays. This method of energy transfer is called <u>radiation</u>.

7　How is radiation different from conduction and convection?

Thermal energy is transferred from the fire to your face by radiation. No substance (solid, liquid or gas) is needed, so the radiation can travel through empty space.

The Sun sends out energy as radiation. This can travel through empty space.

Some of the energy is sent as light rays that we can see. Most of the energy is sent as **infrared** radiation. You cannot see infrared rays, but you can feel them.

8　What types of radiation (rays) does the Sun give out?

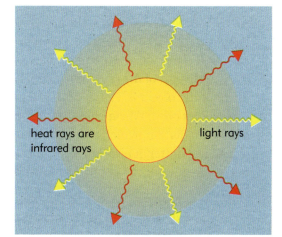

heat rays are infrared rays　　light rays

How can you capture the energy in infrared rays?

A black surface is a good **absorber** of infrared rays. This means that it soaks up infrared radiation very well.

White or shiny surfaces do not absorb infrared rays. They are good at **reflecting** the rays away from them.

9 Look at the pictures below.
Write a sentence to explain each one.

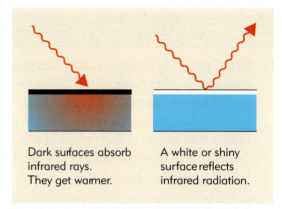

Dark surfaces absorb infrared rays. They get warmer.

A white or shiny surface reflects infrared radiation.

Dark clothes make you feel hot on a sunny day.

Astronauts wear shiny suits for space walks.

The tar on roads can melt in the summer sun.

Houses in hot countries are often white.

What makes a good radiator?

Black surfaces are better at sending out infrared radiation than white or shiny surfaces.

10 What do the probe readings tell you about the temperatures of the motorcycle engines?

11 Explain the difference in temperature between the two engines.

Motorcycle with black engine.

Temperature probe reads 175 °C.

Motorcycle with shiny engine.

Temperature probe reads 200 °C.

How the radiation changes with temperature

All objects which are above the absolute zero of temperature (−273 °C) give out (**emit**) electromagnetic radiation. The <u>rate</u> at which an object radiates energy and the <u>wavelength</u> of the radiation it emits both depend on how hot the object is. Hotter objects emit more radiation than cooler ones. Warm objects emit long-wavelength infrared radiation, which we can feel but not see. As the object gets hotter, it also starts to emit shorter-wavelength radiation, and we can see it glow as well as feel the heat.

12 (a) What types of energy do hot objects send out?
(b) How does the amount of radiation change as objects heat up?

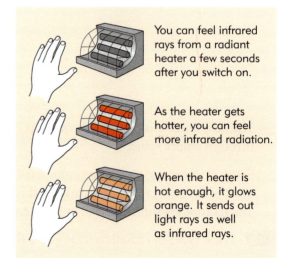

You can feel infrared rays from a radiant heater a few seconds after you switch on.

As the heater gets hotter, you can feel more infrared radiation.

When the heater is hot enough, it glows orange. It sends out light rays as well as infrared rays.

13 The graphs show how the radiation emitted by an object depends on its temperature. Describe, as fully as you can, what the graphs tell you.

(a)
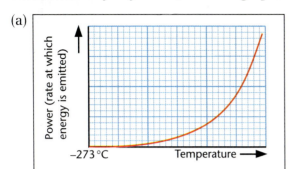

(b)
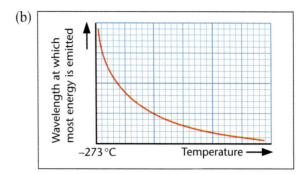

Heating our homes

How is the energy lost?

All buildings lose thermal energy in various ways. This costs money and wastes fuel.

1 Look at the diagram.
(a) In which place is most thermal energy lost?
(b) Write down <u>two</u> places where heat is lost by conduction.
(c) How does the cold air come into the house?

about 2000 J through the **ceilings** (conduction)

up to 4000 J through the **walls** (conduction)

cold air in: 1500 J through **draughts** (convection currents)

1000 J through **floors** (conduction)

1500 J through **window glass** (conduction)

Thermal energy lost each second from a badly insulated house on a cold day.

Keeping the energy in

The thermal energy from the back of a radiator is transferred to the wall by radiation. The thermal energy then moves through the wall by conduction. The diagram shows how to reduce this thermal energy loss.

2 Look at the diagram.
(a) Why does the shiny surface behind the radiator help to keep heat in?
(b) Why is plastic foam used in the air gap?

The most **effective** method is the one that reduces thermal energy loss by the biggest percentage (%).

3 Which method of reducing the thermal energy loss through the wall is the most effective?

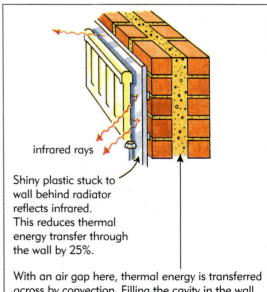

infrared rays

Shiny plastic stuck to wall behind radiator reflects infrared. This reduces thermal energy transfer through the wall by 25%.

With an air gap here, thermal energy is transferred across by convection. Filling the cavity in the wall with plastic foam **insulation** stops convection. The foam is as bad a conductor as air. This reduces thermal energy transfer through the wall by 50%.

Reducing thermal energy loss through a wall.

4 Imagine that <u>all</u> these energy-saving ideas are used in the house shown on page 345.
  (a) Draw a picture of the house and label the thermal energy losses, now that it is insulated.
  (b) What is the total thermal energy loss for the insulated house?
  (c) How does this compare with the uninsulated house?

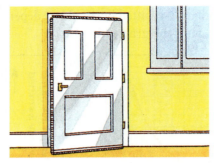

Draught excluders (strips) round doors and windows can save half of the heat lost through draughts.

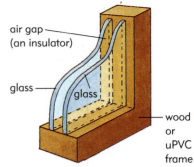

Double glazing can save about half of the thermal energy lost through windows.

Why do foam and fibres make good insulators?

Air, like all gases, is a very poor conductor.
But to use it as an insulator, we must stop it moving about.

Look at the diagrams of foam and fibre **insulation**.

5 Describe how the air is stopped from moving about in the foam and the fibre.

6 Why is it important to stop the air moving about?

Insulating the loft with glass fibre 20 cm thick can save half of the thermal energy lost through the ceiling.

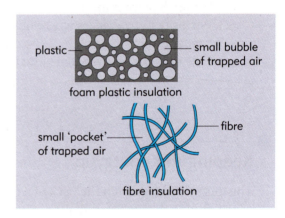

foam plastic insulation

fibre insulation

Saving money by saving energy

We can do lots of things to buildings to reduce thermal energy transfer. But before we spend money on things like double glazing or loft insulation, we need to know how much money this will save on heating bills.

When we are deciding if it is worth spending money on a particular improvement, it is useful to know how long it will take for the improvement to pay for itself. This is called the **pay-back time**. The shorter the pay-back time, the more **cost-effective** the improvement is.

7 Calculate the pay-back time for each of these energy-saving ideas:
  (a) double glazing that costs £1200 and saves £60 per year;
  (b) draught excluders that cost £25 and save £50 per year;
  (c) loft insulation that costs £150 and saves £150 per year.

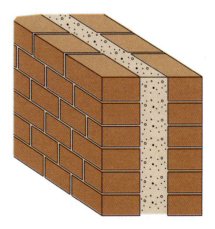

Cavity wall insulation   Costs £450
Saves £75 per year
Pay-back time = £450 ÷ £75
= 6 years

Fifty years ago, most houses in the UK were heated using coal fires. Nowadays, most houses are heated using gas fires or gas central heating.

8 Look at the diagrams below.
(a) Explain why you can save money by using a gas fire instead of a coal fire.
(b) What other advantages do gas fires have over coal fires?

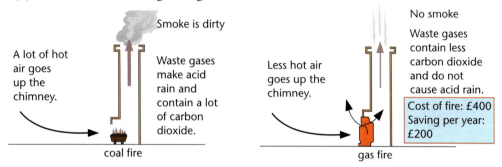

Smoke is dirty

A lot of hot air goes up the chimney.

Waste gases make acid rain and contain a lot of carbon dioxide.

coal fire

No smoke

Waste gases contain less carbon dioxide and do not cause acid rain.

Less hot air goes up the chimney.

Cost of fire: £400
Saving per year: £200

gas fire

It is easy to control the thermal energy output of a gas fire. It is also very easy to turn it on or off.

Some houses are 'all-electric'. This means that electricity is used for all the heating and cooking as well as for lighting, TV sets, etc. So no fuel is burned inside the house.

To keep the air inside the house fresh, it needs to be changed about once an hour. In older houses this happens through cracks, for example around doors or between floorboards. Newer houses can be built with hardly any cracks. They need a ventilation system as shown in the diagram.

9 (a) Explain how a heat exchanger reduces the thermal energy loss from a house.
(b) What would the pay-back time be for the extra cost of reducing cracks and fitting a heat exchanger?
(c) How cost-effective would you say that the heat exchanger is?

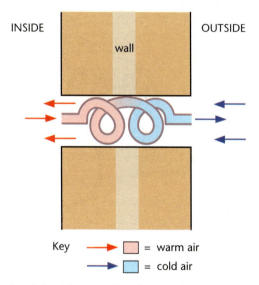

INSIDE   wall   OUTSIDE

Key ⟶ ▢ = warm air
⟶ ▢ = cold air

A 'crack-free' house with a heat exchanger costs about £2000 more to build.
It can save about £400 a year in fuel bills.

Using the Sun's energy to heat your home

If your house gets warm from the Sun, you need less fuel.

This saves fossil fuels like coal, oil and gas. You may also use less electricity.

Even in the winter you can get some benefit from the Sun's energy, although the Sun is lower in the sky.

The diagram shows how a glass window can trap energy from the Sun inside a house.

10 Explain how the glass window is able to do this.

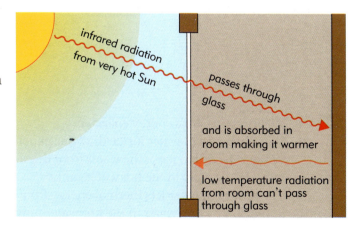

infrared radiation from very hot Sun

passes through glass

and is absorbed in room making it warmer

low temperature radiation from room can't pass through glass

How does a solar panel work?

Some houses have **solar panels** on the roof. These use energy radiated by the Sun to heat water. The diagram below shows how solar panels work.

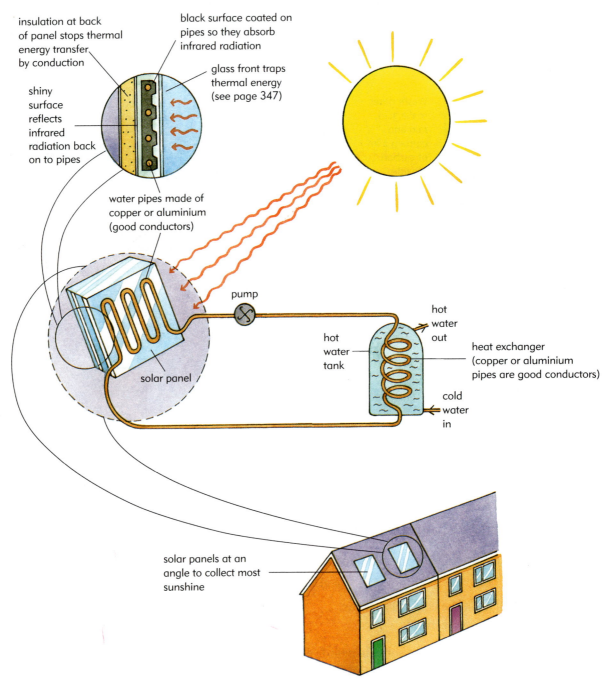

insulation at back of panel stops thermal energy transfer by conduction

black surface coated on pipes so they absorb infrared radiation

glass front traps thermal energy (see page 347)

shiny surface reflects infrared radiation back on to pipes

water pipes made of copper or aluminium (good conductors)

pump

hot water out

hot water tank

heat exchanger (copper or aluminium pipes are good conductors)

solar panel

cold water in

solar panels at an angle to collect most sunshine

11 (a) Where is the Sun's energy collected?
  (b) What happens to water inside the solar panel?
  (c) What is needed to move the water from the solar panel to the pipes in the hot water tank?
  (d) The heat exchanger lets thermal energy from water in the pipe pass to water in the hot water tank. Why is it made of metal?

12 Why is there a shiny surface behind the water pipes in the solar panel?

13 Explain why there is a layer of insulation at the back of the solar panel.

Using electricity

Why is electricity so useful?

Electricity is a form of energy that is used in many places.

We use electricity a lot because it is a very useful sort of energy. Electricity can be easily **transferred** as other kinds of energy. For example, a torch transfers electrical energy as light.

1 Look at the electrical appliances shown in the diagrams. Match the name of each appliance with the form of energy it produces.

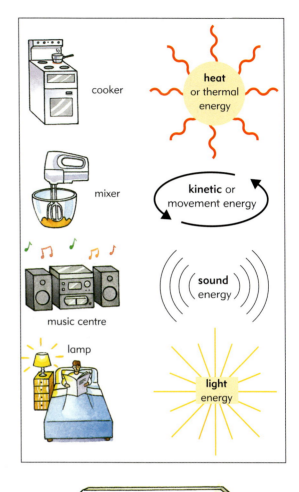

More energy transfers

We design things to get useful energy from electricity. There is usually one main energy transfer we want. For example, a strip light is designed to transfer electrical energy as light.

Here is a way of showing this energy transfer :

strip light

electrical energy ⟶ light

2 Look at the diagrams below. Which piece of equipment carries out each of these energy transfers?
   (a) electrical energy ⟶ movement
   (b) electrical energy ⟶ sound
   (c) electrical energy ⟶ heat

Unwanted energy transfers

Very often you do not transfer all the electrical energy into the sort you want. There are other sorts of energy transfers which are not useful.

For example:

- A light bulb produces more heat than light!
- A drill produces sound as well as movement from the motor.
- An electric motor gets hot as well as moving.

3 The electrical energy supplied to the computer is transferred as four different types of energy.
(a) Make a copy of the diagram and label it to show these <u>four</u> different types of energy.
(b) Which <u>three</u> types of energy transfer is the computer <u>designed</u> to make?
(c) Which energy transfer also happens but isn't really wanted?

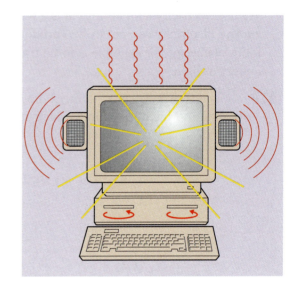

How much electricity do we use?

Electrical appliances transfer energy. How much energy they transfer depends on how long you switch them on for. It also depends on how fast they transfer energy.

Look at these light bulbs.

4 Which bulb is the brightest?

5 Which bulb transfers energy the fastest?

How fast something transfers energy is called its **power**.
Power is measured in **watts** (W).
1000 watts is called a **kilowatt** (kW).

6 Look at the pictures of an electric heater.
Copy and complete the table.

Setting of heater	Power	
	watts	**kilowatts**
low	1000	
medium		2
high		

low = 1 kW   medium = 2 kW   high = 3 kW

How much do electrical Units cost?

Jed lives in a flat. There is a coin meter for his electricity. He put 30p into the meter one evening. He fell asleep for three hours after leaving the heater on 'low'. He had just woken up when the heater went off.

Look carefully at the first two meters.

7 How many Units on the meter did the heater use in three hours on 'low'?

8 How much did each Unit cost?

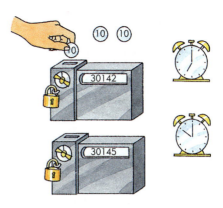

The meter readings show Units used. The difference between the two readings shows how many Units have been used.

It was colder the next night, so Jed put his heater on 'high'. He put 30p into the meter again. A friend arrived to talk to him. He was talking for nearly an hour. When the friend left the heater had just gone off.

9 How many electrical Units did the heater use in one hour on 'high'?

How to work out electrical Units

The amount of electrical energy transferred is worked out by multiplying the power (in kilowatts) by the time (in hours). This gives a **Unit** called the **kilowatt-hour** (kW h).

10 How many Units would an electric heater set to 'medium' use in one hour?

11 Write down the power in kilowatts for:
(a) a light bulb (100 W);
(b) an electric drill (500 W);
(c) a hair dryer (1500 W).
Each piece of equipment is switched on for two hours. How many Units (kilowatt-hours) does each one use?

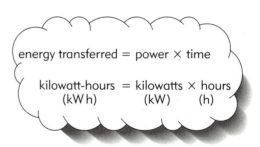

energy transferred = power × time

kilowatt-hours = kilowatts × hours
(kW h)   (kW)   (h)

You may need to change watts into kilowatts before you work out your answers.

100 W = 0.1 kW

500 W = 0.5 kW

1500 W = 1.5 kW

Counting the cost

To work out how much electricity costs, you need to know:

- how many Units (kWh) have been used;
- how much each Unit costs.

total cost for = number of × cost per Unit
electricity used   Units used

12 Copy the table below and complete it. Each Unit costs 10p. The first row has been done for you.

Electrical appliance	Power (in W)	Power (in kW)	Time used (in hours)	Number of Units (kW × hours)	Total cost (Units × 10p)
heater (low)	1000	1	2	1 × 2 = 2	2 × 10p = 20p
heater (medium)			2		
light bulb			2		
electric drill			2		
hair dryer			2		

heater 2000 W = 2 kW (medium)

drill 500 W = 0.5 kW

light bulb 100 W = 0.1 kW

hair dryer 1500 W = 1.5 kW

13 Which appliance costs the most to use?

Working out electricity bills

It is always worth checking electricity bills, but you need information to do this.

Sometimes you may not be at home when the electricity meter reader calls, so you may get an 'estimated bill'.

```
MRS. A. CURRENT                                    MAVISTON
42 WALKER ROAD                                    ELECTRICITY
MAVISTON                                            COMPANY

METER READING      | UNITS   | PENCE   | AMOUNT   | STANDING  | TOTALS
-------------------|         | PER     |          | CHARGE    |
This    | Last     | USED    | UNIT    | £        | £         | £
Time    | Time     |         |         |          |           |
30340E  | 29210C   | 1130    | 10.0    | 113.00   | 14.30     | 127.30
                                                              ----------
        TOTAL EXCLUSIVE OF VAT                                 127.30
                        VAT                                     10.18

THIS BILL IS ESTIMATED. PLEASE COMPLETE THE ENCLOSED PINK CARD FOR AN AMENDMENT.
YOUR CUSTOMER NUMBER  | YOU CAN PHONE US ON  | PERIOD ENDING  | AMOUNT DUE NOW
03 3967 4721 60       | 00136 247            | 12 APR         | £137.48
E against a meter reading means an estimate
C against a meter reading means it is your own reading
```

14 Look at the example bill.
(a) What is the 'This Time' reading on the bill?
(b) What does the 'E' after this reading mean?

15 Which reading did Mrs Current take herself?

Here is the reading on Mrs Current's meter when she was sent the bill.

16 Was the estimated reading right?

17 Did Mrs Current use more or less electricity than it says on the bill?

18 Copy the parts of the bill below. Then fill in all the missing numbers to make it a correct bill.

19 How much less is the new bill?

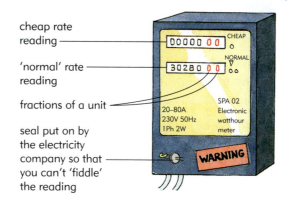

cheap rate reading

'normal' rate reading

fractions of a unit

seal put on by the electricity company so that you can't 'fiddle' the reading

METER READING		UNITS	PENCE PER	AMOUNT	STANDING CHARGE	TOTALS
This Time	Last Time	USED	UNIT	£	£	£
	29210C		10.0		14.30	
	TOTAL EXCLUSIVE OF VAT					
		VAT				9.71
					AMOUNT DUE NOW	

Half-price electricity

After about 11 o'clock at night, most businesses have closed and most people have gone to bed.
This means that much less electricity is needed.

However, some power stations cannot be shut down, and they generate more electricity than is needed. The electricity companies sell 'night-time' electricity at a much cheaper price to persuade people to use more. Storage heaters are an example of appliances that use this cheap electricity.

20 Look at the meter readings. Cheap rate electricity from this company costs 5p per unit. How much would it cost for this electricity on the 'cheap rate' scale?

21 How much would it cost for the same amount of electricity on the 'normal rate' scale (10p per Unit)?

meter reading last time

meter reading this time

Lifting things with electricity

In our homes, we normally use electricity to transfer energy as light, sound, movement or thermal energy. But electricity can also be used to lift things up.

Up and down

Getting the cars up a roller coaster needs energy. The force of **gravity** pulls downwards. So an electric motor pulls the cars to the top against the force of gravity.

Electrical energy from motor pulls cars up.

When the cars get to the top of the slope they are ready to run down the other side. The cars are pulled down by gravity because of their height. We say they have **gravitational potential energy**.

Cars store gravitational potential energy at top.

When the cars are running down the slope they transfer potential energy as movement. This is called **kinetic energy**.

22 Look at the diagrams.
(a) Which way does gravity pull on things?
(b) What type of energy is used to pull the cars up?
(c) What type of energy do the cars store at the top?
(d) What type of energy does this change into as the cars move down the slope?

Cars have transferred half their potential energy to kinetic energy.

Higher still

The cars are then pulled up a slope that is twice as high.

23 How much more gravitational potential energy will the cars have at the top of this slope than at the top of the first slope?

24 How much electrical energy do you think would be needed to pull the cars to the top of this slope? Would it be: the same amount as before, half as much as before, or twice as much as before?

25 How long will it take the same motor to pull the cars up this slope?

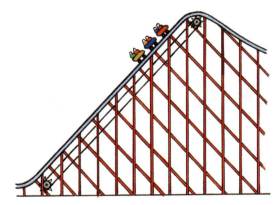

This slope is twice as high as the first, so the gravitational potential energy of the cars at the top will be twice as much. More electrical energy must be transferred to lift the car to the top. The electric motor must be switched on for a longer time.

26 Copy the energy transfer diagram below.
Then fill it in.

| electrical energy | → motor which pulls cars up slope | gravitational _____ energy | → cars moving down slope | _____ energy |

Lift off!

Lifts move people and objects up and down buildings.
They have to work against the pull of gravity to do this.

The diagrams show two different lifts.

The 'standard' lift moves slowly.

The 'express' lift moves faster.
It transfers energy more quickly.

Both lifts need the same **amount** of energy to reach the
same height, but the electric motors in the lifts do not
supply this energy at the same **rate**.

How fast a motor transfers energy is called its **power**.

27 Which electric motor will have to supply energy faster?

28 Which electric motor will have to be the more powerful?

Power is measured in **watts** (W) or in **kilowatts** (kW).

1 kW = 1000 W

29 The standard lift has a 5 kW motor.
The express lift moves twice as fast.
How powerful must the electric motor of the
express lift be?

Power

Athletic power

Athletes train so that their muscles become more powerful. This means that they can transfer energy more quickly. They can then run faster, jump higher or throw further.

1 (a) What can more powerful athletes do?
(b) What does power measure?

Athletes transfer energy. Energy is measured in joules (J).

Who's working hardest?

Look at the weight trainers.

2 Who is working harder?

3 Who has transferred more energy to the weights in one minute?

4 Who is producing more power?

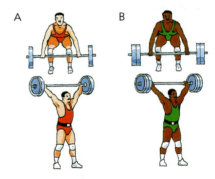

Both lift their weights 10 times in 1 minute. Lifting a bigger weight needs more energy.

Jane lifts the whole of her school bag, loaded with books, on to the bench in one go. Joanne takes out her books one by one and puts them on the bench.

5 Who takes longer to move the books?

6 Who is transferring energy faster?

7 Who is producing more power?

Both girls have the same number of books (of the same mass) in similar bags.

Measuring energy

The unit of energy is the joule (J).

8 Look at the diagrams on the right. Write down two examples of transferring 1 joule of energy.

The 1 watt bulb in this torch transfers 1 joule of energy every second.

1 joule of energy is transferred to the apple when it is lifted on to the table

How much power?

Power measures how fast energy is transferred, so you can
work out power like this:

power   =   energy transferred (joules)
(watts)   time taken (seconds)

(this line means 'divided by')

So 1 watt is 1 joule of energy transferred every second.

watts (W) = joules per second (J/s)

9   Look at the diagram of the person running up stairs.
What is the power produced by this person running up
these stairs?

1000 J

Time to run up stairs
= 2 seconds

How much energy?

You can rearrange the power formula to find out how
much energy an electric lamp or kettle uses.

energy transferred = power × time

Look at the diagrams.

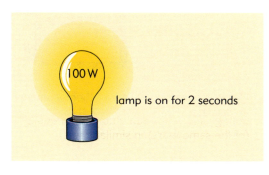

100 W

lamp is on for 2 seconds

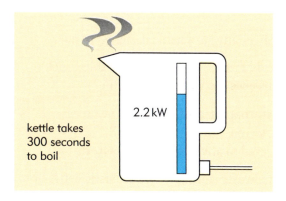

2.2 kW

kettle takes
300 seconds
to boil

10   How much energy has the lamp transferred?

11   How much energy has the kettle transferred?

(Remember: 1 kW = 1000 W)

Efficiency and wasted energy

Which kettle is best?

A hotel owner wants kettles for her guest rooms.
She tests three different kettles. The results are shown
on the right.

1. Which kettle transferred the most energy to boil
   1 litre of water? [1 litre of water is 1 kilogram]

2. Which one would you choose to put in guest rooms if
   you ran the hotel?
   Give a reason for your choice.

3. How was the test made a 'fair test'?

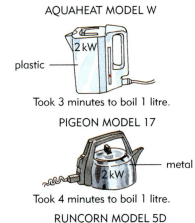

AQUAHEAT MODEL W

plastic — 2 kW

Took 3 minutes to boil 1 litre.

PIGEON MODEL 17

2 kW — metal

Took 4 minutes to boil 1 litre.

RUNCORN MODEL 5D

plastic — 2 kW

Took 3 minutes 10 seconds to boil 1 litre.

Why are some kettles better than others?

An electric kettle is designed to transfer electrical energy
as thermal energy.

We want the energy transferred by the kettle to heat
the water.

4. What energy change is a kettle designed to do?

5. Look at the diagrams. What two other things get
   heated up besides the water in the kettle?

Only the energy that is transferred to the water is
usefully transferred.
Energy transferred to anything else is **wasted**.

6. Which kettle wastes more energy, a plastic kettle or a
   metal one?

7. Look again at the kettles that the hotel owner tested.
   Which kettle usefully transfers the most energy?

The best kettle transfers the most thermal energy to the
water. We say that this kettle transfers energy in the most
efficient way.

8. Which kettle is the most efficient?

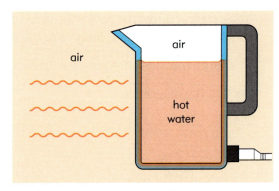

Plastic is a poor conductor of thermal energy. It is
an insulator.

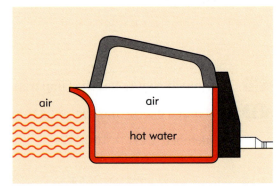

Metal is a good conductor of thermal energy.

Heating food efficiently

An efficient oven transfers as much energy as possible to food.

A microwave oven is very efficient because it only heats up the food. An ordinary oven heats up its surroundings too. An electric cooker uses more energy than the microwave oven to heat the same amount of food. Less energy is usefully transferred in the electric oven.

9 Explain why the microwave oven is more efficient than the ordinary electric oven.

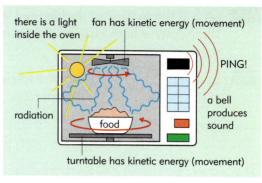

A microwave oven.

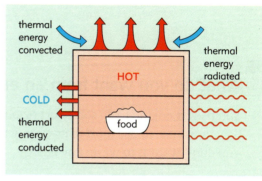

An ordinary electric oven.

Using lamps effectively

A light bulb is not very efficient (see page 360) but we can still use a light bulb effectively. An effective lamp sends light to where we want it.

10 Look at the pictures. Which is the most effective way of using a light bulb:
(a) for reading?
(b) for lighting the whole room?
Give reasons for your answers.

A light bulb sends out light in all directions.

Energy in different forms

When you transfer energy you don't get only the form of energy you want.

Usually other energy transfers also happen. These waste energy and reduce **efficiency**.

11 Look at the TV set.
(a) Which useful forms of energy does it give out?
(b) Which wasted forms of energy does it give out?

12 The TV set is designed to let the waste energy get out easily. How is this done?

13 Look at the pictures of the bike and the skateboard. Write down where friction between moving parts can waste energy.

14 How would you cut down the wasteful energy transfers in the bicycle and the skateboard?

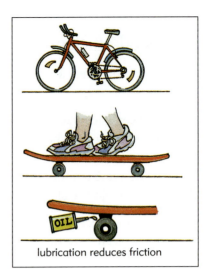

lubrication reduces friction

What do you want from an engine?

You use an engine to do some sort of job for you.
So you want it to use most of its energy in doing that job.
You don't want it to waste any energy.

If an engine is 100 per cent efficient, it means that it is transferring **all** its energy as useful work.

Look at the engines in the diagrams.

15 Which engine is the most efficient?

16 What <u>two</u> sorts of 'waste' energy do all the engines produce?

17 Railways changed from steam engines to diesel engines and then to electric trains.
Why did this happen?

18 Imagine that you used 100 litres of petrol on a long car journey. How many litres would be doing useful work?

% efficiency

about 90%

electric motors

about 35–40%

diesel engines

about 25%

petrol engines

7% or even less

steam engines

Which lamps are most efficient?

A lamp can't produce light without also producing heat. The efficiency of a lamp is the fraction (or percentage) of energy that is transferred as light.

19 Look at the diagrams.
The filament lamp is 4% efficient because it transfers 4% of its energy as light. Write down the efficiency of the fluorescent tube and the street lamp.

20 Write down the three types of lamp in order, starting with the most efficient and ending with the least efficient.

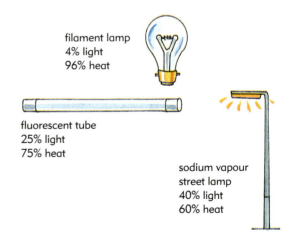

filament lamp
4% light
96% heat

fluorescent tube
25% light
75% heat

sodium vapour
street lamp
40% light
60% heat

Where does the wasted energy go?

Energy doesn't disappear, it just gets changed from one type of energy to another.

Some of the types of energy are what you want, others are just wasted. No energy gets lost – the total amount of energy you end up with is always exactly the same as the amount you started with.

21 Look at the television set. 100 J of energy is put in.
   (a) How many joules are transferred into forms such as heat, light and sound?
   (b) Does any energy disappear?

The arrows on the diagram show how the television set transfers energy. The thickness of each arrow shows how much energy is transferred in that particular way.

The wasted and useful energy always add up to 100%.

22 Copy and complete the energy transfer diagrams below.

(a) Filament lamp

(b) Fluorescent tube

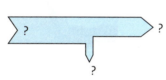

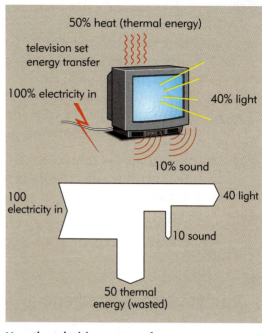

How the television set transfers energy.

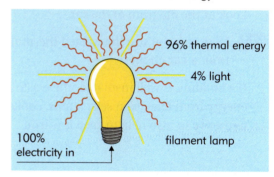

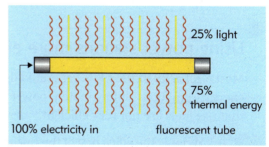

Everything ends up as thermal energy

You use some of the light from the television set to see the picture, but much of the light does not go into your eyes. It is absorbed (soaked up) by everything in the room.

When this happens the absorbed energy makes the room a little bit warmer.
The same thing happens to the sound vibrations.
The energy ends up as thermal energy and spreads out.

This makes it more difficult for you to do anything useful with it.

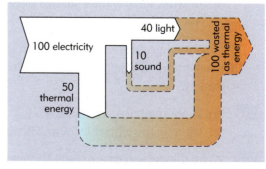

All the energy from the television set gets spread out and wasted as heat in the end.

23 A petrol engine and a steam engine are each given 100 J of energy.
(a) The petrol engine transfers 25 J into movement energy and 10 J into sound energy. How much thermal energy is given out?
(b) The steam engine transfers 7 J into movement energy and 73 J into thermal energy. How much sound energy is given out?

24 Draw a diagram, like the one at the top of this page, to show the energy transfers for the petrol engine. (Use a piece of squared paper.)

25 Which produces more waste thermal energy, a petrol engine or a diesel engine?

Can you use wasted energy for anything?

Sometimes you can make use of 'wasted' energy.

Look at the diagram of the car.

26 How can you use the wasted thermal energy from a car engine?

If a building is well insulated, the waste thermal energy from television sets, lamps, cookers and freezers can also help to keep it warm.

Using power stations more efficiently

Most electricity in the UK is generated using the energy from fuels. Unfortunately, power stations that use fuels are not very efficient.

27 (a) What is the efficiency of the normal power station shown on the diagram on the right?
(b) In what form is most of the energy from the normal power station wasted?

The lower diagram shows how we can use the thermal energy from power stations rather than wasting it.

28 (a) Explain what a 'CHP station' is.
(b) What is the efficiency of a CHP station?

29 Write down two reasons why there are very few CHP stations at present.

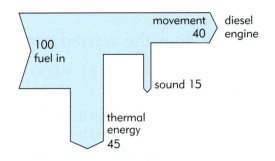

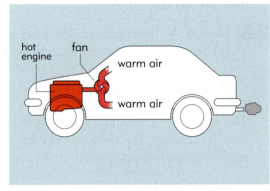

You can use waste thermal energy from a car engine to heat the inside of the car.

Cooling towers: monuments to inefficiency.

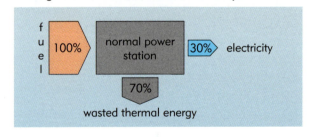

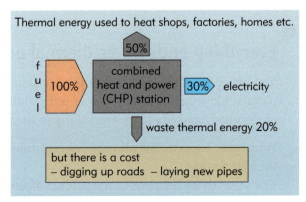

So a CHP system is expensive to build.
It is even more expensive to convert a power station to CHP after it has been built.

Using fuels to make electricity

Electricity gives us a useful form of energy, which we can transfer as many other types of energy.

Electricity has to be generated, and we need an energy source to do this. Because we must generate electricity using some other source of energy, we say that electricity is a secondary energy source.

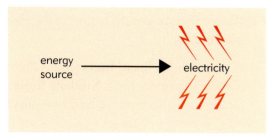

We always need some other energy source to make electricity.

Fuels as energy sources

The four main fuels (**primary** energy sources) we use to generate electricity are coal, oil, gas and nuclear fuel.

Fuels are **non-renewable** energy sources. This means that once you've used them they cannot be replaced. Fossil fuels and nuclear fuels will eventually run out.

1. Look at the bar chart of how long primary energy sources will last.
   (a) How long will coal, natural gas, oil and nuclear fuel last (at present rates of use)?
   (b) Which will last the longest?

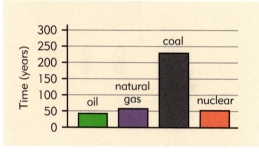

How long fels will last if we use them at the present rate. (Nuclear fuel will last a lot longer if we use it in fast breeder reactors.)

The pie chart shows the energy sources that are used worldwide to generate electricity.

2. (a) Which fuel is used to generate most of the world's electricity?
   (b) What percentage of the world's electricity is generated by gas?
   (c) Which of the energy sources on the pie chart is not a fuel?

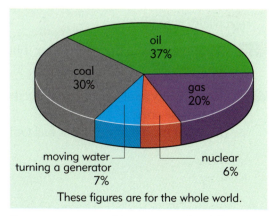

These figures are for the whole world.

Energy sources used to generate electricity.

Using fuels to generate electricity

We use fuels to generate most of the electricity used in Britain.

When we burn **fossil fuels** (coal, oil and gas) they transfer energy as heat. **Nuclear fuels** (uranium and plutonium) also produce heat as atoms are split up.

In a nuclear power station, heat from the reactors is used to make steam. The fuel is used up in a nuclear reaction, but it does not burn.

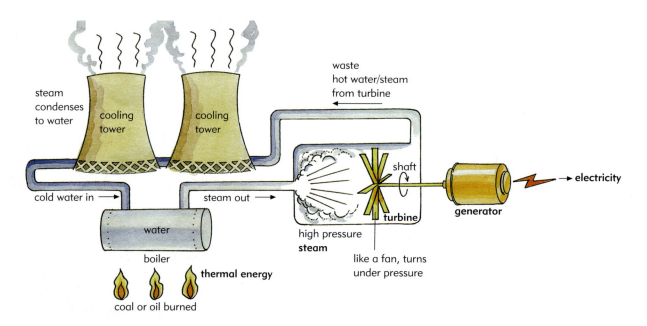

The diagram above shows how the heat (thermal energy) released by a fuel is used to generate electricity.

Thermal energy from fuel heats the water and changes it into steam. The high-pressure steam turns the blades of the turbine. The turbine is connected to the generator by a shaft. As the shaft spins, this spins the generator and electricity is produced.

3 What happens to the steam after it has been through the turbine?

4 (a) Which type of fuel does not need to be burned to release energy?
    (b) What must happen to this fuel for it to release energy?

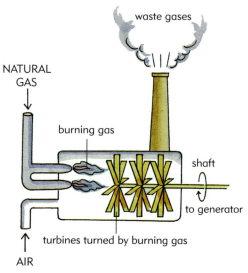

5 Look at the diagram on the right. What is different about a gas-fired power station?

In a gas-fired power station there is no need to use steam. The hot burning gases drive the turbine directly.

What are the problems with using fuels?

When you burn a fuel, **waste** is produced. This can pollute the atmosphere and soil, or may have to be stored until it is safe.

There is only a certain amount of each fuel in the Earth, so they are **non-renewable**. Once they are used up, they are gone **for ever**.

The diagram shows the types of waste from a coal-fired power station.

6   (a) Write down <u>one</u> solid waste that has to be put somewhere.
    (b) Write down <u>four</u> types of waste that go into the atmosphere.

Look at the information in Fact file 1.

7   What are the cooling towers used for?

8   For every 100 tonnes of coal burned in a power station, how many tonnes are usefully transferred as electricity?

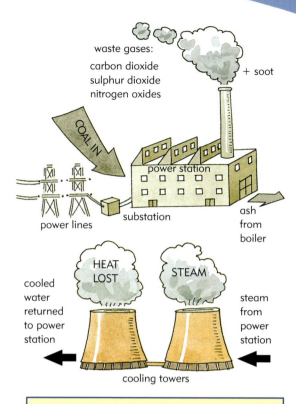

Why do we use fuels to generate electricity?

Fuels are plentiful (at the moment), but before the end of the century they may start to run out.

Power stations that use fuels can generate electricity at any time and in all sorts of weather.

9   Write down <u>three</u> advantages and <u>three</u> disadvantages of using fuels to generate electricity.

Fact file 1
Power stations are only about 30% efficient. Cooling towers change the steam back into water. This water is used again to make steam.

Fact file 2
Power stations that use fuels can be very big and can spoil the countryside. One power station can generate electricity for over 1 million people. Power stations that use fuels produce pollution.

Comparing fuels for electricity

Using fuels always produces waste. These wastes can pollute the air, the water and the soil. We can reduce some of this pollution, but this makes electricity more expensive.

The diagram and the bar charts (on page 366) show some facts about the wastes from different fuels.

10   Look at the diagram.
    (a) What gases can cause acid rain?
    (b) What problems are caused by carbon dioxide gas?
    (c) What forms of solid waste are produced by power stations?

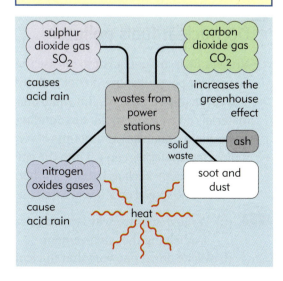

Look at the bar charts.

11 (a) Which type of fuel produces no waste gases and very little solid waste?
(b) Why does this fuel cause a very serious pollution problem?

12 (a) Which fuel produces the most solid waste?
(b) Which fuel – coal, oil or gas – produces the least pollution?

13 (a) Which waste gas increases the greenhouse effect?
(b) Which fuel causes the biggest increase in the greenhouse effect?

14 (a) Which <u>two</u> gases cause acid rain?
(b) Which fuel produces most of the gases that cause acid rain?

15 Look at the amount of carbon dioxide produced by gas. How many times more carbon dioxide do coal and oil produce?

Each bar chart shows the waste produced for every thousand million Units of electricity produced.

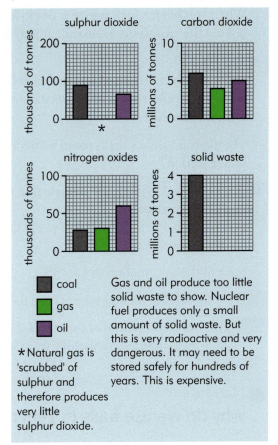

coal

gas

oil

Gas and oil produce too little solid waste to show. Nuclear fuel produces only a small amount of solid waste. But this is very radioactive and very

*Natural gas is 'scrubbed' of sulphur and therefore produces very little sulphur dioxide.

dangerous. It may need to be stored safely for hundreds of years. This is expensive.

How can we cut down air pollution?

Soot and dust are made of tiny particles. These can be removed from chimneys by attracting them using electric charges or by using filters. This makes the electricity a tiny bit more expensive.
Sulphur dioxide can be removed from chimneys by absorbing it in chemicals. This makes the electricity up to 20% more expensive. It is far too expensive to remove the huge amounts of carbon dioxide from the waste gases.

16 (a) How can soot and sulphur dioxide be removed from chimneys?
(b) Why isn't carbon dioxide removed from chimneys?

Do nuclear fuels pollute?

Nuclear fuels are not burnt to release their energy. So nuclear power stations do not pollute the air with carbon dioxide or sulphur dioxide. But small amounts of radioactive materials do escape into the surroundings. Also, if there is an accident, much larger amounts of very dangerous radioactive materials can get into the environment. In a serious accident, such as the one at Chernobyl in 1986, these may be spread over a very large area.

17 (a) Write down <u>two</u> ways that nuclear power stations are helping the environment.
(b) Write down <u>one</u> way that nuclear power stations may harm the environment.

Nuclear power stations produce highly radioactive waste. Some of this can stay dangerously radioactive for thousands of years. So it has to be stored very carefully. This adds a lot to the cost of the electricity generated.

Using renewable energy resources to make electricity

Renewable energy sources

Fuels like wood are <u>renewable</u>. This means that they can be replaced.

Anything that can be grown and used as a fuel is called **biomass**. Other renewable energy sources are **sunlight**, the **wind**, the **waves**, **running water** (hydro-electricity) and the **tides**.

1 Which <u>three</u> ways of generating electricity use moving water?

2 Which renewable energy source:
(a) is used by solar cells?
(b) consists of moving air?

Renewable energy sources will never run out – they are always being replaced.

If you use renewable energy sources you don't have to buy fuel. Also, most renewables do not make harmful waste products. But it is expensive to capture the energy from renewable energy sources.

Moving water

Water is always on the move.

The diagram shows three types of moving water that we can use to generate electricity.

3 Look at the diagram.
Write <u>one</u> sentence to explain each of these:
(a) why the waves move;
(b) what causes the tides;
(c) how the Sun helps rivers to flow.

Rain falls on the land and flows to the sea in streams and rivers. Rain keeps on falling because the Sun keeps on evaporating water from the sea.

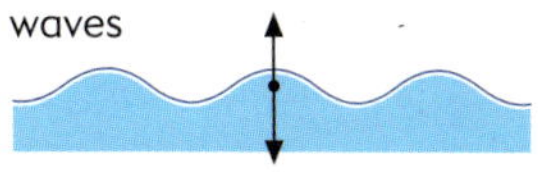

water condenses
Sun
clouds
rain falls
water evaporates
sea
stream
river
waves

waves

The wind produces waves as it blows across the sea. As waves move along, the water moves up and down.

Tides make water flow into estuaries twice a day. Tides keep happening because of the pull of the Moon's gravity as the Earth spins.
tide in

Hydro-electricity

We build **dams** across rivers. These trap a lot of water. We can use this trapped water to generate electricity whenever we need it. Water from one side of the dam flows out through a pipe. The turbine in the pipe spins as the water flows past. This spins a generator, and electricity is produced.

The diagram shows how we do this.

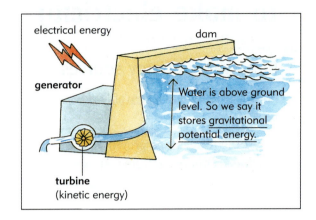

4 (a) What form of energy does the water have behind the dam?
(b) What form of energy does this give the turbine?
(c) What form of energy does the generator give out?

Good and bad news about hydro-electricity

Of all the renewable energy sources, at the moment hydro-electricity is the only one that gives us much electricity. But this is still a very small amount compared to the amount we get from fossil fuels.

> Land has to be flooded to make reservoirs.
>
> Hydro-electric generators can be started up in a few seconds.

5 Look at the information in the Box. Write down <u>one</u> effect of hydro-electricity on the environment.

6 Hydro-electric generators are very useful if more electricity is suddenly needed. Explain why.

Electricity from tides

Tides make water flow into estuaries twice each day. We can trap this water behind a **barrage**. We can then use it to generate electricity when the tide goes out.

7 Where, in Britain, could a large tidal power station be built?

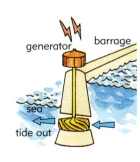

8 Write down <u>two</u> disadvantages of generating electricity from tides.

You can only generate electricity at certain times each day.

Tides vary a lot in height.

Large estuaries like the Severn are the main habitat for wading birds. They find food in the mud. If you flood the estuary there is no mud and no food.

Electricity from waves

The diagram shows a small **wave generator**. No large generators have yet been made to stand rough seas.

9 What drives the turbine in the wave generator shown in the diagram?

rise and fall of water pushes air through turbine

Electricity from wind

Wind is a renewable energy resource. It is a regular part of our weather, which is driven by energy from the Sun. So as long as the Sun keeps shining, the wind will keep blowing.

We have used the energy from the wind for thousands of years. But, during the past 250 years, steam engines, then petrol engines and then electricity took over from wind.

We are now using wind energy again.
This will help to make our stocks of non-renewable energy sources last longer.

The diagram shows a modern **wind turbine**.
This is used to generate electricity.

10 Write down the energy transfers that take place inside the wind turbine.

If the wind is too strong it can damage a wind generator.

11 How is a wind generator protected against this damage?

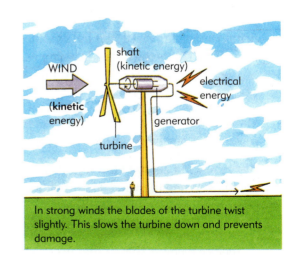

In strong winds the blades of the turbine twist slightly. This slows the turbine down and prevents damage.

Where should we put wind generators?

Large wind generators need a wind of at least 5 metres per second before they generate electricity. So they need to be put in places where there is plenty of wind.

Look at these two maps.

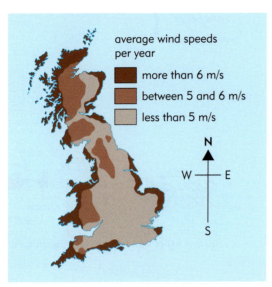

average wind speeds
per year

■ more than 6 m/s

■ between 5 and 6 m/s

☐ less than 5 m/s

N
W — E
S

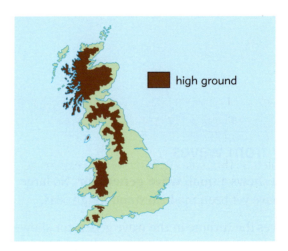

■ high ground

12   Use the maps to help you answer this question.
Where are the best places for wind turbines in Britain?
Explain your choices.

Could wind generators provide all our electricity?

In 1993 there were only 21 wind turbine projects in Britain supplying electricity but by 2000 there were more than 100. This is the fastest growing way of using renewables to generate electricity. But this is still only a small amount of electricity.

Many hundreds of wind generators would be needed to replace just one power station that uses coal.

Also, the wind doesn't blow all the time, even on coasts and hills. When the wind doesn't blow, we need some other way of generating electricity.

13   How many 500 kW wind turbines would be needed to replace a power station that uses coal?

14   Write down <u>two</u> reasons why the wind could not supply all the electricity we need.

Wind farms have lots of wind turbines.
Each wind turbine generates 500 kW to 2 MW.
A coal-fired power station generates about
1300 MW. (1 MW = 1000 kW.)

Wind generators and the environment

People disagree about whether or not wind farms are good for the environment.

15 Write down <u>one</u> advantage and <u>one</u> disadvantage for the environment of using wind farms to generate electricity.

Using the wind does not pollute the air.

People like to visit unspoilt hills and coasts at weekends and holidays.

You can see wind farms for miles around. They can be quite noisy for anyone who lives or works close by.

Solar cells

Solar cells are a very simple way of transferring energy from the Sun as electricity. There is only one energy transfer. Light is transferred directly as electricity.

But one problem is that solar cells are a very expensive way of making each Unit of electricity.

Another problem is that solar cells will work only if the Sun is shining!

16 Write down the energy transfer that takes place in a solar cell.

17 Look at the bar chart. To make each Unit of electricity, which source of energy costs:
(a) the least?
(b) the most?

18 Write down <u>two</u> disadvantages, apart from the cost, of using solar cells to generate electricity.

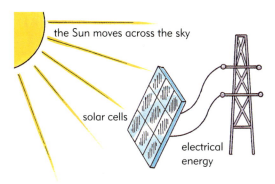

the Sun moves across the sky

solar cells

electrical energy

The solar cells must be at the correct angle to collect the most light energy.

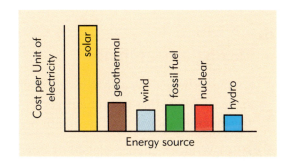

Why use solar cells?

Electricity from solar cells is expensive.
But there are times when they are worth using.

Calculators use only a very small amount of electricity. So you can use a very small solar cell instead of batteries. A solar-powered water pump can be used in remote villages in hot countries. It is often too expensive to put power cables from power stations to remote places.

19 Write down <u>two</u> reasons why solar cells are used in satellites.

20 Where would you use a solar-powered pump? Explain why.

21 Why are solar cells suitable for calculators?

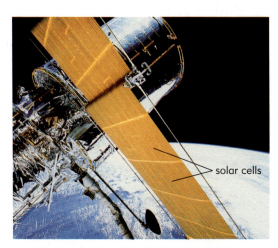

solar cells

Satellites use solar cells. Solar cells weigh less than batteries and will work for many years.

Geothermal energy

'Geo-' means 'from the Earth' and 'thermal' means 'heat'. So **geothermal** energy is heat from the Earth's rocks.

The rocks in the Earth contain radioactive elements like uranium. When these elements decay, the rocks transfer energy as heat. This happens very slowly over billions of years.

There are some places on the Earth where water gets heated up by these reactions, producing steam and hot mud.

22 What do we call steam that spurts out of the ground?

23 Write down the names of <u>four</u> countries where steam from the ground is used to generate electricity.

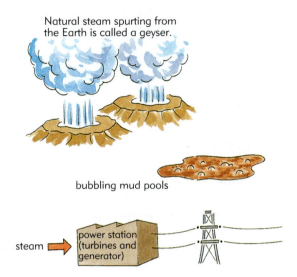

Natural steam spurting from the Earth is called a geyser.

bubbling mud pools

steam → power station (turbines and generator)

Natural steam is used to generate electricity in the USA, Italy, New Zealand and Iceland.

Making a heat mine

Engineers hope to be able to use geothermal energy even in places where there aren't any geysers.
The diagram shows how they might be able to do this.

24 (a) Write a sentence explaining how cold water is changed into steam.
    (b) What is the steam used for?

25 Write down <u>two</u> problems in using geothermal energy to generate electricity.

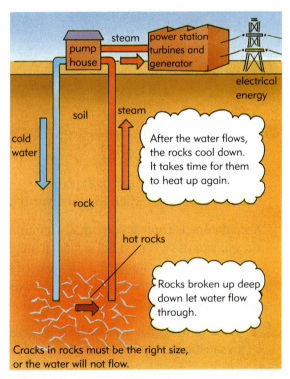

pump house · steam · power station turbines and generator · electrical energy

soil · steam

cold water · rock

After the water flows, the rocks cool down. It takes time for them to heat up again.

hot rocks

Rocks broken up deep down let water flow through.

Cracks in rocks must be the right size, or the water will not flow.

A heat mine.

Should we use renewable energy sources?

Many people think that we should use renewable energy sources instead of non-renewable fuels.

26 Why are renewable energy sources thought to be 'better' for the environment than non-renewable fuels?

But there are problems with using renewable energy sources, as well as advantages. The boxes A to M show some of these.

27 Match each of the renewable energy sources below with the statements in the boxes.

The numbers in the brackets tell you how many boxes you must match with each energy source. You will need to use some of the boxes more than once.

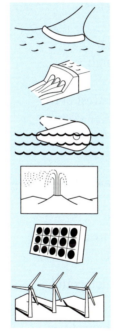

(a) Tidal barrages (4)

(b) Hydro-electric power stations (6)

(c) Wave generators (3)

(d) Geothermal (1)

(e) Solar cells (2)

(f) Wind turbines (2)

28 Which renewable energy source would you choose to generate electricity in these situations:
(a) in a desert region, for pumping water?
(b) in a mountainous region with many streams?
(c) in a country with many volcanoes and geysers?

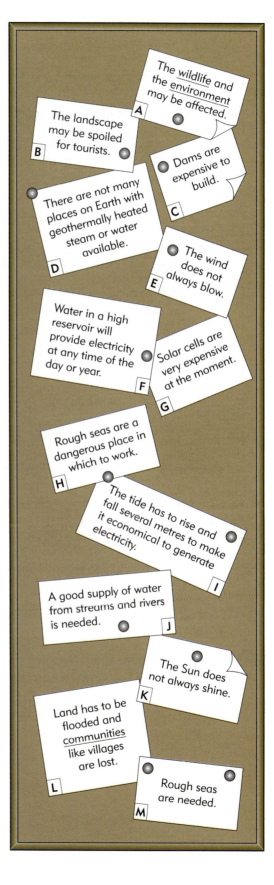

A. The wildlife and the environment may be affected.

B. The landscape may be spoiled for tourists.

C. Dams are expensive to build.

D. There are not many places on Earth with geothermally heated steam or water available.

E. The wind does not always blow.

F. Water in a high reservoir will provide electricity at any time of the day or year.

G. Solar cells are very expensive at the moment.

H. Rough seas are a dangerous place in which to work.

I. The tide has to rise and fall several metres to make it economical to generate electricity.

J. A good supply of water from streams and rivers is needed.

K. The Sun does not always shine.

L. Land has to be flooded and communities like villages are lost.

M. Rough seas are needed.

Choosing the best renewables for the job

Study the map of Arbril Island and read the 'Fact File'
carefully. Then answer the questions.

Fact file on Arbril Island

- Arbril is in the Antarctic Ocean between the
  Falkland Islands and the South Pole.
- There is sunlight for six months of the year
  and then darkness for six months.
- The seas around the island are very rough.
  Sometimes it is impossible to travel to Arbril.
- The tides only rise and fall by less than
  a metre.
- Port Herbert is the main town. It relies
  on fishing as its main industry.
  It is on a sheltered part of the island.
- Soil is poor quality. But rainfall is high
  throughout the year.
- There are hot springs on the south-west of the
  island. They are difficult to reach over very
  high mountains.

Arbril Island

Port Herbert needs a reliable electricity supply.
The old power station is coming to the end of its useful life.
It uses coal, which is very expensive to import.

29 Suggest <u>four</u> renewable energy sources which could
be used.

30 For each energy source, suggest at least <u>one</u> problem to
be solved.

31 (a) Copy the map. Then show on your map where you
would build the new power station for each of your
suggestions.
(b) Give a reason for your choice of site in each case.

How should we generate our electricity?

Higher

Which energy source should we use?

We need to consider several things when deciding which energy source to use. These include how much electricity they can generate, whether the electricity is generated when we need it, how much pollution is caused and the costs of the electricity produced.

There are several different costs to take into account. For example, the energy in the wind is free, so wind generators have no fuel costs. But the energy in the wind is spread out quite thinly. So to generate the same amount of electricity as a coal-fired power station you need a very large number of wind generators. It costs more to build all these wind generators than it does to build a coal-fired power station.

To work out what it costs to produce each Unit (kilowatt-hour, kWh) of electricity, we have to work out:

- the cost of <u>building</u> the system (the capital cost);
- the cost of <u>operating</u> and <u>maintaining</u> the system;
- the cost, if any, of the <u>fuel</u> that is used.

We then need to share this cost between all the Units of electricity the system produces in its lifetime.

1 Look at the diagram.
   (a) For each system, work out the total cost per kWh. Include the building costs, fuel costs and operating costs.
   (b) Which system has:
       (i) the lowest building cost,
       (ii) the lowest fuel cost,
       (iii) the lowest total costs
       for each kWh of electricity?

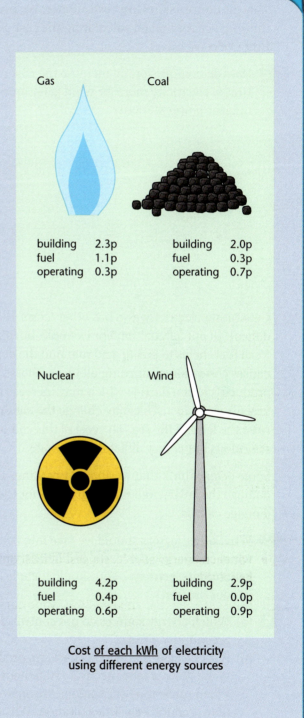

Gas		Coal	
building	2.3p	building	2.0p
fuel	1.1p	fuel	0.3p
operating	0.3p	operating	0.7p

Nuclear		Wind	
building	4.2p	building	2.9p
fuel	0.4p	fuel	0.0p
operating	0.6p	operating	0.9p

Cost <u>of each kWh</u> of electricity using different energy sources

Higher

Electricity generating systems eventually wear out. They then have to be dismantled and removed. This **de-commissioning** costs money. We need to take this into account when calculating the cost of each Unit of electricity.

2 Does de-commissioning increase the cost of each Unit of electricity more for wind generators or for nuclear power stations?

Nuclear power stations don't produce waste gases that pollute the air, but they do produce solid waste that will stay dangerously radioactive for thousands of years. We don't really know how much it will cost to store this safely for such long periods of time. We don't even know if we <u>can</u> do this safely. The power stations themselves contain a lot of radioactive material so dismantling them is expensive.

Generating electricity also has other 'costs' that it is difficult to put a figure on. For example, burning some fossil fuels helps to create acid rain and the damage this causes costs money. Burning all types of fossil fuel produces carbon dioxide which may increase the greenhouse effect. This may change the climate and raise the level of the sea. The cost of this could be enormous but is very difficult to estimate.

Some people think that we should tax things that damage the environment to help to pay for the damage caused.

Many people care about the appearance of their surroundings. They would be prepared to pay more for their electricity to avoid, for example, having lots of wind generators (wind-farms) on most hill-tops.

3 What other factors should we take into account when comparing the 'costs' of different ways of generating electricity? You should include examples in your answer.

4 (a) Which energy sources do not contribute to the greenhouse effect?
   (b) A friend says that wind generators produce no pollution. Write down a more accurate statement.
   (c) When nuclear power stations are operating properly, they produce very little pollution. Without nuclear power stations, we would have regular power cuts in the UK because they generate about a quarter of our electricity. Explain some disadvantages of de-commissioning nuclear power stations.

Making electricity generation more efficient

Some types of power station aren't very easy to shut down. Quite often we generate more electricity than we need. This usually happens during the night. The **surplus** electricity cannot be stored. If we don't want to waste electricity we have to use it in some way. Electrical storage heaters use off-peak electricity during the night. You can buy this for about half the usual price per Unit (see page 353).

5 What do electricity companies do to make sure that surplus electricity they generate during the night isn't wasted?

Another way of using surplus electricity

Surplus electricity can't be stored, but it can be used to pump water uphill to a storage reservoir. The water from this reservoir can then be used to generate electricity when there is a sudden increase in demand. This system is called a <u>pumped storage scheme</u>.

The energy from the surplus electricity is transferred to water as gravitational potential energy.

6 Explain, as fully as you can, the idea behind a pumped storage system.

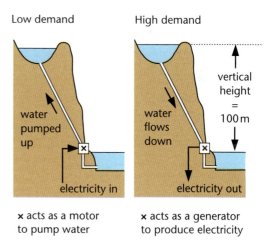

Pumped storage scheme.

Calculating the energy stored by a dam

You can calculate the gravitational potential energy stored in something that is lifted up like this:

$$\begin{array}{ccc} \text{gravitational} & & \text{weight} & & \text{vertical} \\ \text{potential energy} & = & \text{(N, newtons)} & \times & \text{height raised} \\ \text{(J, joules)} & & \text{[10} \times \text{mass in kg]} & & \text{(m, metres)} \end{array}$$

7 One tonne of water is pumped up to the higher reservoir in the pumped storage system shown on the diagram on page 377. Calculate the gravitational potential energy transferred to the water.

Some of the energy from the surplus electricity is, of course, wasted when it is used to pump water up into the upper reservoir (see Box).

Also, not all of the gravitational potential energy of water in the upper reservoir results in electricity when it is used to drive generators.

8 Calculate the efficiency of the pumped storage <u>generator</u> if it generates 750 000 J of electricity when 1 tonne of water falls 100 m.

Despite these energy losses, the overall efficiency of a pumped storage scheme is usually better than 50%.

> The pump uses 1.25 million J of energy to lift the 1 tonne of water 100 m.
>
> So the efficiency of the transfer is:
>
> $$\frac{1 \text{ million J}}{1.25 \text{ million J}} = \underline{0.8} \ (\times 100 = \underline{80\%})$$

Comparing start-up times

The bar chart shows how long it takes to start up different types of power station.

9 Write down the <u>four</u> types of power station in order, starting with the shortest start-up time.

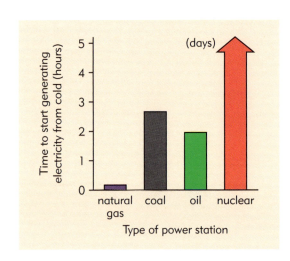

Higher

Matching electricity supply and demand

During each 24-hour period, the demand for electricity rises and falls quite a lot. Sometimes there is a sudden big increase in demand – for example, when many people switch on their electric kettles at the same time. This can happen at the end of a very popular TV programme or at half-time during a televised cup final.

The electricity generating companies have to generate the right amount of electricity to meet these sudden increases in demand. But they don't want to generate this amount of electricity <u>all</u> the time because it isn't all needed so it won't all be sold. To solve this problem, they need to have some generators that they can stop and start very quickly.

10 (a) Which type of power station can be stopped and started most quickly to meet sudden changes in demand?
(b) Which type of power station is least suited to meet sudden changes in demand?

Because nuclear power stations take a very long time to start up and shut down, they are used to produce much of the electricity that is <u>always</u> needed. This is called the **base-load demand**. Other types of power station are started up when they are needed. But there are still times, especially at night, when too much electricity is generated.

Some of this surplus electricity is sold very cheaply to any customers who want it, but some of it is used in pumped storage schemes (see page 377). These can then be used to meet sudden increases in demand during the day.

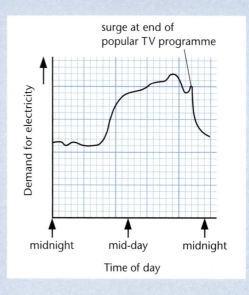

surge at end of popular TV programme

Demand for electricity

midnight   mid-day   midnight

Time of day

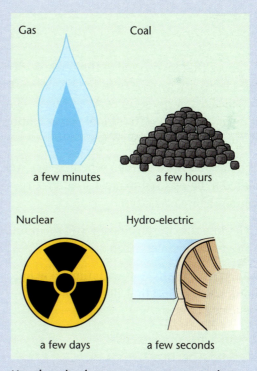

Gas   Coal

a few minutes   a few hours

Nuclear   Hydro-electric

a few days   a few seconds

How long it takes to start up power stations.

Higher

One of the problems with using renewable energy sources to generate electricity is that they are often not very reliable. As you can see from graphs A, B and C, we cannot always depend on these energy sources being available when we need them.

We will almost certainly use renewable energy sources more and more during this century. But because these energy sources are unreliable, we still need to be able to generate a large proportion of the peak demand for electricity using other, more reliable, energy sources.

11. Suppose that by the end of the century, 50% of the electricity generated in the UK comes from renewable energy sources. We would still need almost as many conventional power stations as we do at present. Explain why.

12. Why are pumped storage schemes very well suited to being used to meet sudden surges in the demand for electricity?

13. Calculate the <u>overall</u> efficiency of the pumped storage scheme described on pages 377–378.

14. Match the graphs A–C with the following methods of generating electricity:

   - a wind farm;
   - an array of solar cells;
   - a tidal barrage.

   Give reasons for each of your answers.

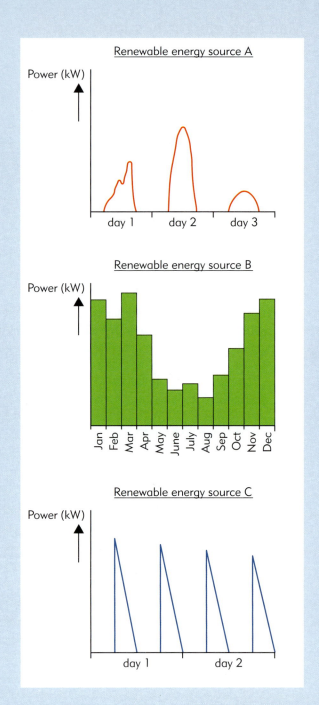

Renewable energy source A

Renewable energy source B

Renewable energy source C

Electricity in the home

Comparing conductors and insulators

Some materials let electricity pass through them very easily. A material like this is called a good **conductor**. You can find out which materials are good conductors. The diagram shows you how.

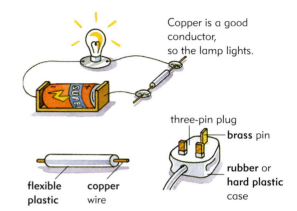

Copper is a good conductor, so the lamp lights.

three-pin plug — **brass** pin

rubber or hard plastic case

flexible plastic   copper wire

1   What happens if you test a good conductor?

The table shows the results of some tests.

Material	Does lamp light?
flexible plastic	no
copper	yes
hard plastic	no
rubber	no
brass	yes

2   Which materials in the table are good conductors?

If you tested your body in the same way, the lamp would not light.

3   What does this tell you about your body?

You can do the same tests with a sensitive electrical meter instead of a lamp. A good **insulator** does not let any electricity pass through it. So the meter stays at zero.

The table shows the results of some tests.

Material	Meter reading
flexible plastic	zero
body (hands dry)	small
body (hands wet)	larger
hard plastic	zero
rubber	zero

4   Which materials in the table are good insulators?

5   Is your body a good insulator?
       Answer as carefully as you can.

Higher

Why metals conduct electricity

The diagrams below explain how a metal such as copper can conduct electricity.

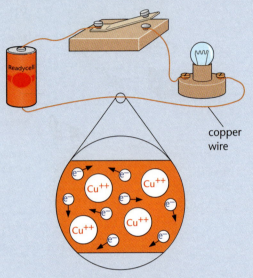

copper wire

Free electrons move about in all directions.

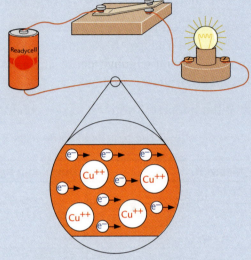

Free electrons drift in one direction.

Key

 copper atom that has lost electrons

e^- 'free' electron that can move anywhere in the metal

A piece of copper does not have an overall electrical charge.
The positive and negative charges <u>balance</u>. It is just that the
free electrons don't belong to any particular atom.

6 Explain <u>in words</u> why a metal like copper is a good
conductor of electricity.

Why can mains electricity kill?

The average voltage of the mains electricity in Europe is
230 volts. This is much higher than the voltage of a
battery. The 230-volt mains can send an electric current
through your body that is big enough to kill you.

7 Make a copy of the table below. Then use the
information from the diagram to complete the table.

Size of current	Effect on body
below 7 mA	
	can't let go; painful but won't kill you
	may kill you
over 30 mA	

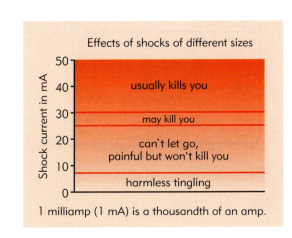

Effects of shocks of different sizes

usually kills you

may kill you

can't let go,
painful but won't kill you

harmless tingling

Shock current in mA

1 milliamp (1 mA) is a thousandth of an amp.

You get a shock if electricity flows through your body to the earth. The size of the shock depends on how easily electricity can pass into and out of your body.

8　Why are you less likely to get a shock that kills you:
(a) if your hands are dry?
(b) if you are wearing shoes or boots with rubber soles?

If you get a shock a current flows...

through your skin into your body...

then through your body...

then through your skin and shoes into the earth

The 230-volt mains can kill you.

Understanding the symbols

Electrical appliances have lots of letters and numbers printed on them. It's important for you to know what these letters and numbers mean. If you don't, you could easily damage the appliances or harm yourself.

What does the V mean?

MODEL 8940
SERIAL No. **B6KK0297**
VIDEO CASSETTE RECORDER
Made in Japan
220/240V ~ 50/60 Hz

Look at the diagram. It shows a base plate, which most electrical appliances have. The information tells you about the electricity supply that the equipment uses.
V stands for **volts**.

9　What voltage supply should the video cassette recorder use?

What does d.c. mean?

The diagram shows a battery pushing a current through a circuit. The electrons always flow in the same direction. This is called a **direct current**, or d.c. for short.

10　Look at the diagram. Describe what happens to the electrons after they leave the negative side (terminal) of the battery.

11　Imagine the battery is connected to the lamp the opposite way round. Draw a diagram to show which way the electrons move now.

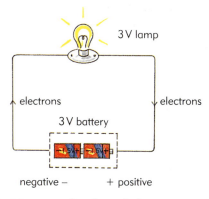

3 V lamp

electrons　　electrons

3 V battery

negative –　　+ positive

An electric current is a flow of electrons.

How can you see the direction of a current?

You can use an **oscilloscope** to draw a graph of the current. The pictures show the current from a battery.

12 Explain how the trace can show:
(a) whether there is a current flowing;
(b) if the direction of the current changes.

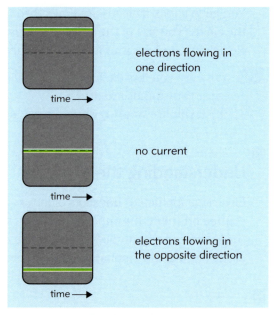

electrons flowing in one direction

time →

no current

time →

electrons flowing in the opposite direction

time →

Oscilloscope pictures of a direct current.

What does a.c. mean?

The circuit diagram shows the same lamp connected to a 3 V a.c. supply. The lamp lights up just as it did with the 3 V battery.

The picture shows how the a.c. supply pushes electrons around the circuit.

13 What does the picture tell you about the a.c. supply?

A current that keeps on changing its direction is called an **alternating current**, or a.c. for short.

14 How does the oscilloscope picture change when it is connected to a 6 V a.c. supply?

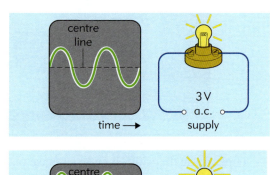

centre line

time →

3 V a.c. supply

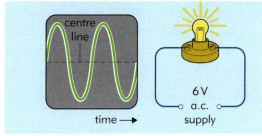

centre line

time →

6 V a.c. supply

What does Hz mean?

The picture shows the graph of an alternating current. A time scale has now been added.

15 How long does it take for the current to change direction and back again (for one cycle)?

16 How many cycles will there be in one whole second?

The number of cycles per second is called the **frequency**. In Europe, mains electricity has a frequency of 50 cycles per second or 50 **hertz** (Hz for short).

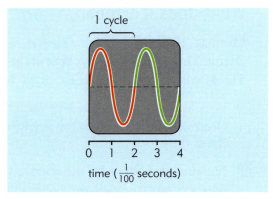

1 cycle

0   1   2   3   4
time ($\frac{1}{100}$ seconds)

The red line shows one complete cycle of the alternating current.

17 The diagrams show the information on the back of two electrical appliances. Explain, as fully as you can, what this information tells you.

| ∼ 220–240 V. 50 Hz | UK |
| 110 V a.c. 60 Hz. | USA |

∼ = a.c.

Why is mains electricity 230 volts a.c.?

The a.c. mains supply often has to be changed to d.c. before it can be used. Lamps and kettles can use a.c. or d.c. but radios and televisions will only work using d.c.

18 Write down the names of <u>two</u> appliances that need d.c. to work.

Also, electricity at 230 volts is high enough to kill. So why do we use an a.c. mains electricity supply of 230 volts?

Why mains electricity is a.c.

Different electrical appliances need different voltages to work properly. It is easy to change the voltage of a.c. using a **transformer**. Transformers do not work with d.c.

To increase the voltage of an a.c. supply you need a **step-up** transformer. To reduce the voltage of an a.c. supply you need a **step-down** transformer.

19 Look at the diagram.
What voltage is needed for:
(a) the personal stereo?
(b) the television?

20 Write down which type of transformer each of these appliances uses.

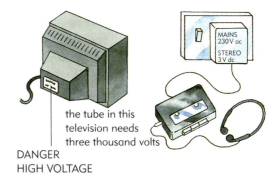

the tube in this television needs three thousand volts

DANGER
HIGH VOLTAGE

MAINS 230 V ac
STEREO 3 V dc

Higher

How transformers work

The diagram shows the main parts of a transformer. It also explains how an alternating current input to the transformer produces an output.

21 Explain how there can be an alternating current in the secondary coil of a transformer even though it is not directly connected to a power supply.

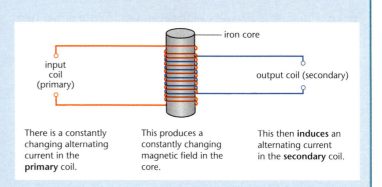

iron core

input coil (primary)

output coil (secondary)

There is a constantly changing alternating current in the **primary** coil.

This produces a constantly changing magnetic field in the core.

This then **induces** an alternating current in the **secondary** coil.

How can you produce a bigger effect?

When we move a magnet into, or out of, a coil, we produce an electric current. This happens because we have produced a voltage across the ends of the coil. We say that we have **induced** this voltage.

5 Look at the diagrams.
Write down <u>three</u> things you can do to induce a bigger voltage.

You can also increase the induced voltage by increasing the area of the coil of wire.

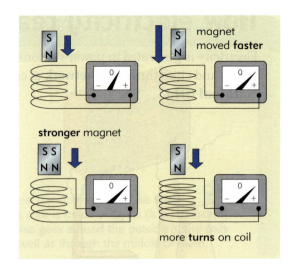

How can you keep on producing a current?

To keep on inducing a voltage and producing a current you can keep on moving a magnet into and out of a coil.

6 Look at the diagram.
Is the electric current produced by moving a magnet into and then out of a coil a.c. or d.c.?

How an a.c. generator works

A more convenient way to get an alternating current is to spin a magnet inside a coil. This is how the generator on a bicycle works.

7 Look at the diagram of a bicycle generator.
Then explain each of the following.
(a) The lights go off when the cyclist stops at some traffic lights.
(b) The lights are brighter when the cyclist is going downhill.

8 If the lamp is taken out of the circuit, it leaves a gap.
(a) Does the generator still produce a voltage?
(b) Explain why no current flows even though there is a voltage generated.

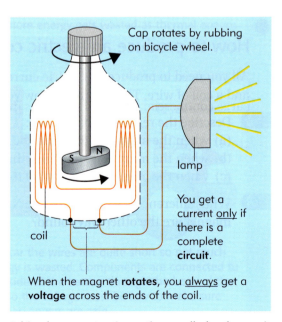

A bicycle generator (sometimes called a dynamo)

Generators in power stations

The generators in power stations are very large.
The rotating magnets need to be very powerful.
These generators use large **electromagnets**, which
work from a d.c. supply.

9 What type of rotating magnet is used in a
large generator?

More about generators

In a small generator used on a bicycle, a magnet
spins between coils of wire.

In the generator shown in the diagram, a coil of wire
rotates in a magnetic field.

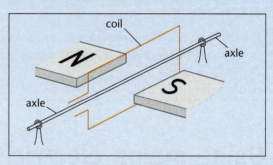

Rotating the coil rather than the magnet causes a
problem if you want to use the induced potential
difference. You can't just connect a circuit across the
ends of the coil because the wires will become
twisted. The diagram shows how this problem
is solved.

10 Describe in words how the potential difference
induced across the ends of a generator coil can be
applied to a circuit.

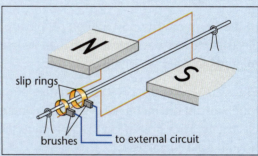

The brushes are made of graphite which
conducts electricity but is soft so it doesn't
wear away the slip rings.

Large commercial generators, such as those used
in power stations, usually have a rotating
electromagnet called a **rotor**. The alternating
current is induced in fixed coils called **stators**.

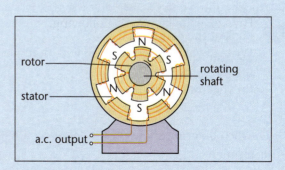

The graph shows the potential difference between the live wire and the neutral wire of the 230-volt mains supply during a small fraction of a second.

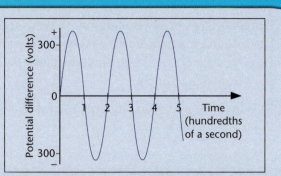

11 Describe how the potential difference between the live wire and the neutral wire varies with time.

The potential difference between the live wire and the earth wire varies in exactly the same way.

12 The graph shows the <u>peak</u> voltage of the mains supply to be about 320 volts.
So why is the mains a <u>230</u>-volt supply?

13 What can you say about the potential difference between the neutral wire of the mains supply and the earth wire?

Using mains electricity safely

In the UK, more than 50 million people use mains electricity every day. Fewer than 100 people are killed by mains electricity in Britain each year.

Plugs and cables are covered with insulators, so we can't touch any of the metal parts that are conducting electricity. They are safe to use as long as they are wired up and used correctly.

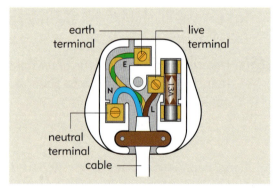

Wiring a three-pin plug

The diagram shows a correctly wired plug.

1 Look at the diagram.
Write down the colour of the wire that goes into:
(a) the earth terminal;
(b) the live terminal;
(c) the neutral terminal.

The next diagram shows how to connect a wire to its terminal.

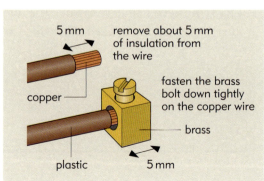

2 Why do you remove some of the plastic insulation from a wire before you connect it?

3 Why do you remove only about 5 mm of insulation from each wire?

Why does a plug have three pins?

The **live** and **neutral** pins of a plug carry the electric current to and from the mains supply.

The **earth** pin is there for safety. It is very important when using electrical equipment that has a metal case. It can help to stop you getting a shock if something goes wrong.

4   The diagram of the three-pin plug shows a cable that has only two wires inside it.
    Which terminals are these wires connected to?

5   The next diagram shows two light fittings.
    One is made of plastic and the other of metal.
    (a) What other difference can you see?
    (b) Why do you think there is this difference?

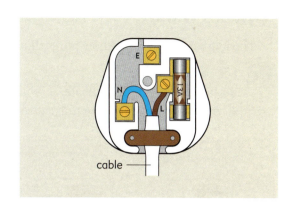

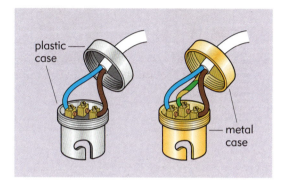

Why is the cable grip important?

The cable connected to a plug often gets pulled.
The **cable grip** means that the whole of the cable takes the strain. Without a cable grip, the copper wires take the strain instead. These are not so strong and can easily break.

6   Would an appliance work if:
    (a) the earth wire broke?
    (b) the live or neutral wire broke?

7   If the earth wire broke, would the appliance be safe to use?

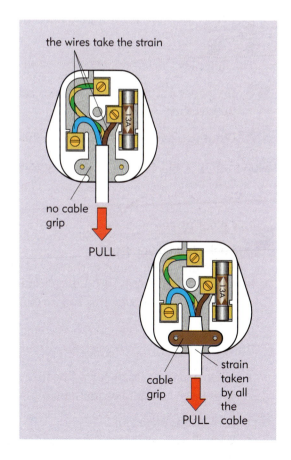

What's wrong?

The following diagrams show plugs that are unsafe.

8 For each diagram:
- say why the plug is unsafe;
- say what you can do to make it safe.

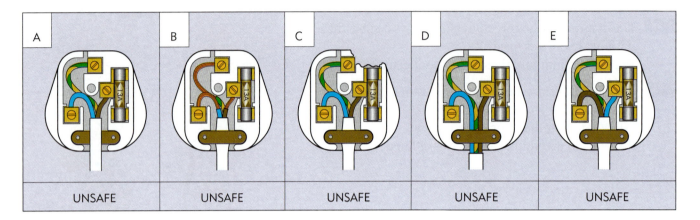

A	B	C	D	E
UNSAFE	UNSAFE	UNSAFE	UNSAFE	UNSAFE

Why do plugs have fuses?

Electricity won't pass through a plug unless it is fitted with a **fuse**.

The diagram shows you why.

9 How does a current get from one end of a fuse to the other?

10 What <u>two</u> parts of a plug does the fuse connect?

11 A plug won't supply a current unless a fuse is fitted. Why not?

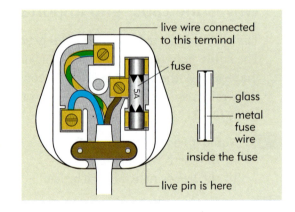

- live wire connected to this terminal
- fuse
- glass
- metal fuse wire inside the fuse
- live pin is here

What are fuses for?

Sometimes the current through a circuit becomes too big. The diagrams show why this can be dangerous.

Electrical appliances can also be damaged by a current that is too big.

The fuse makes a circuit safer. The fuse cuts off the current if the current becomes too big.

12 Look at the diagrams.
Write down <u>two</u> dangerous things that might happen if the current gets too large.

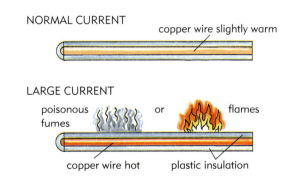

NORMAL CURRENT — copper wire slightly warm

LARGE CURRENT — poisonous fumes or flames, copper wire hot, plastic insulation

How does a fuse do its job?

The diagrams show how a fuse does its job.

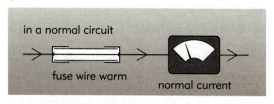

in a normal circuit

fuse wire warm normal current

13　Look at the diagrams.
　　What happens to the fuse wire when the current is:
　　(a) normal?
　　(b) much bigger than normal?

14　(a) What happens to the circuit if the fuse wire melts?
　　(b) What happens to the current?

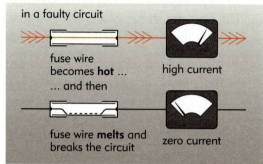

in a faulty circuit

fuse wire
becomes **hot** ...
... and then　high current

fuse wire **melts** and
breaks the circuit　zero current

What size of fuse should I use?

The picture shows the three most common types of fuse.

When you replace a fuse you should use the type
recommended by the makers of the appliance.
The correct type of fuse will already be fitted in the plug
when you buy an appliance. So you should replace a fuse
with one of the same type.

The information on the photograph tells you another way
of choosing the right fuse for an appliance.

A = amperes

A 3 A fuse will
melt if a current
of more than 3 A
passes through it.

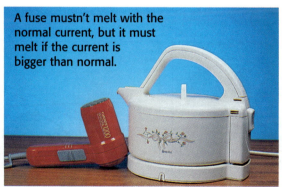

A fuse mustn't melt with the
normal current, but it must
melt if the current is
bigger than normal.

| normal current | 4 A | normal current | 11 A |
| fuse | ? | fuse | 13 A |

Choosing the right fuse.

15　Which type of fuse would you use for the hair dryer?
　　Say why the two other types of fuse would not be right.

Circuit breakers can also be used to do the same job
as fuses. You can read about them on page 406.

Making appliances with metal cases safe

If the **live** wire of the mains supply touches the metal case of an appliance, you could get a bad shock.

To make appliances with metal cases safe, we **earth** them. The diagram shows what then happens if the live wire touches the earthed metal case of an appliance.

16 Explain, in as much detail as you can, why earthing the metal case of a mains appliance makes it safer.

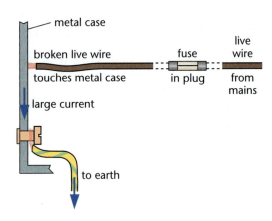

If a broken live wire touches an earthed metal case, a large current flows to earth.
This current flows through the fuse.
More thermal energy is transferred by the fuse so it becomes hotter and melts. This disconnects the live wire from the metal case.

Some 'do's' and 'don'ts' that could save your life

You must have plugged things into the mains thousands of times.
We use mains electricity a lot. So it's very easy to get careless about it.
This is dangerous. You can be killed if you use mains electricity carelessly.

17 An electricity company is making some safety posters for junior schools.
The artist has already drawn the posters. She needs some words to finish them off.
What words would you add to each of the posters below, and on the next page?

18 Think of <u>another</u> thing that it is dangerous to do with mains electricity.
Make a safety poster about it.
Do a drawing, then add the words.

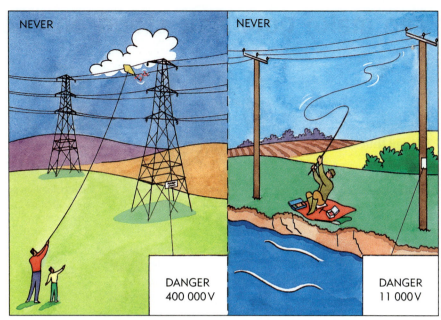

NEVER

NEVER

DANGER
400 000 V

DANGER
11 000 V

NEVER

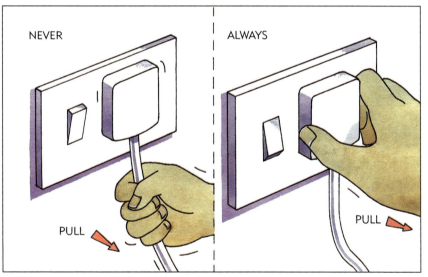

NEVER

PULL

ALWAYS

PULL

NEVER

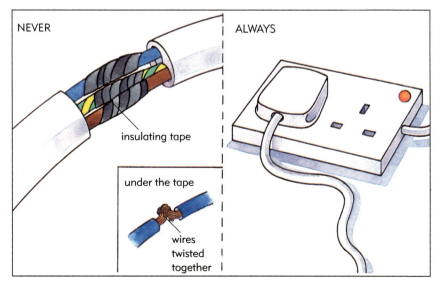

NEVER

insulating tape

under the tape

wires
twisted
together

ALWAYS

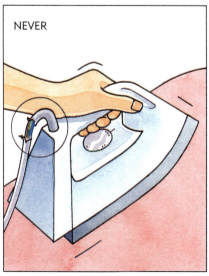

NEVER

Electricity that can make your hair stand on end

Electricity doesn't always flow through circuits.
Electricity can also stay just where it is.
This is called **static** electricity.

You can produce static electricity by rubbing together two different materials. We say that we have **charged** the materials with electricity.

Combing your hair produces static electricity.

Your hairs then push each other away. They repel each other.

1 Combing your hair can produce static electricity.
   (a) Why does this happen?
   (b) Write down <u>two</u> ways in which you can tell when it happens.

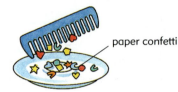

paper confetti

Your comb will also then attract small bits of dust or paper.

Is all static electricity the same?

You can rub two strips of plastic with a cloth.
This charges them with static electricity. The diagrams show what the charged strips of plastic will then do.

2 Look at the diagrams.
   (a) Which types of charged strip attract each other?
   (b) Which types of charged strip repel each other?
   (c) What happens if either strip is near an uncharged, lightweight object?

Charged objects sometimes attract and sometimes repel.

This means that there must be <u>different types</u> of electrical charge.

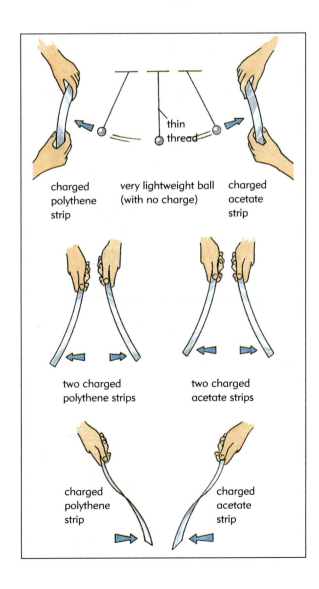

charged polythene strip

very lightweight ball (with no charge)

charged acetate strip

thin thread

two charged polythene strips

two charged acetate strips

charged polythene strip

charged acetate strip

Two types of charge

Two polythene strips that are rubbed with the same cloth must have the same kind of charge. These charges repel each other.

Two acetate strips rubbed with the same cloth must have the same kind of charge. These charges also repel each other.

3 Look at the diagrams.
How can you tell that the acetate strips have the same charge?

Here is a simple way to remember what happens:

> **like** charges **repel**; **unlike** charges **attract**.

The charge you get on a polythene strip when you rub it is called **negative** (−).

The charge you get on an acetate strip when you rub it is called **positive** (+).

4 How can you tell from the diagram that the acetate strip has an opposite charge to the polythene strip?

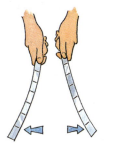
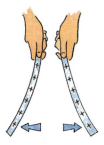

charged polythene strips

charged acetate strips

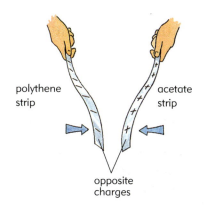

polythene strip

acetate strip

opposite charges

Really making hair stand on end

You can use an electrostatic machine to make hair really stand on end.

5 Explain how this happens, as fully as you can.

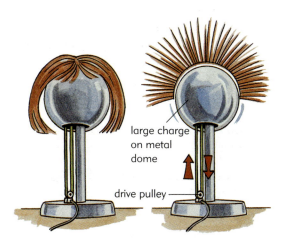

large charge on metal dome

drive pulley

Electrostatic generator. The belt becomes charged as it rubs against the drive pulley. The charge on the belt transfers to the dome.

Why rubbing things together produces electricity

When you rub two different materials together, **electrons** are rubbed off one material and on to the other. These electrons are very tiny. Each electron carries a small electrical charge.

If you rub two pieces of the <u>same</u> material together, electrons don't move.

6 When you rub two different materials together they become charged. Why?

7 What type of electrical charge does each electron carry?

The materials that you rub together must not only be <u>different</u>. At least one of them, and usually both of them, must be an electrical insulator.

Why rubbing polythene gives it a negative charge

The diagrams show what happens when you rub polythene with a cloth.

8 (a) Does the cloth or the polythene gain electrons?
(b) Did the electrons move away from the cloth or the polythene when they were rubbed together?

9 Explain the following, as fully as you can.
(a) Why the polythene ends up with a negative charge.
(b) Why the cloth ends up with a positive charge.

10 The negative charge on the polythene is exactly equal to the positive charge on the cloth. Why is this?

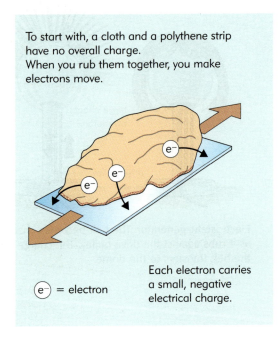

To start with, a cloth and a polythene strip have no overall charge.
When you rub them together, you make electrons move.

e⁻ = electron

Each electron carries a small, negative electrical charge.

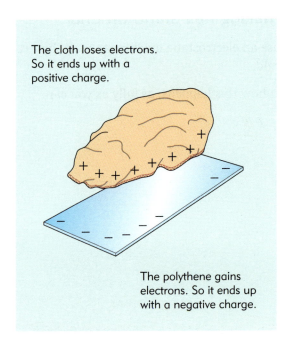

The cloth loses electrons. So it ends up with a positive charge.

The polythene gains electrons. So it ends up with a negative charge.

Why rubbing acetate gives it a positive charge

When you rub a piece of acetate with a cloth, the acetate becomes positively charged.

11 (a) What type of charge do you get on a cloth when you rub a piece of acetate with it?
(b) What happens to electrons as you rub the acetate with a cloth?

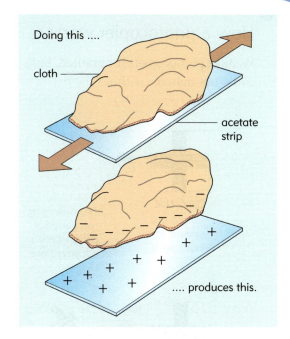

Doing this ....

cloth

acetate strip

.... produces this.

Peeling things apart

Peeling two different materials apart has the same effect as rubbing them together.

Electrons are transferred from one material to the other.

12 Look at the diagram.
Explain which part of the book has:
(a) lost electrons;
(b) gained electrons.

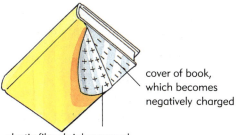

cover of book, which becomes negatively charged

plastic film shrink-wrapped around book, which becomes positively charged

13 When you peel some sticky tape from a roll it often 'tries' to go back on again.
Explain, as fully as you can, why it does this.

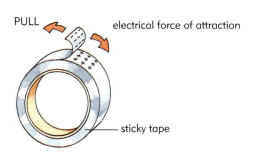

PULL

electrical force of attraction

sticky tape

Making use of static electricity

We can use static electricity in many different ways to do useful jobs.

14 Write down <u>two</u> different ways that static electricity is often used in schools and offices.

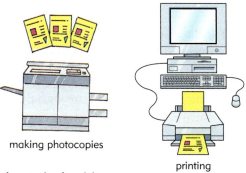

making photocopies

printing

Using static electricity.

How a photocopier works

We take photocopiers for granted, but schools and offices could not manage without them.

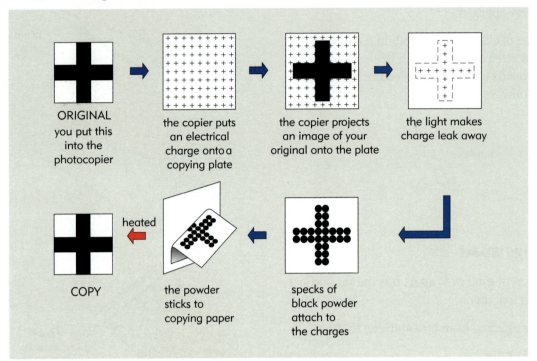

ORIGINAL
you put this into the photocopier

the copier puts an electrical charge onto a copying plate

the copier projects an image of your original onto the plate

the light makes charge leak away

heated

COPY

the powder sticks to copying paper

specks of black powder attach to the charges

There is a copying plate inside the photocopier, which can be charged with electricity. The copying plate loses its charge when light shines on it. Photocopiers use a black powder called **toner**. The toner sticks to charged parts of the plate, and to paper. When the paper is heated, the toner makes a permanent copy.

15 Why is toner made from specks of powder?

How to remove the dirt from smoke

Smoke contains lots of tiny bits of dirt. This dirt falls on houses and gardens. If people breathe in the dirt, it can damage their lungs.

The diagram shows how the dirt can be removed from factory chimneys using a **smoke precipitator**.

The metal antenna and metal grid are positively charged. As smoke particles pass through the grid, they pick up a positive charge too. The metal chimney lining has a negative charge.

16 Explain why the smoke particles are attracted to the metal chimney lining.

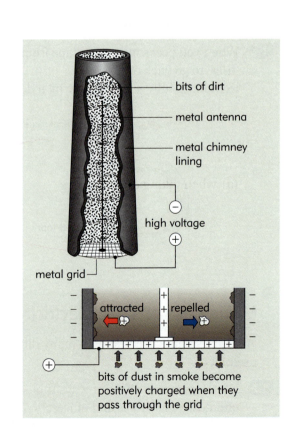

bits of dirt

metal antenna

metal chimney lining

high voltage

metal grid

attracted

repelled

bits of dust in smoke become positively charged when they pass through the grid

How to make a spray hit its target

When a liquid is sprayed out of a nozzle, it becomes electrically charged. Sprays can be made so they charge the droplets as much as possible. The diagrams show how this helps the spray to find its target.

17 Car makers want most of the paint from a spray to end up on a car body. What can they do to the car body to make sure this happens?

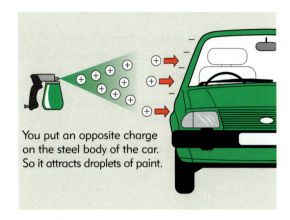

You put an opposite charge on the steel body of the car. So it attracts droplets of paint.

18 Explain why a pesticide spray works better if it charges the droplets of pesticide.

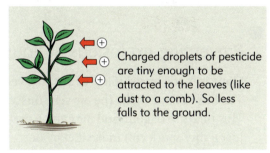

Charged droplets of pesticide are tiny enough to be attracted to the leaves (like dust to a comb). So less falls to the ground.

Danger from sparks

Static electricity can produce sparks. Often these sparks are very small. But sometimes they are huge.

Large sparks such as lightning can kill people. But even quite small sparks can be very dangerous.

19 How is the spark caused:
(a) when taking off a jumper?
(b) from clouds?
(c) from a TV screen?

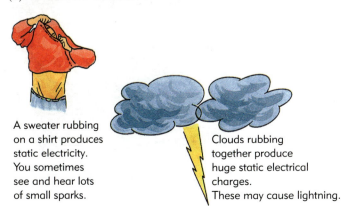

A sweater rubbing on a shirt produces static electricity. You sometimes see and hear lots of small sparks.

Clouds rubbing together produce huge static electrical charges. These may cause lightning.

In a TV tube electrons are fired at the screen. The screen may become charged.

Small sparks may then jump from the screen onto your hand.

Fire hazards with petrol tankers

Petrol is very flammable.
A very small spark can cause a serious fire when petrol or petrol vapour is open to the air.

At filling stations, petrol is transferred from tankers to underground tanks. There must be no sparks while this is being done.

20 In what two ways can petrol tankers produce static electricity?

21 Write down two things that are done to prevent sparks.

A static charge can build up on a petrol tanker because of the tyres rubbing on the road.

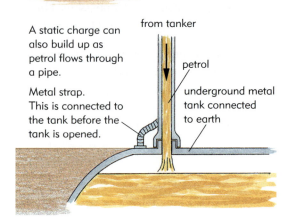

A static charge can also build up as petrol flows through a pipe.

Metal strap. This is connected to the tank before the tank is opened.

from tanker

petrol

underground metal tank connected to earth

Preventing explosions at flour mills

Flour is made by grinding wheat to a very fine powder.
The bits of flour make the air very dusty.
This mixture of flour and air can be very dangerous.
Just a tiny spark can make it explode.

Look at the diagram.

22 There is a danger of a spark between the pipe and the container. Explain why.

23 How can such a spark be avoided?

metal pipe   flour

metal vat

metal strap

conductor to earth

Friction between flour and metal pipe produces static charges.

Solving the problem.

Handle chips with care

People who put electronic chips into circuits have special mats on their workbenches. These mats conduct electricity and are connected to earth. The people sometimes also wear wrist straps connected to earth.

24 Explain, as fully as you can, why these precautions are needed.

An electronic chip is easily damaged by static electricity.

Why does earthing work?

To earth something, you must connect it to the earth with a **conductor**. The conductor allows electrons to flow through it. This is a current of electricity.
The diagram shows how earthing works.

25 Look at the diagrams.
When an object is earthed, which way do electrons flow:
(a) if it is negatively charged?
(b) if it is positively charged?

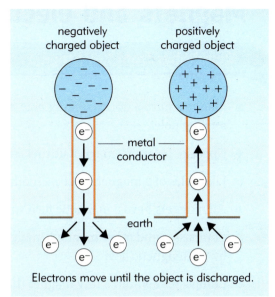

negatively charged object   positively charged object

metal conductor

earth

Electrons move until the object is discharged.

Higher

Potential differences in static electricity

Batteries and power supplies can supply an electric current because they have a potential difference across their terminals.

You also get a potential difference between a charged object and the earth. The graph shows how this potential difference varies with the size of the charge on the object.

26 Describe the relationship between the size of the electric charge on an object and the potential difference between the object and the earth.
[Make use of all the information from the graph.]

The diagrams show what can happen if the potential difference between an object and the earth becomes big enough.

27 Explain why an electrically charged body can create a spark.
[Make use of all the information from the diagrams.]

Potential difference (volts)

Charge (coulombs)

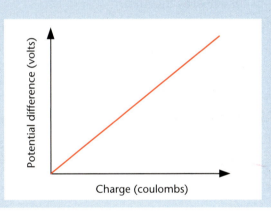

copper wire

spark

electric charge flows to earth

earth

The air between the copper wire and the sphere is normally a very good insulator.

If there is a very high voltage, air will conduct. This causes a spark.

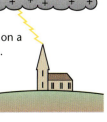

The same thing happens on a large scale with lightning.

In this case, electrons flow from the earth to neutralise the charge in the cloud.

Magnets and electromagnets

Remember from Key Stage 3

- Magnets have a north-seeking pole and a south-seeking pole.

- Opposite poles of a magnet attract each other.

- Like poles of a magnet repel each other.

- All magnets have a magnetic field surrounding them.

- Compasses and other magnetic materials are affected by the magnetic field.

- Compasses contain a tiny magnet that is free to spin. The north end of the magnet is a pointer.
  The pointer points north unless there is a magnet nearby. Otherwise the pointer points to the south pole of a nearby magnet.

- Electromagnets can be switched on or off.

- You can make one by passing an electric current through a coil of wire. The coil has a north-seeking pole at one end and a south-seeking pole at the other end.

- Changing the current's direction changes the direction of the magnetic field.

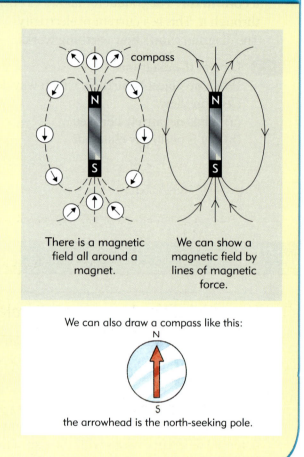

There is a magnetic field all around a magnet.

We can show a magnetic field by lines of magnetic force.

We can also draw a compass like this:

the arrowhead is the north-seeking pole.

How to make a magnet move a copper wire

A magnet only attracts certain metals. It has no effect on things made from copper. But it <u>does</u> have an effect on a copper wire if an electric current is flowing through it. The diagram shows this effect.

1. Which way does the copper roll when the current is flowing?

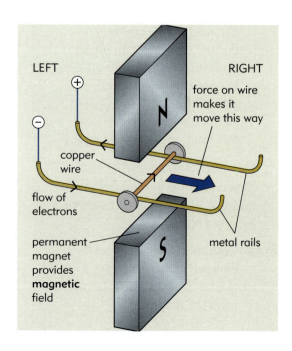

What happens if you reverse the current?

You can reverse the current by swapping the + and – connections to the power supply.
The diagram shows what happens then.

2 (a) Which way will the wire roll?
(b) What happens to the forces on the wire when the current is reversed?

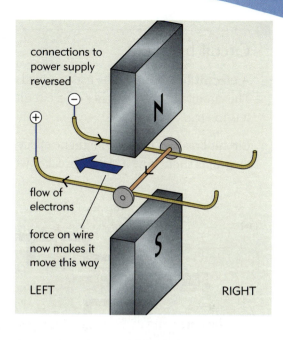

connections to power supply reversed

flow of electrons

force on wire now makes it move this way

LEFT   RIGHT

What happens if you change the poles of the magnet round?

The next diagram shows what happens if you swap round the poles of the magnet.

3 What is the effect of swapping the poles of the magnet?

4 What do you think will happen to the direction of the force on the wire if you reverse the current and also swap round the poles of the magnet?

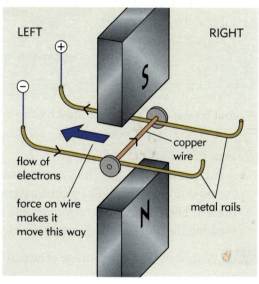

LEFT   RIGHT

copper wire

flow of electrons

force on wire makes it move this way

metal rails

How can you increase the force on the wire?

The diagrams below show two ways of increasing the size of the force that acts on the wire.

5 What are these <u>two</u> ways?

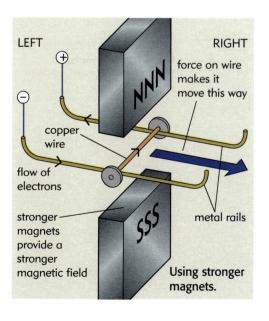

LEFT   RIGHT

copper wire

force on wire makes it move this way

flow of electrons

stronger magnets provide a stronger magnetic field

metal rails

Using stronger magnets.

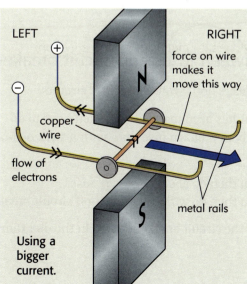

LEFT   RIGHT

copper wire

force on wire makes it move this way

flow of electrons

metal rails

Using a bigger current.

Circuit breakers

A **circuit breaker** can be used instead of a fuse.
The circuit breaker switches off a circuit if the current flowing through
it is too big.

Circuit breakers use the magnetic effect of an electric current. The
diagrams show how a circuit breaker works.

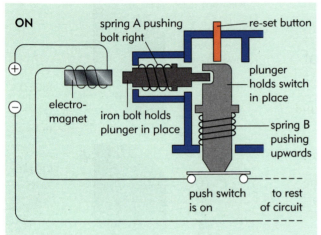

A circuit breaker with a normal current passing through.

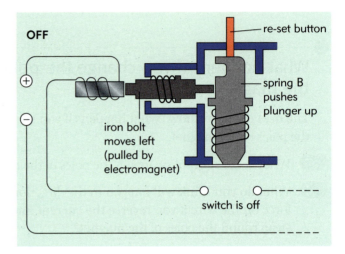

This is what happens if the current passing through the
circuit breaker is too big.

A current that is bigger than normal makes the electromagnet
stronger than normal. The electromagnet pulls the iron bolt more to
the left. The plunger can now move up. The push switch goes to 'off'.
Pushing the plunger down resets the circuit breaker.

6 Write down <u>one</u> advantage of circuit breakers.

A difference between fuses and circuit breakers

Fuses and circuit breakers both switch off faulty circuits.

Once the faulty circuit has been mended, you then have to do
something different about a blown fuse than with a switched-off
circuit breaker.

7 (a) What must you then do about a 'blown' fuse?
  (b) What must you then do with a switched-off circuit breaker?

8 How do you re-set the circuit breaker shown in the diagrams?

Electric motors

An electric motor uses electricity to make things move.
Many of the electrical appliances that we use in our homes
have an electric motor inside them.

9 Write down the names of <u>eight</u> household appliances
that have an electric motor inside them.

10 Look at the diagram.
(a) What can you say about the direction of the
current in the top and the bottom of the coil?
(b) What can you say about the direction of the force
felt by the top and the bottom of the coil?
(c) Describe how the coil moves when a current
is flowing.

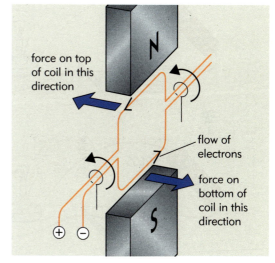

The forces on the top and bottom of the coil
are in opposite directions, so the coil turns ↻
(anticlockwise).

Making a more powerful motor

A motor with only <u>one</u> coil is very weak.
To make it stronger you need to have <u>lots</u> of coils of wire.

11 Write down at least <u>two</u> other ways of making the
motor more powerful.

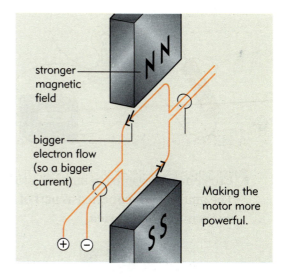

Making the
motor more
powerful.

Circuits

- A current will flow if there is a complete circuit <u>and</u> a voltage across the circuit.

- Components are things like lamps and buzzers that can be connected in the circuit.

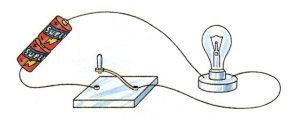

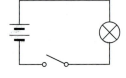

Switch open (off).
No voltage across lamp.
No current through lamp.
Lamp doesn't light.

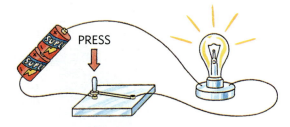
PRESS

Switch closed (on).
A voltage across lamp.
A current through lamp.
Lamp lights.

flow of electrons

- **Voltage** is measured in volts using a **voltmeter**. The voltmeter is connected across components – making an extra loop in the circuit. We say it is connected **in parallel**.

2.4 V

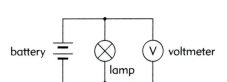
battery   lamp   voltmeter

- **Current** is measured in amperes using an **ammeter**. The ammeter is connected as part of the circuit. We say it is connected **in series**.

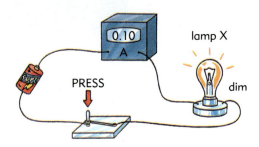
0.10 A   lamp X   PRESS   dim

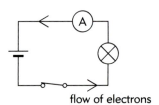
flow of electrons

- **Resistance** measures how much voltage is needed to push the current through a component. A bigger resistance means that a bigger voltage is needed for the same current.

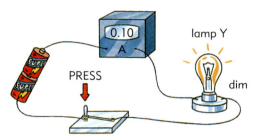

lamp Y

PRESS

dim

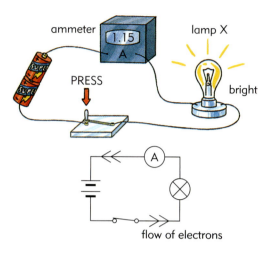

ammeter

lamp X

PRESS

bright

flow of electrons

1 Which lamp has the bigger resistance, lamp X or lamp Y? Explain why.

Circuit diagrams

You can show an electric circuit by drawing a <u>picture</u> of what it looks like. But it is a lot easier, and much more convenient, to draw **symbols** for each component in a circuit. For example:

instead of drawing a lamp

you can draw this symbol

2 Look at the diagram.
Write down the correct symbol for:
(a) a voltmeter
(b) an ammeter
(c) a lamp
(d) a switch (open)
(e) a switch (closed)
(f) a battery
(g) a fuse

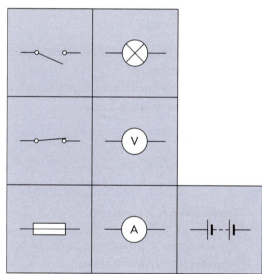

Drawing and interpreting circuit diagrams

You should be able to draw a circuit diagram from looking at a circuit.

3 Draw the circuit diagram for each of these <u>two</u> circuits.

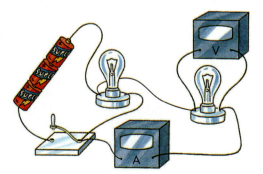

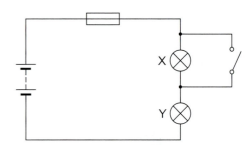

You should be able to understand, from a circuit diagram, how components are connected.

4 Look at the diagram.
   (a) What are components X and Y?
   (b) What is connected across X?
   (c) What other components are included in the circuit?

Energy transfer in circuits

When a lamp lights up, it is transferring energy to its surroundings.

The diagram shows how the lamp is able to do this.

5 What is actually moving round a circuit when a current flows?

6 (a) What happens as an electric current flows through a resistance?
   (b) Where does the energy that is transferred by a lamp filament come from?

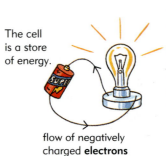

The cell is a store of energy.

The lamp filament has a resistance. So when a current is pushed through it gets hot. Energy is transferred to the surroundings as **heat** (thermal energy) and light.

flow of negatively charged **electrons**

Transferring energy

The wattage, or **power**, of an electrical appliance tells you how fast it transfers energy.
Power is measured in **watts** (W for short).
A 60-watt lamp transfers 60 joules of energy a second.

7 How fast does a 100-watt lamp transfer energy?

Electrical push is called voltage.
Voltage is also called potential difference.

You can work out the power of an electrical appliance like this:

power = potential difference (p.d.) × current
(watts)   (volts)   (amperes)

8 Look at the diagrams of electrical appliances.
   In each case work out the missing figure.

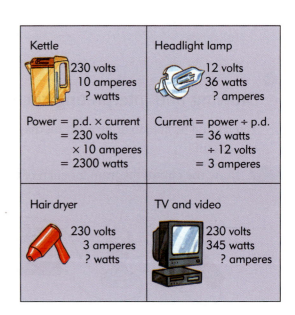

Kettle
230 volts
10 amperes
? watts

Power = p.d. × current
= 230 volts
× 10 amperes
= 2300 watts

Headlight lamp
12 volts
36 watts
? amperes

Current = power ÷ p.d.
= 36 watts
÷ 12 volts
= 3 amperes

Hair dryer
230 volts
3 amperes
? watts

TV and video
230 volts
345 watts
? amperes

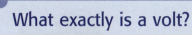

What exactly is a volt?

An electric current is a flow of **charge**. When the charge flows through a resistance, energy is transferred.

The voltage or potential difference across a resistance is a measure of the energy that is transferred when a charge flows through it. If the flow of 1 coulomb of charge transfers 1 joule of energy, then the potential difference is 1 volt.

$$\begin{array}{c}\text{energy}\\(\text{joules, J})\end{array} = \begin{array}{c}\text{potential difference}\\(\text{volts, V})\end{array} \times \begin{array}{c}\text{charge}\\(\text{coulombs, C})\end{array}$$

But energy transferred = p.d. × current × time.
This is because charge can also be written as current × time.

So we can write that energy transferred = p.d. × current × time.

However, $\dfrac{\text{energy transferred}}{\text{time}}$ = power, so

$$\begin{array}{c}\text{power}\\(\text{watts, W})\end{array} = \begin{array}{c}\text{p.d.}\\(\text{volts, V})\end{array} \times \begin{array}{c}\text{current}\\(\text{amperes, A})\end{array}$$

Example 1
The potential difference between a charged object and the earth is 25 000 V. When the object is earthed, one millionth of a coulomb of charge flows. How much energy is transferred?

$$\begin{aligned}\text{Energy transferred} &= \text{p.d.} \times \text{charge}\\ &= 25\,000 \times 0.000001\\ &= \underline{0.025\,\text{J}}\end{aligned}$$

Example 2
When 1000 C of charge flows, 400 kJ of energy are transferred. What is the potential difference that made the charge flow?

Energy transferred = p.d. × charge

$$\begin{aligned}\text{So p.d.} &= \frac{\text{energy transferred}}{\text{charge}}\\ &= \frac{400\,000}{1000}\\ &= \underline{400\,\text{V}}\end{aligned}$$

---

9 Calculate the following. [In each case start with the formula you are using. Show all your working.]
   (a) Calculate the energy transferred when a potential difference of 12 V is used to make 500 C of charge flow.
   (b) Calculate the potential difference used if the flow of 240 C of charge transfers 9600 J of energy.

10 Explain why the power of an electrical appliance (in watts) is given by the formula:

$$\begin{array}{c}\text{p.d.}\\(\text{volts})\end{array} \times \begin{array}{c}\text{current}\\(\text{amperes})\end{array}$$

Cells and batteries

Many things need electricity to work. Some of these things plug in to the mains electricity. The diagram shows some things that work from 'batteries'.

11 What is the proper name for what we often call a battery?

12 Different things need different numbers of cells to work properly.
How many cells are needed by each of the things shown in the diagram?

For a cycle without a dynamo, the cells in a cycle lamp provide the push that forces an electric current through the lamp. The lamp needs two cells connected in series (in line) to provide enough push to make it light properly.

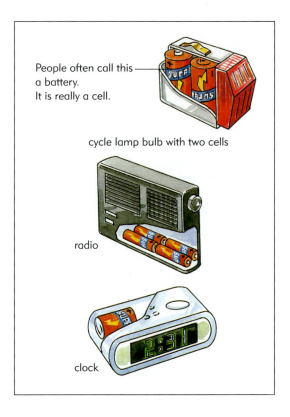

People often call this a battery.
It is really a cell.

cycle lamp bulb with two cells

radio

clock

The lamp lights normally.

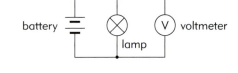

battery   lamp   voltmeter

13 To make the cycle lamp light normally, you need two cells.

What voltage do these two cells provide?

14 The diagram shows two cells connected the opposite way round.
What voltage do these two cells give? Give a reason for your answer.

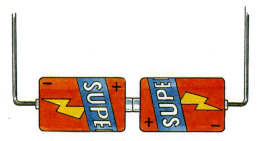

These cells are pushing in opposite directions. Their voltages cancel out.

Cycle lamp with one cell.

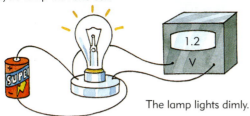

The lamp lights dimly.

Cycle lamp with three cells.

The lamp lights very brightly but does not last long.

15 Look at the <u>three</u> cycle-lamp circuits.
What is the voltage provided by:
(a) 1 cell?
(b) 2 cells?
(c) 3 cells?

16 What pattern do you see in the voltages when you connect cells together?

17 What voltages do the cycle lamp, radio and clock from question 11 need to work properly?

What is a battery?

A battery is made up of two or more cells joined together. When you connect cells in series you can add their voltages to get the total.

18 What voltage does each cell in the car battery provide?

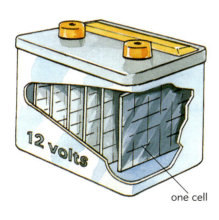

one cell

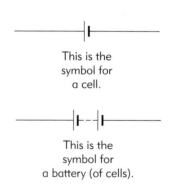

This is the symbol for a cell.

This is the symbol for a battery (of cells).

Higher

Transmitting mains electricity efficiently

You can get a higher <u>voltage</u> out of a transformer than you put into it. But you can't get more <u>energy</u> out of it than you put in.

The rate at which energy is transferred to and from the transformer is given by:

$$\frac{\text{power}}{\text{(watts, W)}} = \frac{\text{p.d.}}{\text{(volts, V)}} \times \frac{\text{current}}{\text{(amperes, A)}}$$

The power can't become bigger, so if the voltage <u>increases</u> 100 times for example, the current must <u>decrease</u> 100 times.

19 At a power station the voltage is increased from 25 000 to 400 000 V. What happens to the current?

The smaller the current the less the power that is wasted in the cables. The reduction in the wasted power is <u>more than</u> in proportion to the reduction in the current (see Example 1).

20 How many times smaller are the power losses in transmission cables when the voltage is increased from 25 000 to 400 000 V?

Another advantage of using a high voltage to transmit electricity is that there is a much smaller drop in voltage along the power line (see Example 2).

21 (a) Calculate the available p.d.:
    (i) if a p.d. of 230 V is used to send a current of 100 A through power lines with a resistance of 0.5 Ω;
    (ii) if a p.d. of 23 000 V is used to send a current of 1 A (i.e. to send energy at the same rate) through the same power lines.
  (b) Calculate the percentage of the original p.d. that is still available in (i) and (ii).

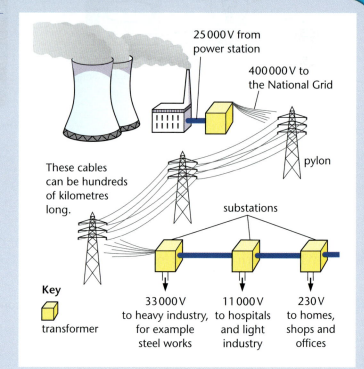

25 000 V from power station

400 000 V to the National Grid

pylon

These cables can be hundreds of kilometres long.

substations

Key

transformer

33 000 V to heavy industry, for example steel works

11 000 V to hospitals and light industry

230 V to homes, shops and offices

Example 1
If you step <u>up</u> the voltage 10 times, you step <u>down</u> the current 10 times.
This reduces the power loss 10 × 10 = <u>100 times</u>.
In other words, by stepping up the voltage 10 times, you reduce the power loss to 1 % of what it was before.

Example 2
A p.d. of 12 V is used to send a current of 3 A along wires whose total resistance is 0.1 Ω.
The voltage drop along the cables is given by:

  p.d. = current × resistance
       = 3 × 0.1
       = <u>0.3 V</u>

So the <u>available</u> p.d. is 12.0 − 0.3 = 11.7 V

This is $\dfrac{11.7}{12} \times 100\%$

i.e. <u>97.5% of the original p.d.</u>

Higher

Even if we use a very high voltage to transmit electricity, there will still be <u>some</u> power loss in the cables. There are also power losses in the transformers that are used to step down the voltage at the consumer's end. These transformers are in electricity **substations**. Each substation supplies 230 V mains electricity to many homes. Large transformers, however, are very efficient and waste very little energy.

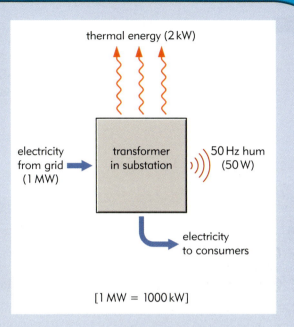

thermal energy (2 kW)

electricity from grid (1 MW) → transformer in substation → 50 Hz hum (50 W)

electricity to consumers

[1 MW = 1000 kW]

22 (a) In what <u>two</u> ways does a transformer wastefully transfer energy?
　(b) Calculate the efficiency of the transformer shown in the diagram.

23 The mains supply in your home is 230 V. Suppose it was transmitted at this voltage.
　(a) How much bigger would the current be in the transmission cables?
　(b) How many times greater would the power losses be?

Different types of circuit

A lamp lights brightly when you connect it as shown in the diagram below.

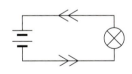

In all of the diagrams

——→

shows the flow of electrons.

You can connect another lamp to the same battery so that both lamps light brightly. This diagram shows how you can do this.

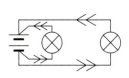

The full voltage is applied across each lamp. A current flows separately through each lamp. We say the lamps are connected <u>in parallel</u>.

1. Are the lamps connected in series or in parallel?

2. Why do both bulbs light brightly?

3. Suppose one lamp breaks or you take it out.
   How will this affect the other lamp?

The two lamps in the diagram are connected to the two cells by separate wires. But this isn't the only way to connect things in parallel.

The lamps in these circuits are also connected in parallel.

current to both lamps through the same wire

current to just one lamp separately

Connecting things in parallel

Connecting wires let electricity flow through them very easily. They have hardly any resistance.

Look at the two circuits. Each lamp is connected directly to both sides of the battery. In each circuit the two lamps are connected in parallel. The current to both lamps flows from the battery through the same wire. Then the current splits and flows through each lamp separately. Then the current joins back up and flows back to the battery through the same wire.

When lamps are connected in parallel there is exactly the same voltage across each lamp. So if one lamp has a smaller resistance, a bigger current will flow through it.

4. Look at the diagram.
   (a) What is the current through lamp X?
   (b) What is the current through lamp Y?
   (c) Which lamp has the bigger resistance?
   Give a reason for your answer.

5. (a) What is the <u>total</u> current supplied by the battery to the two lamps? This is shown by meter A_1.
   (b) How does this current compare to the <u>separate</u> currents through lamp X and lamp Y?

In a parallel circuit, the total **current** from the supply is the same as the currents through the separate branches added together. It is the **sum** of these separate currents.

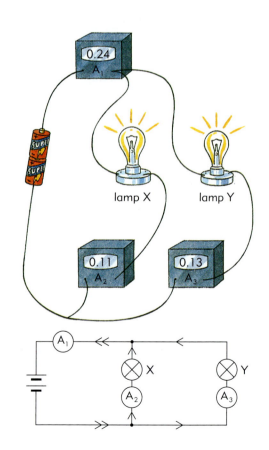

lamp X   lamp Y

Safety when connecting things in parallel

When we plug things into the mains supply we are connecting them in parallel. But it isn't safe to take more than 13 A of current from one mains socket. So we need to be very careful about what we plug into it.

6 Use the table to help you. Would it be safe to connect these appliances in parallel?
(a) room heater and large TV
(b) lamp and CD player
(c) small TV and hair dryer

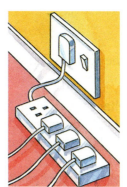

Parallel connections to the mains.

How much current?	
room heater	12 A
hair dryer	3 A
large TV	2 A
small TV	1 A
CD player	1 A
lamp	0.5 A

Different ways of connecting circuits

In our homes, lights, power sockets and appliances are usually connected in parallel.

7 Write down two reasons for connecting things to the mains supply in parallel.

In the diagram the switch for each lamp is connected differently. It is connected in series with its lamp.

8 The diagram shows an important component connected in series with all of the lamps and switches in the circuit. What is this component?

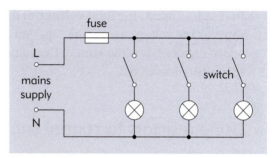

The lights in a house are connected in parallel. This means:
• they all get the full 230 volts
• they can all be switched on and off separately.

Connecting things in series

The lamp in the first diagram is shining brightly. The second diagram shows what happens when you connect another lamp in series with it.

9 What can you say about the current flowing in different parts of a series circuit?

10 (a) What happens to the brightness of the lamps when the second lamp is connected?
(b) Explain, as fully as you can, why this happens.

When lamps are connected in series their total **resistance** is the same as their separate resistances added together. It is the **sum** of their separate resistances.

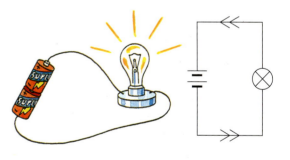

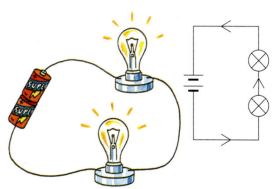

The same current flows through one lamp and then through the other. So we say they are connected in series. The two lamps are dimmer than with one lamp by itself. So we know that:
• the current is smaller
• the resistance of the two lamps is bigger than the resistance of one lamp

The diagram shows the voltages across the cells and the lamps in a series circuit.

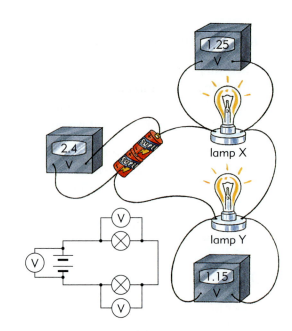

11 (a) What is the voltage across lamp X?
    (b) What is the voltage across lamp Y?
    (c) Which lamp needs a bigger voltage to push the same current through?
    (d) Which lamp has the higher resistance?

12 (a) What is the voltage across the two cells?
    (b) How does this voltage compare with the separate voltages across the lamps?

In a series circuit, the total **voltage** of the supply is the same as the voltage across the separate components added together. It is the **sum** of the voltages across the separate components.

Using resistors to control currents

We can put a resistor in series with a component. This increases the total resistance of the circuit and so reduces the current through the component. The diagram shows an example.

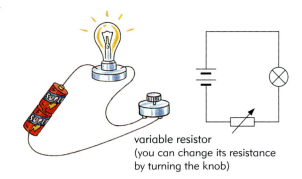

variable resistor
(you can change its resistance by turning the knob)

13 The lamp starts to glow when there is 0.8 V across it. This happens when there is 1.7 V across the variable resistor. What is the voltage across the two cells?

Resistance

The current through a lamp depends on the voltage you apply across it. It is the same for any other electrical component.

You can use the circuit shown in the diagram to find out how current changes with voltage.

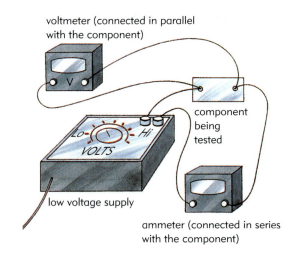

voltmeter (connected in parallel with the component)

component being tested

low voltage supply

ammeter (connected in series with the component)

Current–voltage graph for a resistor

Different components give different results. You can show these results on a **current–voltage graph**.

The graph shows how the current through a resistor changes when you change the voltage across it. The graph is like this only if the resistor does not get hot. So you must keep the current through the resistor quite small.

1 Describe, as carefully as you can, how the current through the resistor changes as you increase the voltage.

The resistance of the resistor stays the same provided its temperature does not change.

It sometimes matters which way round you connect a component to the supply. You can check this by connecting it the other way round.

2 Look at this graph.
   (a) What changes when you connect the resistor the opposite way round?
   (b) What stays the same?

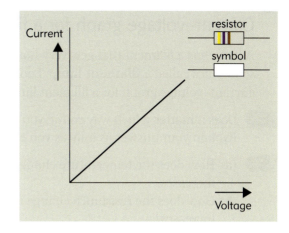

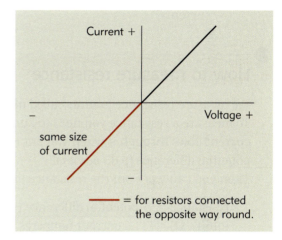

Current–voltage graph for a diode

Some circuits use a component called a **diode**. The current–voltage graph for a diode shows that it behaves differently from a resistor.

3 Does it matter which way round you connect the diode? Explain your answer as fully as you can.

4 Write down <u>one</u> other difference between the diode and a resistor.

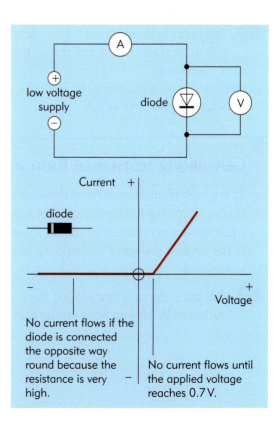

419

Current–voltage graph for a filament lamp

A lamp has a filament that gets very hot. A lamp is sometimes called a **filament lamp**. Look carefully at the current–voltage graph for a filament lamp.

5 Does it matter which way round you connect the lamp? Explain your answer as fully as you can.

6 (a) How does the temperature change as the current increases?
(b) How does the resistance change as the current increases?
(c) How does the shape of the graph change at higher currents?

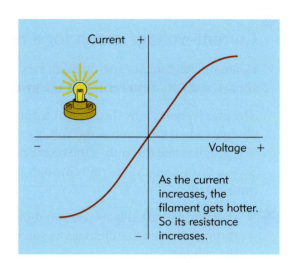

As the current increases, the filament gets hotter. So its resistance increases.

How to measure resistance

We measure resistance in units called **ohms** (symbol, Ω). To measure a resistance you need to know how big a current flows through it when you put a particular potential difference (p.d.) across it.
Then you can work out the resistance like this:

$$\frac{\text{resistance}}{\text{(ohms, }\Omega)} = \frac{\text{potential difference (volts, V)}}{\text{current (amperes, A)}}$$

The example shows how to use this formula.

7 The current through a 12 V car headlamp is 3 A. Calculate the resistance of the filament when the lamp is operating.
[Start by writing down the formula. Show all your working.]

Example

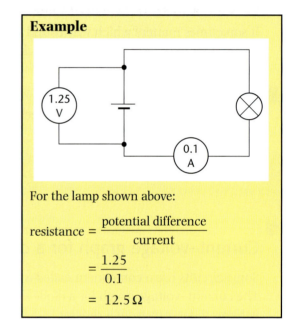

For the lamp shown above:

$$\text{resistance} = \frac{\text{potential difference}}{\text{current}}$$

$$= \frac{1.25}{0.1}$$

$$= 12.5\,\Omega$$

Calculating resistance from a graph

This graph shows the current through a **resistor** that stays at a constant temperature. For a resistor at a constant temperature, the current increases in proportion to the applied voltage. This happens because the resistance is constant.

8 Calculate the resistance using the pairs of values indicated by broken lines on the graph.

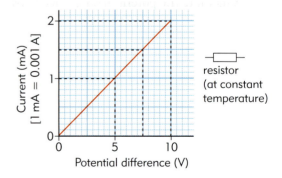

resistor (at constant temperature)

The graph is a straight line through the origin (0, 0). This means that the current is directly proportional to the voltage. This happens because the resistance stays the same.

Using resistance to calculate potential difference

The formula:

$$\frac{\text{resistance}}{\text{(ohms, } \Omega)} = \frac{\text{potential difference (volts, V)}}{\text{current (amperes, A)}}$$

can be rearranged to give

$$\frac{\text{potential difference}}{\text{(volts, V)}} = \frac{\text{current}}{\text{(amperes, A)}} \times \frac{\text{resistance}}{\text{(ohms, } \Omega)}$$

If you know:
- the resistance of a component in a circuit and
- the current flowing through the component

you can use this version of the formula to calculate the potential difference across the component.

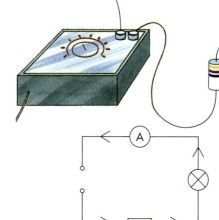

47 Ω resistor

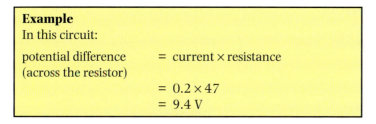

> **Example**
> In this circuit:
>
> potential difference = current × resistance
> (across the resistor)
>
> = 0.2 × 47
> = 9.4 V

9 The lamp in the circuit has a resistance of 12 Ω.
(a) Calculate the potential difference across the lamp.
(b) What is the potential difference across the power supply?

The final diagram shows a circuit that can be used to get different voltages from the same d.c. supply.

10 Calculate the potential difference between X and Y when the current through the resistor is 0.001 A.

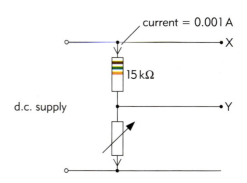

current = 0.001 A
• X
15 kΩ
d.c. supply
• Y

This circuit can be used to split up the potential difference of the d.c. supply.
It is called a <u>potential divider</u>.

Resistances that change

The resistance of an ordinary resistor stays the same, provided its temperature doesn't change. But in a lamp, the temperature of the filament <u>does</u> change, so its resistance changes too.

Some resistors are made so that we can change the resistance when we want to. These are called **variable resistors**. Some other electrical components are designed to change their resistance when there are changes in their surroundings.

The resistance of a filament lamp

The graph shows the current through the filament of a lamp when different voltages are applied across it.

11 (a) Make a copy of the current–voltage graph for the lamp.
 (b) Underneath your graph, work out the resistance of the filament:
   (i) when the p.d. across it is 0.5 V;
   (ii) when the p.d. across it is 1.0 V.
 (c) What happens to the resistance of the filament when the p.d. across it increases?
 (d) Why does this happen?

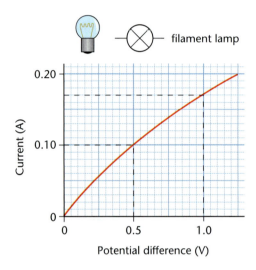
filament lamp

The filament of a lamp becomes hot because it <u>resists</u> the current flowing through it.
The graph becomes less and less steep. This means that the current through the filament does not increase as fast as the potential difference across it. This happens because the resistance of the filament increases as it gets hotter.

Variable resistors

You can change the resistance of a variable resistor by turning the knob or sliding the slider.

12 Write down some examples of where you often use variable resistors in your everyday life.

13 In this circui, the resistance of the variable resistor is reduced.
 (a) How will this affect the current through the 15 kΩ resistor?
 (b) What will happen to the p.d. between X and Y?

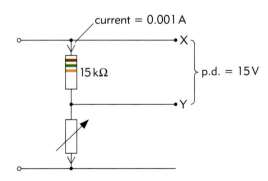

Thermistors

A **thermistor** is designed so that its resistance changes a lot when the temperature changes.

14 (a) Sketch the resistance–temperature graph for the thermistor.
   (b) Describe how the resistance of the thermistor changes with temperature.

15 Write down <u>two</u> uses for thermistors.

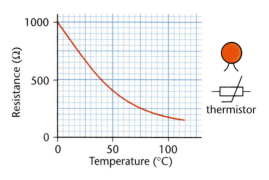

Because their resistance changes a lot with temperature, thermistors are used in electrical thermometers and thermostats.

Light dependent resistors (LDRs)

A **light dependent resistor** (LDR) is designed so that its resistance depends on the brightness of the light that falls on it. We call this the **light intensity**.

16 (a) Sketch the resistance–light intensity graph for the light dependent resistor (LDR).
   (b) Describe how the resistance of the LDR changes with the light intensity.

17 Write down <u>two</u> uses for LDRs.

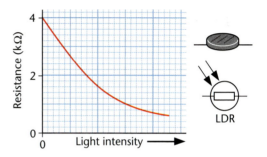

LDRs are used in light meters and in circuits which switch on lights automatically when it gets dark.

Using electricity to split things up

Most solid substances won't conduct electricity.
But some substances will conduct electricity if you melt them or dissolve them in water.

1 Which solid substances will conduct electricity?

2 Look at the diagrams.
   Write down which chemicals are:
   (a) insulators;
   (b) conductors.

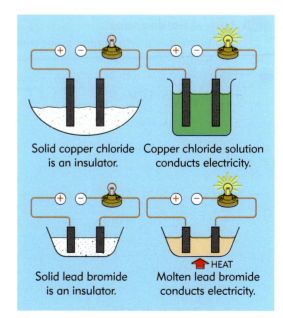

Solid copper chloride is an insulator.　Copper chloride solution conducts electricity.

Solid lead bromide is an insulator.　Molten lead bromide conducts electricity.

What happens to things when you pass electricity through them?

When you pass an electric current through a metal it gets warmer. The same happens with graphite.

But something else also happens when you pass electricity through a melted or dissolved substance. The substance gets split up into simpler substances. This is called **electrolysis**.

3 Look at the diagrams.
(a) When copper chloride solution is electrolysed, what chemical is produced at:
  (i) the positive (+) electrode;
  (ii) the negative (−) electrode?
(b) When molten lead bromide is electrolysed, what chemical is produced at:
  (i) the positive (+) electrode;
  (ii) the negative (−) electrode?

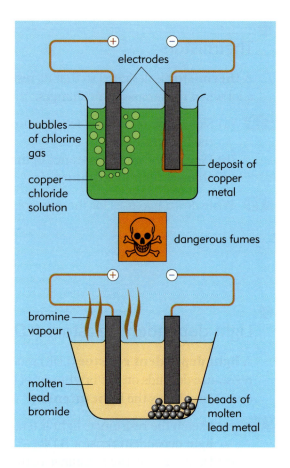

How does electrolysis work?

Some substances are made of electrically charged particles called **ions**. When the substance is melted (molten) or dissolved, these ions can move about. These are the substances that you can split up by electrolysis.

4 Look at the diagrams.
(a) Write down which ions move towards the positive electrode.
(b) Write down which ions move towards the negative electrode.

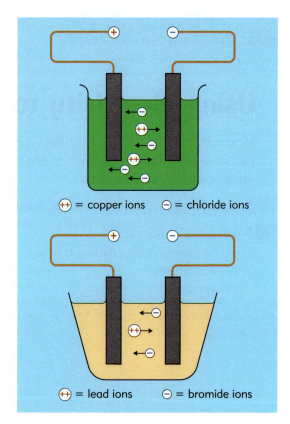

What is an electric current?

An electric current is always a flow of electrical charges. In solids, like copper wires and most electrical components, an electric current is a flow of negatively charged electrons. But during electrolysis, different sorts of charged particles called **ions** carry the current through the liquid.

5. (a) What particles form the current in the wires?
   (b) What particles form the current in the liquids?
   (c) What can you say about the charges of these particles?

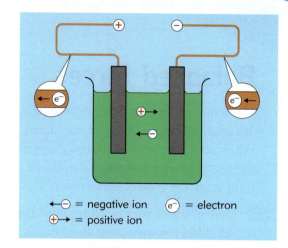

←⊖ = negative ion   (e⁻) = electron

⊕→ = positive ion

Higher

A closer look at electrolysis

When an electric current passes through copper chloride solution, copper and chlorine are produced.

6. What two factors does the amount of copper or chlorine produced during electrolysis depend on?

The amount of charge that flows while an electric current is switched on also depends on the size of the current and how long it is on. We measure charge in units called **coulombs** (C, for short).

The amount of charge that flows is given by:

$$\underset{\text{(coulombs, C)}}{\text{charge}} = \underset{\text{(amperes, A)}}{\text{current}} \times \underset{\text{(seconds, s)}}{\text{time}}$$

So the mass (or volume) of copper and chlorine produced during the electrolysis of copper chloride is also directly proportional to the number of coulombs of charge that has flowed.

7. Work out the charge that flows when a current of 0.8 A is switched on for 12 minutes. [Start by writing down the formula. Show all your working.]

8. During an electrolysis, 20 cm³ of hydrogen gas is produced. How much hydrogen will be produced if a current twice as big is used for five times as long?

9. During an electrolysis, a current of 0.5 A flows for 2 hours and deposits 8.0 g of silver.
   (a) How much charge has flowed?
   (b) How much silver will be deposited if a current of 0.2 A flows for 1½ hours?

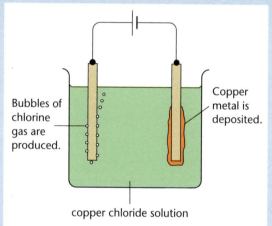

Bubbles of chlorine gas are produced.

Copper metal is deposited.

copper chloride solution

The amount (mass or volume) of copper and chlorine produced is proportional to:
- the size of the current;
- the time that the current flows.

Example 1
If you use a current that is half as big but for ten times as long, you will produce ½ × 10 = 5 times as much copper and chlorine.

Example 2
During an electrolysis, a current of 1.5 amperes is used for 5 minutes.

$$
\begin{aligned}
\text{charge (C)} &= \text{current (A)} \times \text{time (s)} \\
&= 1.5 \times (5 \times 60) \\
&= \underline{450\,C}
\end{aligned}
$$

[Remember to change the time to seconds.]

11 Forces

Balanced forces

You might think that things stay still because no forces are acting on them. A tug of war shows that this isn't so. When both teams pull with <u>equal force</u> in <u>opposite directions</u>, the forces cancel out. When the forces are **balanced**, the rope does not move. The rope is stationary.

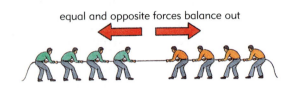
equal and opposite forces balance out

1 Look at the picture of a dumb-bell being held still.
   (a) What force is acting downwards on the dumb-bell?
   (b) What force is acting upwards on the dumb-bell?
   (c) Why does the dumb-bell stay still?

lifting force of arm muscles

weight of dumb-bell

Why you don't fall through the floor

Your weight is always pushing downwards against the floor. The floor holds you up, so it must be pushing upwards on you. The two forces are equal in size but act in opposite directions. So the forces balance and you stay where you are.

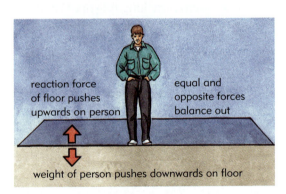
reaction force of floor pushes upwards on person

equal and opposite forces balance out

weight of person pushes downwards on floor

People don't fall through floors.

2 Look at the picture of a book on a table. Explain, as fully as you can, why the book doesn't fall through the table.

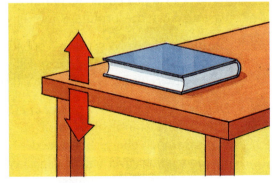
Books don't fall through tables.

How can a helicopter hover?

When people are lifted from a boat, the helicopter has to keep very still. This is difficult because the weight of the helicopter is always pulling it downwards.

3 (a) What force is acting upwards on the helicopter?
(b) Explain how the helicopter can keep still.

4 What would happen to the helicopter if:
(a) the uplift was greater than the weight?
(b) the weight was greater than the uplift?

uplift force of rotor

weight of helicopter and rescue cage

Using forces to keep still

5 Look at the pictures of stationary objects.
For each one:
(a) copy the picture;
(b) mark on the forces that are acting on the object;
(c) label the forces that you have marked on the picture;
(d) explain why the object keeps still.

Balanced forces

holding up a shopping bag

child standing on bathroom scales

helium balloon against the ceiling

holding a dog back

Why do objects slow down?

If a fish stops pushing itself forwards it will slow down and stop. The same thing happens with a car – if the engine stops, the car slows down. This happens because of the force of **friction** when things move through air or water.

6 (a) What force acts in the opposite direction to the moving fish and car?
(b) How does this force change the car's and the fish's motions?

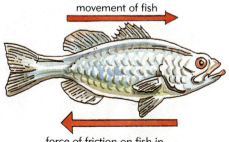

movement of fish

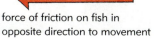

force of friction on fish in opposite direction to movement

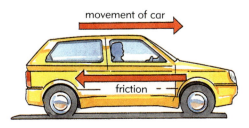

movement of car

friction

The force of friction slows down the fish and the car.

Forces all around us

Whenever an object moves through air or water, there will be a force of friction acting on it in the opposite direction. This force makes the object slow down.

7 (a) Copy the diagrams. On each one, mark the direction of movement and the direction of the force of friction.

(b) Why do all the objects slow down?

a cyclist who has stopped pedalling

a canoeist who has stopped paddling

a shuttlecock flying through the air

Balanced forces

When a fish is moving, a force of friction acts against it. To move at a steady speed its swimming force must **balance** the friction force.

8 (a) What happens if the friction force is greater than the swimming force of the fish?

(b) What must the fish do to speed up?

9 A car is moving along a level road at a steady speed. How big is the driving force of the car compared to the friction force?

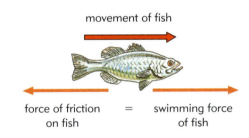

movement of fish

force of friction   =   swimming force
on fish   of fish

Forces balance so the fish moves at a steady speed.

Saving fuel

The friction force of the air is called **air resistance**. If there is a lot of air resistance, a greater force is required to move a car forwards. This means more fuel is used by the engine. Car designers try to produce **streamlined** shapes that make the air flow smoothly around the car.

10 Why are cars made in a streamlined shape?

Getting in shape for victory

Chris Boardman won a gold medal for cycling in the 1992 Olympics. His bicycle was designed to reduce friction with the air. He could then reach a higher speed using the same pedalling force.

11 Look at the picture. Write down <u>five</u> ways in which friction was reduced.

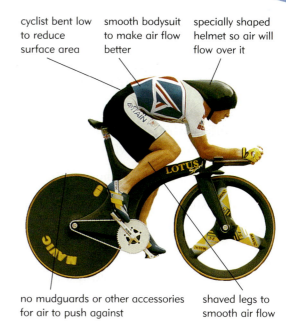

cyclist bent low to reduce surface area

smooth bodysuit to make air flow better

specially shaped helmet so air will flow over it

no mudguards or other accessories for air to push against

shaved legs to smooth air flow

Making the most of air resistance

Air resistance is also called **drag**. It can be very useful if you actually want to slow down. The space shuttles use this idea. As the shuttle lands it opens a parachute to help the braking. The parachute provides a bigger surface for more air to push against.

12 Why do you think the shuttle needs a large parachute?

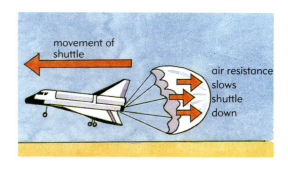

movement of shuttle

air resistance slows shuttle down

Using friction

You don't just get friction when things move through air or water. Friction also acts when solid surfaces slide across each other. This is because parts of the two surfaces catch on each other. When a surface is moving one way, the friction force acts on it in the <u>opposite</u> direction.

There is less friction between smooth surfaces.

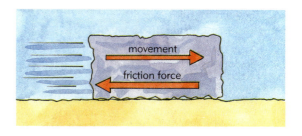

movement

friction force

1 (a) What stops you from slipping every time you take a step forwards?
(b) Why is it difficult to walk on an icy pavement?

Effects of friction

Rub your hands together quickly for a few seconds and they will feel warm. Friction causes surfaces to heat up.

Friction causes objects to <u>wear away</u>. Tiny pieces break off the surface when it rubs against another surface. This can happen quickly or slowly, depending on the materials.

Look at the pictures and answer the following questions.

2 Why should you never slide down a rope?

3 Why does a match light when you strike it?

4 Friction forces make clothes wear away.
   (a) Which parts wear away first in shoes?
       Give a reason for your answer.
   (b) Why do elbows in jumpers wear away first?

5 A pencil eraser loses tiny pieces of rubber each time you use it. Why does this happen?

rope

skin

friction between rope and skin

chemical on end of match that will burn easily

sandpaper

friction between match and sandpaper on matchbox

MATCHES

Friction against a moving surface makes it slow down. This is how brakes for bicycles and cars work. A car needs a big force to stop it, so a lot of heat is produced. Car brake pads are made from a special material that can stand the heat.

Look at the diagrams and answer the following questions.

6 (a) What happens to the brake blocks when the cyclist pulls the brake lever?
   (b) How does this slow the bike down?

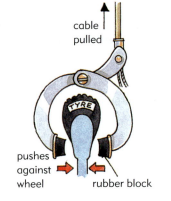

cable pulled

TYRE

pushes against wheel

rubber block

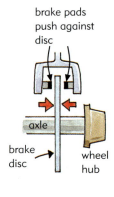

brake pads push against disc

axle

brake disc

wheel hub

Large stone blocks produce a lot of friction if they are dragged along the ground. This makes moving them very slow work, and causes wear to the block and to the ground. Rollers between these two surfaces reduce the friction.

rock and rollers

large friction between stone and ground

rollers turn around instead of scraping along the ground

Why cars can't stop instantly

If someone steps out in front of a car, it takes time for the driver to react. This is called **reaction time**. The distance the car travels during the reaction time is called the **thinking distance**.

7 Look at the diagram. Why does it take time to react?

8 Look at the table.
   (a) What is the thinking distance when travelling at 30 miles per hour?
   (b) What happens to the thinking distance if the speed is doubled?

When the driver presses the brake pedal, it takes time for the brakes to slow the car down. During this time the car travels a distance called the **braking distance**.

9 Look at the table.
   (a) What is the braking distance for a speed of 30 miles per hour?
   (b) What happens to the braking distance if the speed is increased?
   (c) How is the **stopping distance** calculated?

10 After drinking alcohol people may feel perfectly normal but actually their reactions are much slower. Why is it a bad idea for people to drive after drinking alcohol?

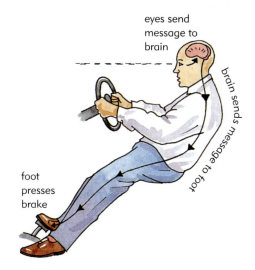

eyes send message to brain

brain sends message to foot

foot presses brake

Making a quick stop.

Stopping distances on dry roads (30 miles per hour is 48 kilometres per hour).

Speed in miles per hour	Thinking distance in metres	Braking distance in metres	Stopping distance in metres
20	6	6	12
30	9	14	23
40	12	24	36
50	15	38	53
60	18	55	73
70	21	75	96

Why do tyres have tread?

You need good tyres to stop quickly. Tyres can grip the road only if they are touching it. They lose their grip when the road is wet. The tread on a tyre is designed to push away the water. In dry conditions the tread doesn't help. In dry weather, racing cars use tyres with no tread.

11 Why do racing drivers stop to change their tyres when it starts raining?

There is friction between tyres and the road, which makes the tyres grip the road. When you brake there is friction between the brakes and the wheels, which slows the wheels down.

12 Look at the diagrams.
(a) What job does friction do when the car is driving along?
(b) What are the <u>two</u> jobs done by friction when the car brakes?
(c) Which friction force is larger when the car brakes?
(d) What is happening when the car skids?
(e) Which friction force is larger when the car skids?

How to reduce skidding

There is less chance of skidding when there is a lot of friction between the tyres and the road. That is why there are often very rough road surfaces at crossings and junctions.

An icy road has very little friction. It is very difficult to stop a car quickly without skidding on an icy road.

13 Explain the following as fully as you can.
(a) Why is it safer to have a rough road surface before a pedestrian crossing?
(b) Why must a driver brake very carefully when there is ice on the roads?

Speed limits

A sky-diver jumps out of an aeroplane. The graph shows what happens to her speed as she falls.

14 Describe how the sky-diver's speed changes after she first jumps from the aeroplane.

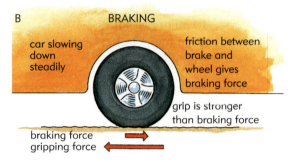

A — DRIVING ALONG
car moving at steady speed
friction between tyre and road gives grip
braking force — zero
gripping force ←

B — BRAKING
car slowing down steadily
friction between brake and wheel gives braking force
grip is stronger than braking force
braking force →
gripping force ←

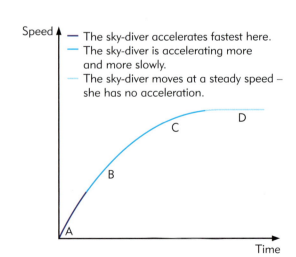

C — SKIDDING
too much braking force locks the wheel
braking force is stronger than grip causing car to **skid**
skidding car is difficult to control
braking force →
gripping force ←

Speed
— The sky-diver accelerates fastest here.
— The sky-diver is accelerating more and more slowly.
— The sky-diver moves at a steady speed – she has no acceleration.

A B C D Time

Why does the sky-diver stop accelerating?

When the sky-diver stops accelerating we say that she has reached her **terminal velocity**.

The sky-diver accelerates when there is an <u>unbalanced</u> force acting on her.

The diagram shows the forces acting on the sky-diver at points A, B, C and D on the graph.

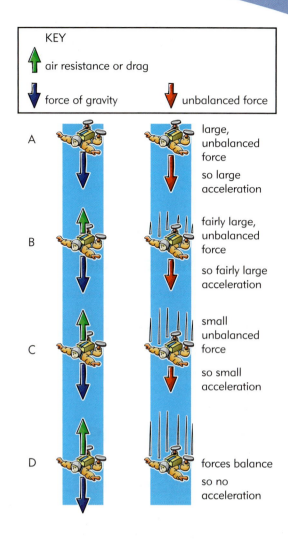

KEY

↑ air resistance or drag

↓ force of gravity   ↓ unbalanced force

A — large, unbalanced force so large acceleration

B — fairly large, unbalanced force so fairly large acceleration

C — small unbalanced force so small acceleration

D — forces balance so no acceleration

15 (a) Which force is strongest at A?
    (b) How does air resistance change as the sky-diver falls faster?

16 How does her speed change when the forces are:
    (a) unbalanced?
    (b) balanced?

Well before she reaches the ground, the sky-diver opens her parachute.

17 Explain why opening her parachute gives her a much smaller terminal velocity.

Large area gives more air resistance. So this balances gravity at a much lower speed.

Cruising along the motorway

The diagrams show two cars moving at a steady speed along a flat stretch of motorway.

18) If the cars are driving at a steady speed, what can you say about their frictional forces and their driving forces?

19) You need a much bigger driving force to keep a car going at 70 miles per hour than at 50 miles per hour. Explain why.

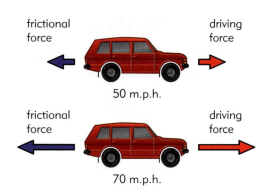

frictional force   driving force

50 m.p.h.

frictional force   driving force

70 m.p.h.

Falling through liquids

There is a lot more friction when things fall through liquids than when they fall through air. So things falling through liquids have a much smaller terminal velocity.

20) Look at the diagram. What does it tell you about the terminal velocity of the two marbles?

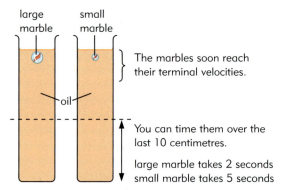

large marble   small marble

The marbles soon reach their terminal velocities.

oil

You can time them over the last 10 centimetres.

large marble takes 2 seconds
small marble takes 5 seconds

Moving along

Remember from Key Stage 3

How to calculate speed

You can work out speed like this:

$$\frac{\text{speed}}{\text{(metres per second)}} = \frac{\text{distance travelled (metres)}}{\text{time taken (seconds)}}$$

[On your calculator: distance ÷ time]

> **Example**
> On a motorway, a car travels 300 metres in 10 seconds.
>   distance travelled = 300 metres (m)
>   time taken = 10 seconds (s)
>   speed = ?
>   so speed = $\dfrac{300}{10}$ = 30 metres per second (m/s)

1) Look at the examples in the pictures and work out the missing item for each one.

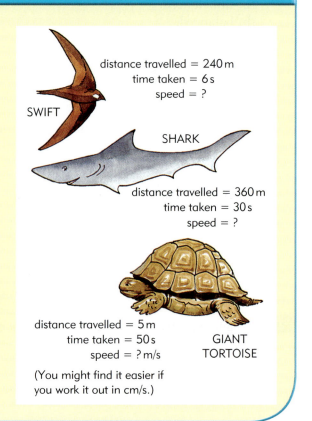

distance travelled = 240 m
time taken = 6 s
speed = ?

SWIFT

SHARK

distance travelled = 360 m
time taken = 30 s
speed = ?

distance travelled = 5 m
time taken = 50 s
speed = ? m/s

GIANT TORTOISE

(You might find it easier if you work it out in cm/s.)

Showing movement on a graph

This graph shows distance travelled against time taken, so it is called a **distance–time graph**. You can read from the graph the distance travelled in a period of time. For example, it takes 4 seconds to travel 20 metres.

2 What speed is this?

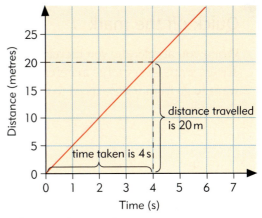

A distance–time graph.

Jane's journey

This example shows how a journey can be described on a distance–time graph. Look at it carefully and use it to answer the questions.

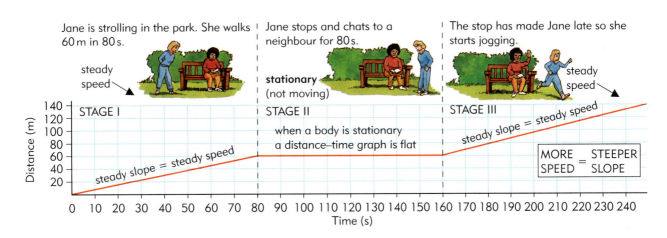

3 Look at stage I of the journey, where Jane is walking.
   (a) How many metres does Jane walk?
   (b) How long does it take Jane to walk this distance?
   (c) What is her speed for this part of her journey?

The **slope** of the graph tells you about the **speed**.

4 (a) What is Jane's speed when she chats (stage II)?
   (b) Describe the shape of the graph for this stage.
   (c) Write down a rule for telling when something is stationary on a distance–time graph.

5 (a) Which part of the graph is steeper, (I) or (III)?
   (b) Which stage of the journey is faster, (I) or (III)?
   (c) What is the connection between speed and slope on a distance–time graph?

Quick off the mark

When you accelerate, you increase your speed. Racing drivers want a very large **acceleration**. This means they want to go from a slow speed to a fast speed in a very short time.

At the start of a race this car can accelerate up to 50 metres per second in just 5 seconds. (50 m/s is more than 100 miles per hour.)

How to calculate acceleration

The racing car in the photograph takes 5 seconds to reach a speed of 50 metres per second. So every second its speed increases by 10 metres per second. This is its **acceleration**. You can work out acceleration like this:

$$\text{acceleration} = \frac{\text{change in velocity}}{\text{time taken}}$$

(On your calculator: change in velocity ÷ time)

So for the racing car,

$$\text{acceleration} = \frac{50 \text{ m/s}}{5 \text{ s}} = 10 \text{ m/s}^2$$

Don't worry about the word 'velocity' here. For the moment, think of it as another word for speed.

The change in velocity is measured in metres per second (m/s). The time taken is measured in seconds (s). So the units of acceleration are metres per second, per second. We call these **metres per second squared** (m/s²).

6 Look at the examples in the pictures and work out the missing items in each one.

Rocket being launched

change in velocity = 600 m/s
time taken = 20 seconds
acceleration = ?

Cheetah hunting

velocity at start = 0
new velocity = 20 m/s
time taken = 2 seconds
change in velocity = ?
acceleration = ?

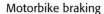

Horse overtaking another

velocity at start = 15 m/s
new velocity = 23 m/s
time taken = 4 seconds
change in velocity = ?
acceleration = ?

Motorbike braking

velocity at start = 20 m/s
new velocity = 0
time taken = 10 seconds
change in velocity = –20 m/s
acceleration = ?

Showing acceleration on a graph

The slope of a **velocity–time graph** tells you how the speed or velocity is changing.

Look at the velocity–time graph for a cyclist.

7 How does the graph show a steady speed?

8 What is the cyclist's speed:
(a) to begin with?
(b) after she accelerates?

9 How does the graph show an acceleration?

10 How does the graph show that the cyclist slows down quicker than she speeds up?

A steeper slope on a velocity–time graph means a bigger acceleration (or deceleration).

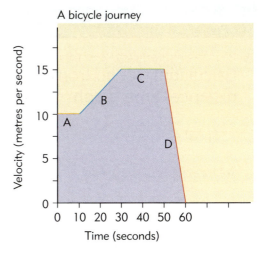

A bicycle journey

First the cyclist is at a steady speed (A) then she slowly accelerates to a higher speed (B). She travels at this higher speed for a while (C) then she quickly slows down (decelerates) and stops (D).

Why do we need another word for speed?

The diagram shows a ball being thrown into the air. When the ball comes back down it is moving at the same **speed** as you threw it up. But it is moving in the opposite **direction**. We say it has a different **velocity**.

11 (a) How much has the velocity of the ball changed?
(b) What is its acceleration?

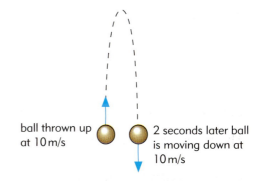

ball thrown up at 10 m/s

2 seconds later ball is moving down at 10 m/s

The speed of the ball has not changed. The velocity of the ball has changed by 20 m/s.

Higher

Calculating speed from a distance–time graph

$$\text{Speed} = \frac{\text{distance}}{\text{time}}$$

So, for part A of the graph:

$$\text{speed} = \frac{20}{5} = \underline{4 \text{ m/s}}$$

To calculate the speed for other parts of the graph, you need to use the slope or **gradient** of the graph. Example 1 shows you how to do this for part B of the graph.

12 (a) Make a copy of the graph but mark the changes for part C instead of part B.
(b) Calculate the speed for part C of the graph. [Show your working.]

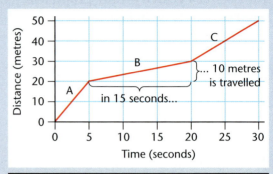

Example 1
In part B of the graph:

$$\text{speed} = \frac{\text{distance}}{\text{time}} = \frac{10}{15} = \underline{0.67 \text{ m/s}}$$

A note about speed and velocity
The velocity of an object includes the <u>direction</u> that it is travelling as well as its <u>speed</u>.

Velocity–time graphs, however, are about objects moving in the same direction along a <u>straight line</u>. This means that you don't have to worry about the difference between speed and velocity. You can think of velocity as just another word for speed.

Calculating acceleration from a velocity–time graph

$$\text{Acceleration} = \frac{\text{change in velocity}}{\text{time taken}}$$

So, for part P of the graph:

$$\text{acceleration} = \frac{2}{2} = \underline{1 \text{ m/s}^2}$$

To calculate the acceleration for other parts of the graph, you need to use the slope or gradient of the graph. Example 2 shows you how to do this for part Q of the graph.

13 (a) Make a copy of the graph but mark the changes for part R instead of part Q.
(b) Calculate the acceleration for part R of the graph. [Show your working.]

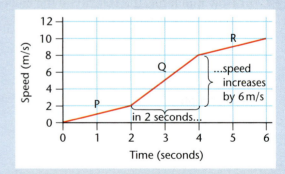

Example 2
In part Q of the graph:

$$\text{acceleration} = \frac{\text{change in velocity}}{\text{time taken}}$$

$$= \frac{6}{2} = \underline{3 \text{ m/s}^2}$$

Higher

What if the graphs are curved?

Distance–time graphs are not always made up of straight lines. This graph, for example, is curved. The line on the graph is getting less and less steep. This tells you that the object is slowing down.

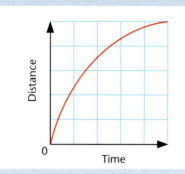

14 How does the gradient change as the object slows down?

Velocity–time graphs can also be curved. This means that the acceleration is changing.

Calculating distance travelled from a velocity–time graph

The **area** beneath a velocity–time graph tells you the distance that an object has travelled.
The box explains why.

Example 3 shows how you can work out from the graph on the right the distance travelled in the first 5 seconds.

> **Using area to calculate distance**
> Area below graph = average height × width
> But average height = average speed
> and   width = time
> So
>   area below graph = average speed × time
>   = distance travelled

> **Example 3**
> For 0–5 seconds:
> area below graph = $\dfrac{20 + 30}{2} \times 5 = \underline{125\ metres}$

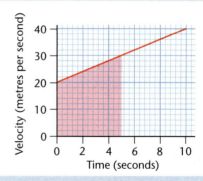

15 Make a copy of the graph, but mark the area under the graph for between 5 and 10 seconds. Then calculate the distance travelled during this period. [Show your working.]

16 What is happening to the gradient on the graph?
What is happening to the speed?

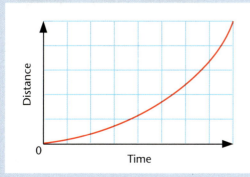

17 What is happening to the gradient on the graph?
What is happening to the acceleration?

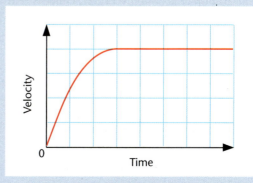

Forces and mass

Now that Malcolm is older, he isn't so active and he has started to put on weight. Malcolm wants to lose weight. He reads in a magazine that people weigh less on Mars. He wishes he could go there.

Though Malcolm would **weigh** less on Mars, his body is still exactly the same. There is just as much of him. He still has exactly the same **mass**.

ON EARTH
bathroom
scales read 90

ON MARS
bathroom
scales read 36

What is mass?

Mass tells you how much there is of something. A 10 kg bag of potatoes contains twice as much potato as a 5 kg bag, so it has twice the mass. We measure the mass of things in **kilograms** (kg).

1. Malcolm says his weight is 90 kg. This is wrong. What should he say?

What is weight?

When you drop something it falls downwards. That is because the force of **gravity** pulls it. The size of this force on an object is called **weight**. Objects with more mass have more weight. That is because there is more mass for the force of gravity to act on. We measure forces in **newtons** (N).

2. (a) What is the force that causes weight?
   (b) In which direction does this force act?

3. Try to find out why we measure weight in newtons.

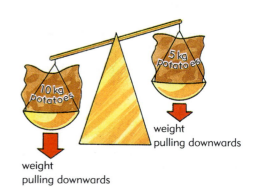

weight
pulling downwards

weight
pulling downwards

More mass means more weight.

What's the difference?

If two objects are in the same place, the one with more mass also has more weight. But mass and weight aren't the same thing. You will have the same mass wherever you are because you have the same amount of body. Your weight might not always be the same.

4 Look at the picture of astronauts. How much do the astronauts weigh? Explain your answer.

In space the force of gravity can be very, very small …

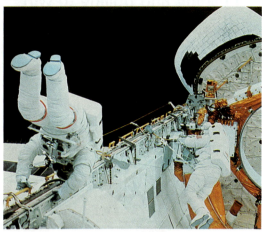

… so astronauts have hardly any weight.

Working out weight

Weight depends on how much mass an object has. It also depends on the force of gravity. The diagrams show the weights of two objects on Earth.

5 What is the weight (in newtons) of a 1 kg mass on Earth?

We say that Earth's force of gravity is 10 newtons per kilogram (10 N/kg). We can work out the weight of an object as follows:

$$\frac{\text{weight}}{\text{(newtons)}} = \frac{\text{mass}}{\text{(kilograms)}} \times \frac{\text{force of gravity}}{\text{(newtons per kilogram)}}$$

6 Look at the examples in the pictures and work out the missing numbers.

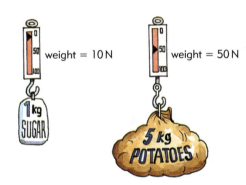

weight = 10 N

weight = 50 N

mass = 6 kg
weight = ?

mass = 60 kg
weight = ?

mass = 0.5 kg
weight = ?

mass = ? kg
weight = 12 000 N

Malcolm goes planet hopping

Malcolm has a mass of 90 kg. On Earth the force of gravity is 10 N/kg. Therefore, his weight is $90 \times 10 = 900$ N.

7 Copy the table. Complete the table to show what Malcolm's weight would be in different parts of the solar system.

	Malcolm's mass (kg)	Force of gravity (N/kg)	Malcolm's weight (N)
Earth	90	10	900
Moon	90	1.6	
Mars		4	
Jupiter		23	

Pushing and pulling to win

Sumo wrestlers push against each other. When both wrestlers push with the same force nothing happens. The forces balance and cancel each other out. To win, one wrestler must push the other off the mat. He must push harder. The forces must be **unbalanced**.

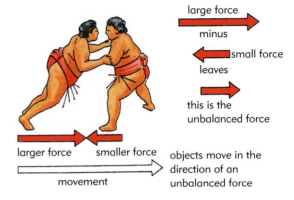

large force

minus

small force

leaves

this is the unbalanced force

8 Look at the diagram of a tug-of-war. Which team will win? Give a reason.

larger force   smaller force   objects move in the direction of an unbalanced force

movement

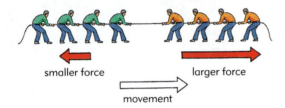

smaller force   larger force

movement

In fact, any stationary object will start to move when an unbalanced force acts on it. The movement is always in the direction of the unbalanced force.

Speeding up a bobsleigh

A force is needed to speed up a bobsleigh. The bigger the force, the faster the bobsleigh speeds up.

A bigger force produces a bigger **acceleration**.

9 Look at the picture of the bobsleigh.
How does the team start the sleigh? Explain as fully as you can why it is started in this way.

Slowing down a bicycle

Slowing something down also needs a force. A bicycle has front brakes and rear brakes. You could slow down a bicycle by just using the rear brake.

10 (a) What difference does it make if you use both brakes?
(b) What difference does it make if you squeeze the brakes harder?
(c) What is the effect of a bigger braking force?

Accelerating trucks and trains

The bigger the mass of an object, the harder it is to make it speed up or slow down. Look at the picture of three lorries. The lorries have identical engines so each lorry produces the same force.

11 (a) Which lorry has the smallest mass?
(b) Which lorry speeds up the fastest?
(c) Which lorry speeds up most slowly?
(d) How does mass affect the time taken to reach the same speed?

12 (a) If all the lorries are moving at the same speed, which one can slow down in the shortest time?
(b) How does mass affect the time taken to slow down?

13 Work out the accelerations of the three lorries.

> ### Remember
>
> $$\text{acceleration} = \frac{\text{change in speed (metres per second)}}{\text{time taken (seconds)}}$$
>
> The answer is in metres per second squared.

mass = 5 tonnes
can slow down
quickly

takes 5 seconds
to go from 0 to 5
metres per second

mass = 10 tonnes
takes longer to slow
down

takes 10 seconds
to go from 0 to 5
metres per second

mass = 20 tonnes
takes much longer
to slow down

takes 20 seconds
to go from 0 to 5
metres per second

Look at the picture of four trains. Each engine [🚂] is identical and produces the same force.

14 (a) Why does train Q accelerate faster than train P?
(b) Why does train R accelerate more slowly than train P?
(c) Why does train S have the same acceleration as train P?

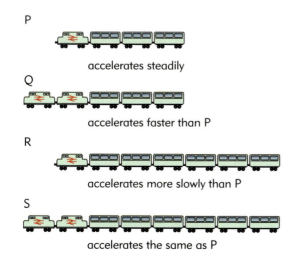

P

accelerates steadily

Q

accelerates faster than P

R

accelerates more slowly than P

S

accelerates the same as P

Experiments in space

Space scientists need to send a spaceship to just the right place. So they need to know just how much acceleration the rockets will produce. A spaceship is also a good place to measure the accelerations produced by different forces.

15 Why is a spaceship a good place to do experiments on force and acceleration?

In space, friction and gravity don't mess up our experiments with forces.

How force affects acceleration

The diagrams show some experiments in a spaceship. The same 1 kg mass is accelerated using different forces.

16 (a) How much acceleration does a 1 N force give to the 1 kg mass?
(b) How much acceleration does a 2 N force give?
(c) How does the acceleration of an object change if the force is doubled?

Changing the force

speed increases by 1 m/s every second
acceleration is 1 m/s^2 (metre per second squared)

1 N force

speed increases by 2 m/s every second
acceleration is 2 m/s^2

2 N force

speed increases by 3 m/s every second
acceleration is 3 m/s^2

3 N force

How mass affects acceleration

A 1 N force is then used to accelerate different masses.

17 (a) How much acceleration do you get when the mass is twice as big?
(b) How much acceleration do you get when the mass is three times as big?
(c) How does the acceleration of an object change if the mass is doubled but the force stays the same?

Changing the mass

speed increases by 1 m/s every second
acceleration is 1 m/s^2

1 N force

speed increases by $\frac{1}{2}$ m/s every second
acceleration is $\frac{1}{2}$ m/s^2

1 N force

speed increases by $\frac{1}{3}$ m/s every second
acceleration is $\frac{1}{3}$ m/s^2

1 N force

Higher

A force of 1 N acting on a mass of 1 kg produces an acceleration of 1 m/s^2.

18 (a) Copy and complete the table using the results of all <u>six</u> experiments from questions 16 and 17.

Force (N)	Mass (kg)	Acceleration (m/s^2)	Mass × Acceleration
1	1	1	1

(b) What do you notice about the first and last columns in the table?

The force is always the same as the mass times the acceleration.

$$\frac{\text{force}}{\text{(newtons)}} = \frac{\text{mass}}{\text{(kilograms)}} \times \frac{\text{acceleration}}{\text{(metres per second squared)}}$$

19 Look at the examples in the pictures. Use the formula to work out the missing items. The first one is done for you.

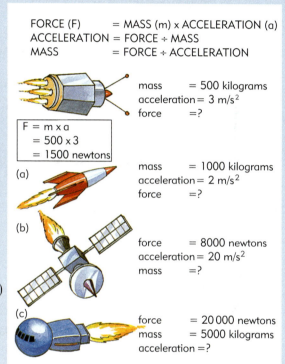

FORCE (F)   = MASS (m) x ACCELERATION (a)
ACCELERATION = FORCE ÷ MASS
MASS   = FORCE ÷ ACCELERATION

mass   = 500 kilograms
acceleration = 3 m/s^2
force   =?

F = m x a
= 500 x 3
= 1500 newtons

(a) mass   = 1000 kilograms
acceleration = 2 m/s^2
force   =?

(b) force   = 8000 newtons
acceleration = 20 m/s^2
mass   =?

(c) force   = 20 000 newtons
mass   = 5000 kilograms
acceleration =?

Meanwhile, back on Earth …

All objects on Earth are pulled by **gravity**. This makes them fall with the same **acceleration** unless friction from the air interferes.

Try dropping a paper clip and a boot together, from the same height. They should hit the floor together even though the boot has a lot more mass. The boot has more mass than the paper clip and so it feels more gravitational force. However, one thing does not change – that is the force per unit of mass.

The gravitational force on 1 kilogram is

20 10 newtons. What acceleration does this produce?

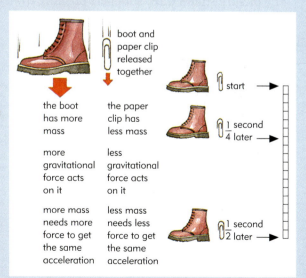

boot and paper clip released together

the boot has more mass   the paper clip has less mass

more gravitational force acts on it   less gravitational force acts on it

more mass needs more force to get the same acceleration   less mass needs less force to get the same acceleration

start

$\frac{1}{4}$ second later

$\frac{1}{2}$ second later

21 A golf ball and a table tennis ball are almost the same size but the mass of the golf ball is many times greater than the mass of the table tennis ball.

When they are dropped from a height of 10 metres above the ground, the golf ball reaches the ground before the table tennis ball.
(a) What would you expect to happen if the Earth's gravity were the only force acting on the balls?
(b) What other force acts on the balls when they are moving? [See pages 433–434 if you need help.]
(c) Why does this other force have a bigger effect on the table tennis ball than on the golf ball?

Work and energy

You move a shopping trolley by applying a force to it. When you push the trolley you transfer **energy** to it. We say that you are doing **work**. Work, like energy, is measured in **joules** (J).

The picture shows three shoppers pushing trolleys to their cars.

1 (a) Who does more work, Ravi or Mandy? Explain why.
(b) Who does more work, Mandy or Jeroen? Explain why.

Ravi pushes with a <u>small</u> force for a <u>short</u> distance

Mandy pushes with a <u>big</u> force for a <u>short</u> distance

Jeroen pushes with a <u>big</u> force for a <u>long</u> distance

The amount of work done depends on how much force is used and what distance is moved. In fact, we can calculate it like this:

$$\frac{\text{work done}}{\text{(joules)}} = \frac{\text{force}}{\text{(newtons)}} \times \frac{\text{distance moved}}{\text{(metres)}}$$

2 Look at the pictures and work out the missing numbers. The first one is done for you.

WORK DONE = FORCE x DISTANCE
DISTANCE = WORK DONE ÷ FORCE
FORCE = WORK DONE ÷ DISTANCE

model train
force = 5 newtons
distance moved = 3 metres
work done = force x distance
= 5 x 3
= 15 joules

crane
force = 7000 newtons
distance moved = 6 metres
work done = ?

barrow
force = 300 newtons
distance moved = ?
work done = 12 000 joules

weight-lifter
force = ?
distance moved = 2 metres
work done = 1400 joules

Energy on the move

To make a trolley move you must transfer energy to it. The trolley then has what we call **kinetic energy** (movement energy). Objects with a lot of kinetic energy are more difficult to stop. Look at the pictures of runaway trolleys and answer the following questions.

3 (a) Which trolley is easier to stop, P or Q?
(b) Which trolley is easier to stop, R or S?

4 What <u>two</u> things does kinetic energy depend on?

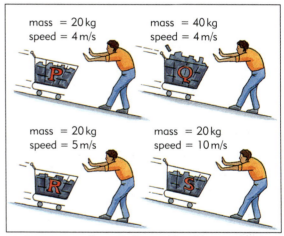

mass = 20 kg
speed = 4 m/s
P

mass = 40 kg
speed = 4 m/s
Q

mass = 20 kg
speed = 5 m/s
R

mass = 20 kg
speed = 10 m/s
S

Runaway trolleys. Trolleys with a big mass and a big speed have a lot of kinetic energy.

Friction is a waste of work

When you first push a shopping trolley it starts to move. But you then have to keep on pushing to keep it moving at the same speed. You are then doing work against the force of friction. All the work you do ends up as **heat** (thermal energy).

5 What form of energy does work against friction end up as?

At first, work done gives the trolley kinetic energy.

acceleration

pushing force

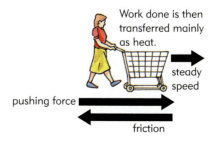

Work done is then transferred mainly as heat.

steady speed

pushing force

friction

Working your way up the stairs

When you walk up stairs you are doing work. You are lifting your own weight so you are working against gravity. The distance you move your weight is the height of the stairs.

6 How much work must the person in the diagram do against gravity to climb the stairs?

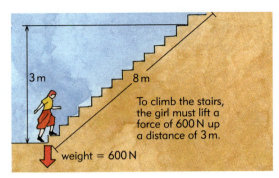

To climb the stairs, the girl must lift a force of 600 N up a distance of 3 m.

weight = 600 N

Climbing stairs.

Higher

A closer look at kinetic energy

You can work out the kinetic energy of a moving object using this formula:

$$\text{kinetic energy (joules, J)} = \frac{1}{2} \times \text{mass (kilograms, kg)} \times \frac{[\text{speed}]^2}{[(\text{m/s})]^2}$$

The example shows how to work out the kinetic energy of a ball that is thrown up into the air.

7 A car of mass 750 kg is moving at a speed of 20 metres per second. Calculate its kinetic energy. [Start by writing out the formula. Show your working.]

Example

speed 20 m/s

mass 0.1 kg

$$\text{kinetic energy} = \frac{1}{2} \times \text{mass} \times [\text{speed}]^2$$
$$= \frac{1}{2} \times 0.1 \times [20]^2$$
$$= \underline{20 \text{ joules (J)}}$$

Why fast-moving things can do a lot of damage

In most collisions, the kinetic energy of the objects that collide is transferred to their internal structure. This can bend or break the objects. It also usually makes them warmer.

The more kinetic energy a moving object has, the more damage it can do in a collision. It can do this damage to itself, to whatever it collides with, or to both. An example of this is explained in the diagram.

8 The speed limit in some residential streets is now 20 miles per hour rather than 30 miles per hour. How many times less kinetic energy has a car travelling at 20 mph than it has at 30 mph?

A stone from a catapult travels at 100 m/s.

A bullet from a gun travels at 1000 m/s.

Kinetic energy ∝ (speed)^2
[∝ means 'proportional to'.]
The bullet has 10 times the speed.
So it has (10)^2 times = 100 times as much kinetic energy (for the same mass).

What goes up usually comes down

If you throw a ball straight up into the air, it gradually slows down, stops, and then falls back to the ground. The diagram shows what is happening to the energy of the ball as it does this.

9 Copy and complete the table.

	Type of energy the ball has (kinetic or potential)
When you first throw the ball up from the ground	
When the ball is at its highest point	
When the ball reaches the ground again	
When the ball is halfway up (or down)	

At the top of its flight, all of the ball's energy is gravitational potential energy.

When it is first thrown, all the ball's energy is kinetic energy.

When the ball reaches the ground all its energy is kinetic energy again.

Escaping from the Earth

If you could throw the ball fast enough, it would <u>never</u> lose all its kinetic energy. So it would escape from the Earth altogether. The speed that is needed for this to happen is called the **escape speed**.

10 Use the information from the diagram to answer the following questions.
(a) What is the escape speed from the Earth's surface?
(b) Quite a lot of hydrogen escapes into the Earth's atmosphere from industrial processes. But the atmosphere contains hardly any hydrogen at all. Suggest a reason for this.

11 A car and its driver have a combined mass of 1 tonne (1000 kg).
The car is travelling at 30 miles per hour (48 km/h).
Calculate the kinetic energy of the car.

12 A bullet has a mass of 20 g and has 2.5 kJ of kinetic energy when it leaves the barrel of the rifle.
Calculate the speed of the bullet.

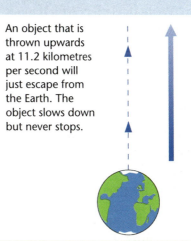

An object that is thrown upwards at 11.2 kilometres per second will just escape from the Earth. The object slows down but never stops.

How fast do gas molecules move?

The average speed of hydrogen molecules in the Earth's atmosphere is greater than 11.2 km/s.
The average speed of oxygen and nitrogen molecules in the Earth's atmosphere is less than 11.2 km/s.

Looking at the skies

We get days and years because of the way the Earth moves.

Why does it go dark at night?

The Sun shines non-stop, sending its light out into space in all directions.

The light that hits the Earth makes it daytime on one side.

On the other side, facing away from the Sun, there is no light. It is dark there so it is night.

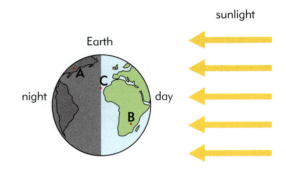

1 (a) Is it night or day at the places marked A and B on the diagram?

(b) What is happening at place C?

The spinning Earth

Places move from the dark side into the light because the Earth spins. It turns around once every 24 hours. So every 24 hours we have a period of dark then light (night then day).

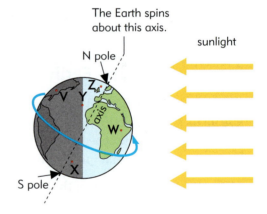

2 (a) Which places are in daylight in this picture? (Write the letters which mark the places.)

(b) Which places are in night?

(c) Which place is going from night into day (dawn)?

3 The Earth spins about a line called an axis. What do we call the ends of this axis?

Patterns in the sky

The stars in the night sky make patterns called **constellations**. These patterns do not change from year to year. A famous example is Orion the Hunter, which you can see in the evening sky in mid-winter. People imagine the stars joined up to make shapes.

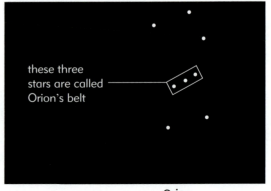

Orion

Shifting patterns

Because the Earth spins, we do not always look out at the same constellations. The patterns seem to spin around us each night as the Earth turns.

4 Look at the diagram. Does the sky seem to be turning clockwise or anticlockwise?

5 Which star do the other stars seem to be turning around?

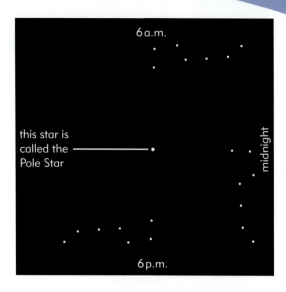

What is a star?

Our Sun is an ordinary star. It is a huge ball of very hot, glowing gas. In the centre of the Sun, hydrogen is turned into helium in nuclear reactions that give off vast amounts of energy. This energy is radiated from the Sun as light and heat.

6 What is the temperature of the Sun:
    (a) in the centre?
    (b) at the surface?

7 Why is it dangerous to look straight at the Sun?

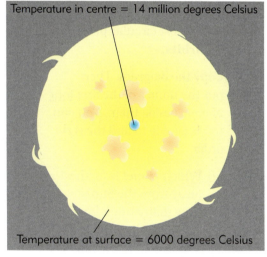

The Sun produces so much light that looking at it is dangerous. It is so bright it can damage the sensitive parts of your eye.

Why do other stars look so faint?

The stars that you can see in the night sky are made from hot gas, like the Sun. They all look faint, but some are actually much bigger and brighter than the Sun. They only look so faint because they are so far away.

8 The star Sirius is actually 26 times brighter than the Sun. Explain why it looks so much fainter.

Sirius is the bright star at the bottom left.

How far away is the nearest star?

The Sun is the nearest star to Earth.

Light travels <u>very</u> fast, but it still takes light about 8 minutes to reach us from the Sun.

Light from the next nearest star, Proxima Centauri, takes about 4 years to reach us.

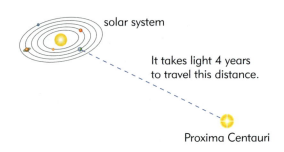

solar system

It takes light 4 years to travel this distance.

Proxima Centauri

9 How many times further away from us is Proxima Centauri than the Sun?

What is a planet?

Planets are much smaller than the stars. They go around the Sun in almost circular paths called **orbits**.

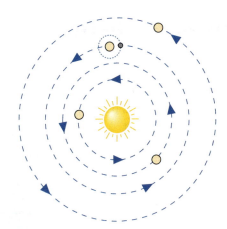

10 Copy the diagram. Then put the names of the first <u>four</u> planets beside them. You can use the table to find the right order.

11 (a) Which planet is furthest away from the Sun?
(b) How many times further away is it than the Earth?

Planet	Distance from Sun (compared to Earth's)
Mercury	0.39
Venus	0.72
Earth	1
Mars	1.5
Jupiter	5.2
Saturn	9.5
Uranus	19.2
Neptune	30.1
Pluto	39.4

How can you see a planet in the night sky?

Planets look like very bright stars in the night sky, but they do not shine with their own light. They only reflect light from the Sun back to us. This diagram shows how sunlight hits Jupiter and reflects back to us on Earth. This makes Jupiter look bright against the darkness of the sky. It also looks brighter than the much more distant stars behind it.

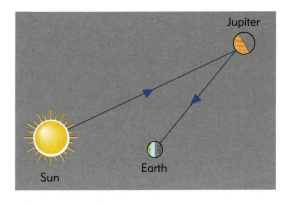

Jupiter

Earth

Sun

12 You can see Venus from place X on the Earth in the early morning.

Copy the diagram and draw arrows to show how sunlight is reflected off Venus so that we can see it.

Venus

Sun

X
night side

Earth

Wandering planets

Planets do not keep fixed positions amongst the background patterns of the stars. They slowly wander about past these constellations. This is because the planets are much nearer than the stars. Also the planets, including the Earth, are all moving in their orbits about the Sun.

13   Copy the diagram. Show where you think the planet might be in another 20 days.

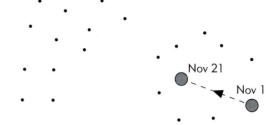

The planets appear to move slowly through the constellations. (Through a <u>telescope</u>, planets look bigger than stars.)

Orbiting the Sun and Earth

To understand why the planets move in orbits we need to understand a bit more about gravity.

The force of gravity

Any two objects pull each other together with a force called **gravity**. Two milk bottles on a doorstep attract each other. We do not notice the force because it is a million times smaller than the tiniest brush of a feather. The force is only big enough to feel if one of the objects is very big, like the Earth or one of the other planets.

The Earth has a million, million, million, million times more mass than books or bottles so we can feel the pull of its gravity.

1   What happens to the pull of the Earth's gravity if you get further from the Earth?

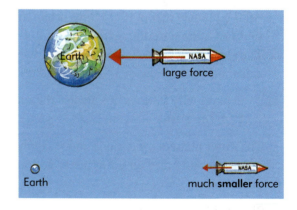

Forces of gravity in the solar system

Every planet in the solar system pulls on every other planet. But the distances between them are so large that for most of them the force is felt only faintly. The Sun attracts all the planets too. Look at the forces acting on the planet Mercury.

2   (a) Which planet is closer to the Sun?
      (b) Which planet feels the stronger pull towards the Sun?
      (c) Why is the pull of gravity towards the Sun?

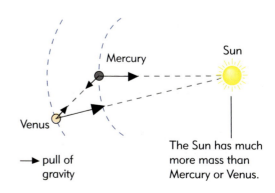

The Sun has much more mass than Mercury or Venus.

Why the planets orbit the Sun

The Sun has an enormous pull of gravity on everything in the solar system. Its pull acts even across the great distances to the outer planets.

This is because the Sun has a very large mass: a thousand times bigger than Jupiter's mass and 330 000 times bigger than Earth's.

Each planet would fall straight towards the Sun if it was not moving sideways at high speed. The Earth travels at 30 km/s along its path. This is exactly the right speed to keep it moving in a circle. We call this an **orbit**.

3 Look at the diagram. Which direction does gravity act on each planet?

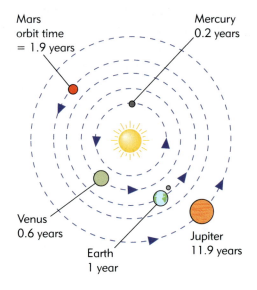

The orbits of the planets are not quite circular

All the planets except Pluto have orbits that are almost (but not quite) circular, with the Sun at the centre. The circles are very slightly squashed, so they are called **ellipses**. Pluto's orbit is very elliptical.

4 Explain why Pluto is not always the most distant planet from the Sun.

The outer planets orbit much more slowly than the inner ones and also have further to go to get around their orbit once.

5 How does the orbit time change as planets are further from the Sun?

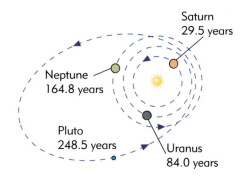

Satellites

The Moon orbits the Earth.
It is sometimes called the **satellite** of the Earth.

The first **artificial satellite** was sent into orbit around the Earth in 1957. Since then many satellites have been put into Earth orbit. We use these satellites to do many different jobs.

What makes satellites go round the Earth?

The force of gravity pulls a satellite down towards Earth. But when satellites are put up into space they are given a sideways speed so that as they fall they also move sideways. This makes them move in a curve. If a satellite has just the right speed for its height, it will move in an orbit around the Earth.

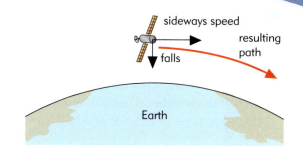

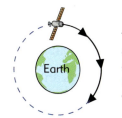

The path keeps curving around and so the satellite moves in an orbit around the Earth.

Making things move in a circle

You can tie a small object to a piece of string and whirl it around your head. The object travels in a circle, just like a satellite. You can feel a force in the string. This force pulls the object towards the centre of the circle.

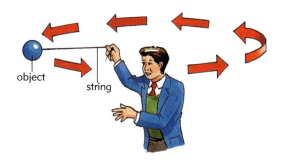

6 (a) What <u>two</u> things keep the object moving in a circle around your head?
   (b) What do you think would happen if you cut the string?

7 What provides the inwards force needed to keep a satellite moving in a circle?

the object is trying to move this way

the string pulls this way

so the object moves in a circle

Are orbits always a circle?

Most satellites orbit the Earth in a path that is very close to a circle. Some are put into orbits that are slightly squashed circles called **ellipses**.

8 Copy the diagram. Then draw on it <u>two</u> arrows to show the force of gravity on the satellite in both the places marked.

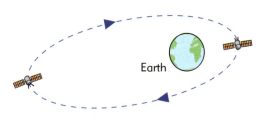

An elliptical orbit.

What can you see from a satellite?

Some satellites carry cameras and infrared sensors. They can take pictures of cloud patterns, which are used to predict the weather. They can also be used for spying, taking pictures of airfields and harbours.

Satellites orbit above the atmosphere, so they get a very clear view of the stars and the rest of the Universe.

9 Write down <u>one</u> reason why you can see the Universe better from a satellite than from the ground.

10 Write down <u>three</u> things an astronomer could see better from a satellite than from Earth.

> With a telescope on a satellite:
> - you can see fainter stars;
> - you can see more distant galaxies;
> - you can see more detail on the planets.

What else can we use satellites for?

If a satellite is at exactly the right height and speed, it will stay above the same place on the surface of the Earth. Satellites like this are used to send telephone messages and television programmes around the world.

11 How is a message sent from A to B on the Earth?

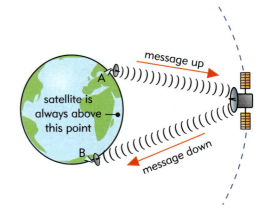

satellite is always above this point

message up

message down

Orbits of satellites

To stay in its orbit at a particular distance, a satellite must move at a particular speed around a planet.

The orbit that a satellite is put into depends on the job that you want the satellite to do. The diagram shows the two main types of orbit that are used.

12 (a) What type of satellite is usually put into a polar orbit?
(b) Why is this the best orbit to use?

13 (a) What type of satellite is usually put into a geostationary orbit?
(b) Why is this the best orbit to use?

Weather satellite in a polar orbit. All the Earth is seen in 24 hours as the Earth spins.

orbit time $1\frac{1}{4}$ hours

Spin time = 24 hours

Equator

orbit time 24 hours

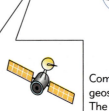

Communications satellite in a geostationary orbit. The satellite orbits the Earth at the same rate as the Earth spins. So the satellite is always in the same place in the sky when you look at it from Earth.

Orbits of planets

To stay in its orbit at a particular distance, a planet must move at a particular speed around the Sun.

The table shows the orbit periods of the planets and their average distances from the Sun.

14 (a) What happens to the orbit period when the distance of a planet from the Sun doubles?
[Hint: compare Saturn and Uranus.]
(b) How many times further from the Sun does a planet have to be for its orbit period to double?
[Hint: compare Earth and Mars.]

Planet	Average distance from Sun (Earth = 1)	Time taken for one orbit (Earth = 1)
Mercury	0.4	0.24
Venus	0.7	0.62
Earth	1.0	1.0
Mars	1.5	1.9
Jupiter	5.2	11.9
Saturn	9.6	29.5
Uranus	19.2	84.1
Neptune	30.1	165
Pluto	39.5	249

We have been using the <u>average</u> distances of planets from the Sun because the orbits of planets are not, in fact, circles. They are an oval shape called an **ellipse**.

The orbits of most of the planets are, however, quite close to being circular. They are only slightly elliptical.

15 Which planet has a much more elliptical orbit than the others?

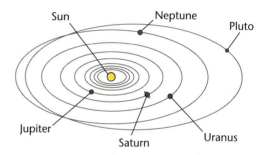

Orbits of comets

Comets are balls of ice and rock, usually a few kilometres in diameter. The diagrams show what the orbit of a comet is like and when you can see a comet.

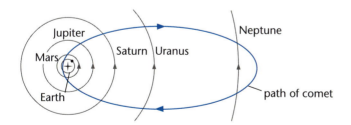

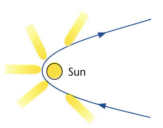

Comets have very large, very elliptical orbits. They can take tens, or even thousands, of years to make one orbit. During the most distant parts of their orbits, comets move more slowly.

You can see comets when they are closest to the Sun. Energy from the Sun melts the ice, and the 'solar wind' makes the vapour into a tail.

16 (a) Describe the orbit of a comet.
(b) Explain <u>when</u> you can see a comet and why you can see it.

The Universe

Our Sun is just one of many billions (thousands of millions) of stars in a galaxy. Our galaxy is just one of billions of galaxies in the Universe.

Some galaxies are beautiful spirals, like our galaxy. Others are round or cigar-shaped.

1 Look at the photograph.
How many galaxies can you see?

This shows part of the Virgo Cluster. It takes light about 40 million years to reach us from these galaxies.

How big is the Universe?

Stars in a galaxy are often millions of times further apart than the planets in the solar system.

Galaxies are often millions of times further apart than the stars inside a galaxy.

Astronomers often tell us how far away things are by saying how long it takes the light from them to reach us.

2 What is the name of our galaxy?

3 Where in our galaxy is the solar system?

4 How long does it take light to reach us from:
(a) the Sun?
(b) the nearest other star?
(c) the Virgo Cluster?

5 Light travels at 300 000 kilometres per second.
Work out how far away the Sun is from the Earth.

Our Sun is in a galaxy called the Milky Way, about two-thirds of the way out from the centre. It takes light about 100 000 years to cross the Milky Way.

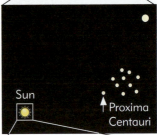

It takes four years for light to travel to the Sun from the next nearest star.

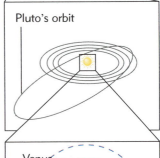

It takes light about 10 hours to cross the solar system.

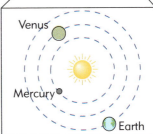

It takes light 500 seconds to travel from the Sun to the Earth.

How are stars made?

Galaxies don't just contain billions of stars. There is also a lot of gas and dust between the stars. The diagrams show how this gas and dust can form new stars.

6 (a) What materials are stars formed from?
(b) How does gravity help a star to form?
(c) What happens when enough material is pulled together?

There are clouds of dust and gas in galaxies.

The force of gravity between the bits of dust and the particles of the gas pulls them together. This takes a very long time because the force is very tiny.

If enough material gets pulled together it gets very hot and starts to shine. A new star is born. The star can make the gas around it glow.

How long do stars last?

Stars like the Sun do not burn for ever. They are so hot in the centre that hydrogen gas is turned into helium gas by a nuclear reaction called **fusion**. Vast amounts of heat and light are produced in this reaction. This is what makes stars shine brightly. Our Sun has been doing this for about 5 billion years, and it is only about halfway through its life.

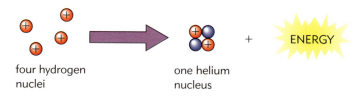

four hydrogen nuclei   one helium nucleus   +   ENERGY

7 Why can't the Sun keep shining for ever?

8 For how much longer do scientists think the Sun will continue to shine?

Is there life elsewhere in the Universe?

The Earth is just one of several planets orbiting the Sun. The Sun is just one of billions of stars. Once people knew this, they wondered if there was life on other planets.

At the beginning of the twentieth century, an American astronomer called Percy Lowell claimed to have discovered canals on Mars. This was evidence, he thought, that there must be intelligent life on Mars.

9 We now know that Lowell's canals came as much from his imagination as from his observation.
What does this tell us about science and scientists?

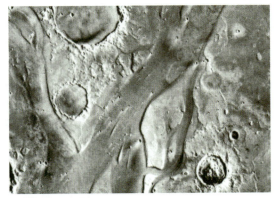
The surface of Mars.

Could there be life on Mars?

Scientists have observed Mars through telescopes, from spacecraft and from Martian landers. So they know that there are no large living organisms on Mars.

To decide whether there might be microscopic forms of life (microorganisms) on Mars, scientists look for evidence of **water** and of changes to the **atmosphere**.
For evidence of past life they can look for **fossils**.

Water
Most scientists think that life is very unlikely to exist unless water is present in its liquid state.

There is now very little, if any, water on the surface of Mars. Even if there were, it would be in the form of ice.

But the landscape of Mars shows that it has been eroded in the past by flowing water. There might be liquid water deep below the surface of Mars.

10 Where, on Mars:
(a) might life have existed in the past?
(b) might life exist today?

Atmosphere
Living things normally change the atmosphere around them. The Earth's atmosphere, for example, contains far more oxygen than it would if there were no living organisms.

An experiment was done by a robot from the Viking lander in 1976. It put Martian soil into a closed container and added water plus various nutrients. It then measured changes in the gases above the soil.

Gases in atmosphere	Earth (no life)	Earth (actual)	Mars
carbon dioxide	98%	0.03%	95%
nitrogen	1.9%	79%	2.7%
oxygen	trace	21%	0.13%

The only changes in the gases were what you would expect from purely chemical reactions with the soil.

11 Is the idea of life on Mars supported by:

(a) the percentages of different gases in the Martian atmosphere?

(b) the Viking experiments?

Fossils from Mars?

Scientists think that some meteorites found in Antarctica came from Mars. In 1996, the American space agency, NASA, claimed that there was fossil evidence of bacteria in one of these meteorites.

12 Give one reason <u>for</u>, and one reason <u>against</u>, the claim that the Martian meteorite contains fossils.

During the 1980s and early 1990s, the American public had lost interest in NASA and its funding had been reduced.

13 Why might the space scientists have decided to make such a controversial announcement in 1996?

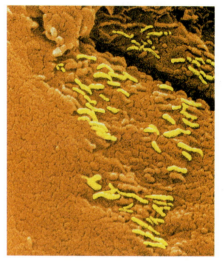

The forms inside the Martian meteorite are the same shape as some bacteria on Earth. But they are 100 times smaller.
Similar forms are found in rocks on Earth, but scientists don't know whether or not they are fossilised mini-bacteria.

Where else might there be life?

The most likely place, besides Mars, to find life in the solar system is on Europa, one of the satellites ('moons') of Jupiter. Pictures from spacecraft tell us that there is a lot of ice on the surface of Europa.

14 (a) Why do scientists think that there might be life on Europa?

(b) How might this life be detected by a lander without having to drill through the very thick ice?

Many of the billions of stars in the Universe are likely to have planets. So it's possible that there is life on some of these.

But these planets are so far away that we're not likely to find out if there's life there unless there's intelligent life that can send us signals. [See pages 462–463.]

There are changing patterns of cracks in the ice and very few meteorite craters on the surface of Europa. This suggests that there may be liquid water beneath the ice which sometimes wells up through the cracks.

SETI – the search for extra-terrestrial intelligence

Scientists have long thought that many other stars besides the Sun would have planets. But it is only recently that they have obtained definite evidence for such planets.

Because there are billions of stars in our own galaxy alone, it is very likely that there will be life of some kind on planets around some of these other stars.
However, we are unlikely to be able to detect such life unless there are intelligent beings who have developed the technology to be able to send us signals.

15 (a) Why do scientists <u>think</u> that there is probably life elsewhere in our galaxy?

(b) Explain, as fully as you can, why scientists will never <u>know</u> that there's life around other stars unless there is <u>intelligent</u> life?

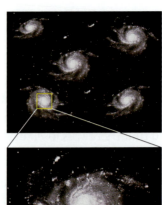

There are billions of galaxies...

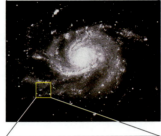

...each with billions of stars...

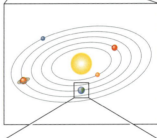

...many of which have planets...

...some of which have conditions suitable for life.

How other intelligent life might detect us?

For the past 100 years, humans have used radio waves to carry information between different places on Earth. We say that we have used radio <u>signals</u>.

More recently, we have sent TV and telephone signals up to satellites and back down to Earth again using microwaves.

Some of these signals would travel out of the solar system to the stars beyond. They would, however, be very weak by the time they reached even the nearest star.

Since the 1960s, signals have also been deliberately sent out into space in the hope that there might be intelligent life-forms out there with the technology to pick them up.

16 Explain, as fully as you can, how intelligent beings elsewhere in the galaxy might know about us.

But the planets around other stars are much too far away for us to explore. So we'll only know if there are living things there if they send out a signal that they're there.

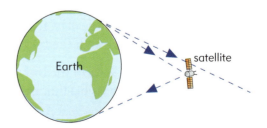

Microwaves sent up to satellites will also travel out into space. So aliens could eavesdrop on our telephone calls and TV programmes.
The signals would, however, be very weak by the time they reached an alien planet.

How we might detect intelligent life elsewhere

Just as other intelligent life-forms might detect microwaves that we have (accidentally or deliberately) sent out into space, so we might detect signals that <u>they</u> have sent out.

In 1967, a Cambridge astronomer called Antony Hewish and a young researcher called Jocelyn Bell picked up a strange 'signal' on their radio-telescope (see the graph). In the following weeks, they found several other sources of similar signals, each with its own frequency.

17 (a) Why do you think that Hewish and Bell thought, at first, that they had received signals from intelligent aliens? Why did they change their minds?
 (b) Where <u>did</u> the pulses of microwaves come from?

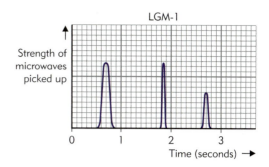

LGM-1

Strength of microwaves picked up

Time (seconds)

At first, Hewish and Bell thought that they had received signals from some extra-terrestrial beings. So they called their signals LGM-1, LGM-2 etc. (LGM = little green men).
Astronomers later realised that the signals came from rotating neutron stars. These stars are called pulsars.

SETI continues

Scientists continue to use radio-telescopes to search for information-carrying signals from space. Many individuals help with this search. They hook up their personal computers (PCs) to the observatories whose dishes collect the microwaves that reach Earth from elsewhere.

18 What would an information-carrying signal probably be like?

19 Why do scientists look mainly for microwaves with wavelengths between 18 cm and 21 cm?

Even if a genuine signal from space were detected, there would still be a HUGE problem in trying to find out what it meant!

> **What sort of signal are we looking for?**
> The very regular pulses from a pulsar would not carry any information. Signals sent by extra-terrestrials would be irregular, and would probably be repeated over and over again.

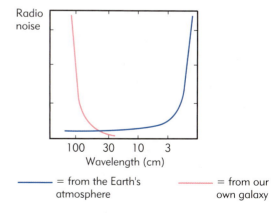

Radio noise

Wavelength (cm)

——— = from the Earth's atmosphere ——— = from our own galaxy

There is least background noise for microwaves of wavelength about 20 cm. There are several million separate frequencies in this range which need to be monitored.

How it all began

The life history of a star

The Sun is an average star that is in the **stable** period of its life. It has been more or less like it is now for the past 5 billion years, and it will stay the same for another 5 billion years. The diagrams show why.

When most of the Sun's hydrogen has been converted into helium, the Sun will expand into a **red giant**. This will be about ten times the present diameter of the Sun. The inner planets will be vaporised. Only the rocky cores of the outer planets will remain.

1 Why is the Sun at present in the middle of a very long stable period?

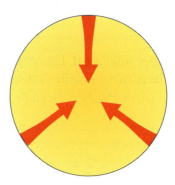

A star has a very big mass, so the force of its own gravity tends to make it collapse. This makes the core of the star hot enough for nuclear fusion reactions to occur. The bigger the mass of the star, the faster these reactions are.

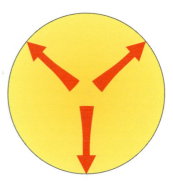

The energy released by the nuclear fusion reactions produces a very high temperature. This creates a pressure which tends to make the star expand.

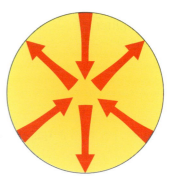

During the main period of a star's life, the gravitational forces and the outward pressure forces are balanced. So the star is stable.

Part of the red giant that will eventually form from the star shown above.

What happens after the red giant stage depends on the mass of the star.

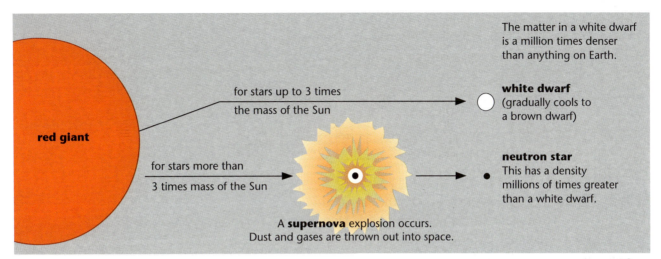

for stars up to 3 times the mass of the Sun

red giant

for stars more than 3 times mass of the Sun

A **supernova** explosion occurs. Dust and gases are thrown out into space.

The matter in a white dwarf is a million times denser than anything on Earth.

white dwarf (gradually cools to a brown dwarf)

neutron star This has a density millions of times greater than a white dwarf.

2 Describe, as fully as you can, what will happen to the Sun starting from about 5 billion years in the future. [You will also need information from page 464.]

3 What would be different about the Sun's future if it were about four times as massive as it actually is?

Did you know?

The mass of 1 teaspoonful of matter:
- from the Earth's crust is about 15 g
- from the Earth's core is about 50 g
- from a white dwarf is about 5 tonnes
- from a neutron star is about 500 million tonnes.

Higher

Second generation stars and black holes

The 'big bang' that started our Universe about 15 billion years ago created just three types of atom (elements).

The main element that was created was hydrogen. Some helium was also created and a much smaller amount of lithium.

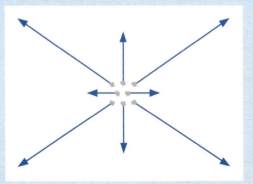

The 'big bang' created mainly hydrogen atoms, some helium atoms and far fewer lithium atoms.

Higher

These atoms were spread out very thinly through space. In some places huge numbers of atoms are gradually pulled together by the gravitational attraction between them. This is how stars are produced.

During the stable period of a star, the energy that is needed to keep the star stable is produced by nuclear fusion reactions. Hydrogen atoms join together (fuse) to produce helium atoms.

When much of the hydrogen in a star has become helium, the hydrogen → helium fusion reaction can no longer maintain the star in its stable state. It then becomes a **red giant**.

Different nuclear fusion reactions, inside a red giant, produce bigger and bigger atoms up to the size of iron atoms. The biggest atoms produced are iron atoms. Fusion reactions for making atoms that are bigger than iron atoms do not release energy and do not occur in a red giant.

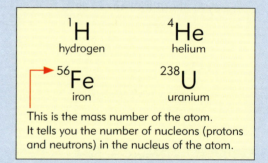

^{1}H hydrogen ^{4}He helium

^{56}Fe iron ^{238}U uranium

This is the mass number of the atom. It tells you the number of nucleons (protons and neutrons) in the nucleus of the atom.

Eventually, the fusion reactions inside a red giant start to slow down. If the star was massive enough to begin with, the red giant then collapses very rapidly to produce a **supernova explosion**.

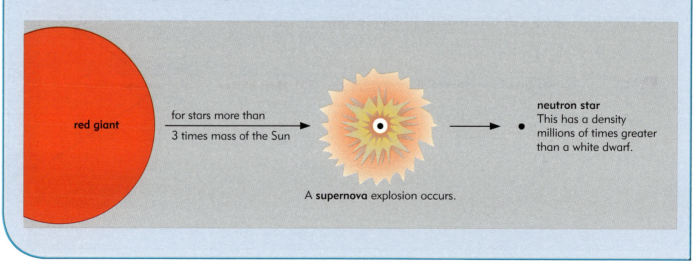

red giant

for stars more than 3 times mass of the Sun

A **supernova** explosion occurs.

neutron star
This has a density millions of times greater than a white dwarf.

Higher

During a supernova explosion, enough energy is available for fusion reactions to produce atoms bigger than iron atoms. In fact, atoms of all the naturally occurring elements are normally formed, including uranium which has the biggest and most massive atoms.

When dust and gases are thrown into the surrounding space from a supernova explosion, atoms of all the elements are spread out amongst the hydrogen and helium atoms that have been there since the 'big bang'.

Whenever these atoms are pulled together by their gravitational attraction into large enough clumps, new stars are produced.

These stars contain not only hydrogen and helium but also material from supernova explosions. So we call them **second generation stars**. They contain atoms of all the elements.

After the supernova explosion from a very massive star, a **black hole** may be left behind, rather than a neutron star.

The matter in a black hole is so dense that nothing – not even light and other forms of electromagnetic radiation – can escape from its gravitational pull.

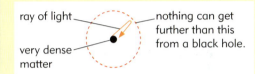

Because no radiation is emitted from a black hole, you cannot observe it directly. But matter which is spiralling into a black hole emits X-rays and we can sometimes detect these.

4 What elements would you expect to find in the following stars?
   Explain your answer in each case.
   (a) A first generation star during its stable phase.
   (b) The red giant from (a) towards the end of that phase of its life.
   (c) A second generation star

5 Explain what a black hole is, how is it produced and why it is called a 'black hole'.

6 The solar system contains all the naturally occurring elements, up to and including uranium.
   What does this tell you about the solar system?

7 The table shows how long stars of different masses exist as stable stars like the Sun.
   (a) What pattern do these figures show?
   (b) Can you suggest a possible reason for the pattern?
   (c) How would each star listed in the table end up?

Mass of star (Sun = 1)	Life as a star like the Sun (billions of years)
3.0	0.5
1.0	10
0.5	200

How did the Universe begin?

During the 1920s, the astronomer Edwin Hubble discovered that distant galaxies are moving away from us. He also discovered that the further away galaxies are from us, the faster they are moving away.

This suggests that, at one time, all the matter in the Universe was in the same place and that the Universe began with a huge explosion. This explosion is called the 'big bang'.

8 (a) Explain what is meant by the 'big bang' theory.
    (b) Why do astronomers think that the Universe began in this way?

By measuring the speeds of many galaxies and their distances from us, astronomers have calculated when the 'big bang' happened. They think that it was probably about 15 billion years ago (see Box below).

Astronomers measure the huge distances to stars and galaxies in **light-years**. A light-year is the distance that light travels in a year. Light travels 300 million metres every <u>second</u>, so it travels a very long way in a whole year.

A galaxy that is 1.6 billion light-years away is travelling away from us at about one-tenth of the speed of light.

$$\text{speed} = \frac{\text{distance}}{\text{time}}$$

So

$$\text{time} = \frac{\text{distance}}{\text{speed}}$$

$$= \frac{1.6 \text{ billion}}{0.1}$$

$$= 16 \text{ billion years}$$

The 'big bang' theory of the beginning of the Universe

To begin with, all the matter of the Universe was in one place.

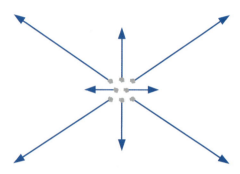

Then a huge explosion sent matter flying out in all directions. The particles of matter were spread out a long way from each other.

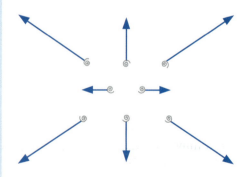

Gravity pulled the matter together in some places to make galaxies of stars. These galaxies are still moving away from each other. The further away from each other galaxies are, the faster they are moving apart.

How do we know the Universe is expanding?

Astronomers examine the light from a star or a galaxy by splitting it up into a **spectrum**. They can tell from these spectra that distant galaxies are moving away from us.

The box explains how they do this.

The spectra of stars and galaxies have patterns of dark lines. These are the 'fingerprints' of different elements.

For example, in the yellow part of the spectrum of sunlight there is a pair of dark lines. These are due to sodium in the Sun's atmosphere.

In the spectrum of the light from a distant galaxy, the lines of particular elements such as sodium are shifted towards the red end of the spectrum. This is called a **red-shift**. It tells us that the galaxy is moving away from us.

A more distant galaxy has a bigger red-shift. This tells us that it is moving away faster.

[Note: There are lots of dark lines, spread right across the spectrum. Different lines correspond to different elements.]

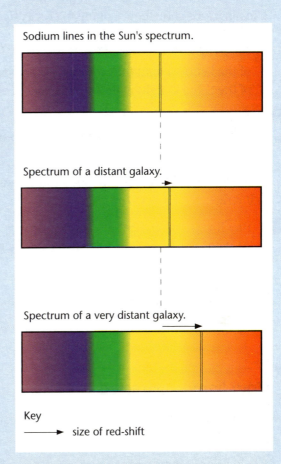

Sodium lines in the Sun's spectrum.

Spectrum of a distant galaxy.

Spectrum of a very distant galaxy.

Key

→ size of red-shift

9 (a) Which way is a galaxy moving if the light from it shows a red-shift?
   (b) What does the size of a red-shift tell us?

10 The Universe may continue to expand for ever. Or it may eventually stop expanding and start to collapse again. This would end in a 'big crunch'. Explain each of these possibilities in terms of kinetic and potential energy.

11 The light from a few nearby galaxies shows a slight <u>blue-shift</u>. What does this tell us?

Hint for question 10
When bodies move apart against the force of gravity they slow down. So kinetic energy decreases. There is a corresponding increase in their gravitational potential energy.

12 Waves and radiation

Reflection and refraction

How light is reflected

A mirror that is flat is called a <u>plane</u> mirror.
The diagrams show how a plane mirror **reflects** light.

Look at the angle with the mirror:

- of the beam of light that comes from the lamp;
- of the beam of light after it is reflected.

1 Look at the diagrams.
What can you say about the angle for beams that arrive at, and reflect from, the mirror?

Light is reflected from a plane mirror at the same <u>angle</u> as it strikes the mirror.

2 Ray X is going straight towards a mirror.
At what angle will it be reflected from the mirror?
Give a reason for your answer.

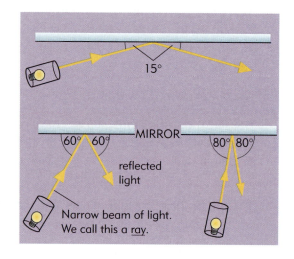

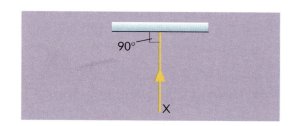

How light is refracted

Light can be **refracted** when it passes from one substance into another.

The diagrams show you more about the way light is refracted.

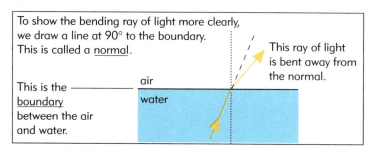

To show the bending ray of light more clearly, we draw a line at 90° to the boundary. This is called a <u>normal</u>.

This is the <u>boundary</u> between the air and water.

air

water

This ray of light is bent away from the normal.

3 (a) When light travels from water into air, which way is it bent (refracted)?
(b) When light travels from air into water, which way is it bent (refracted)?

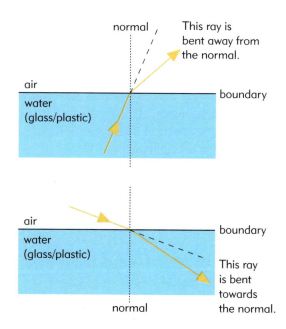

normal — This ray is bent away from the normal.

air

water (glass/plastic)

boundary

air

water (glass/plastic)

boundary

This ray is bent towards the normal.

normal

Rays that aren't refracted

The diagrams show what happens to rays of light when they strike a boundary head-on at 90° (along a normal).

4 What happens to light as it crosses a boundary along the normal?

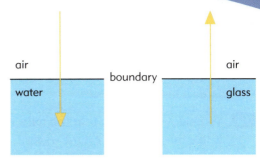

These rays of light strike the boundary at 90°. They do not change direction.

How sound is reflected

It isn't only light that can be reflected.
You can also reflect **sounds**.

4 (a) What sort of surface does sound reflect well from?
(b) What angle does the sound reflect from the surface at?

The ticks sound loudest when the tubes are at the same angle to the surface.

How sound is refracted

Sound can also be bent or refracted, just like light. This happens when sound travels across the boundary between two different substances.

6 (a) What is used to focus light on a screen?
(b) What is used to focus sound at a microphone?

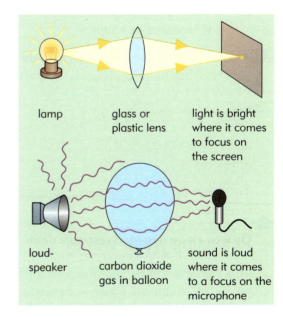

Water waves

You can see waves on lakes and on the sea.
The wind makes these waves.
But we can look at water waves more easily if we make them in a tank.
The diagram shows how you can do this.

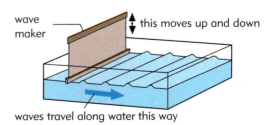

Reflecting water waves

The diagram shows what happens if you send water waves towards a hard, flat surface.

7 Write down <u>one</u> way that the reflection of water waves is like the reflection of light waves.

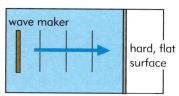

wave maker

hard, flat surface

the lines show the top of each wave

reflected waves

Your friend sends some water waves at 90° to a hard, flat surface.

8 What will happen to these waves after they have hit the surface?

wave maker

hard, flat surface

these waves are going straight at the hard, flat surface

Making the water shallower

The diagrams show what happens to water waves when the water suddenly gets shallower.

9 (a) What happens to the <u>distance</u> between the water waves in the shallower water?
(b) Why does this happen?

10 What happens to the <u>direction</u> of these water waves when they move into the shallower water?

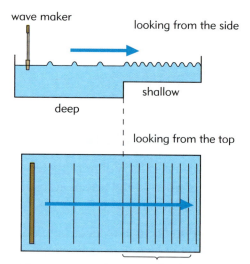

wave maker

looking from the side

shallow

deep

looking from the top

The waves get closer together here. This is because they travel slower.

Refracting water waves

This diagram shows what happens if you send water waves into shallower water at an angle.

11 What <u>two</u> things happen to water waves as they move into shallower water at an angle?

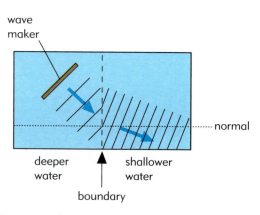

wave maker

normal

deeper water

shallower water

boundary

The waves change direction. They are refracted.

Light waves and sound waves

Water waves are reflected and refracted just like light and sound. So scientists think that light and sound are also waves.

Light and sound are refracted because they travel faster through some substances than through others.
For example, light travels slower through glass or plastic than it does through air.

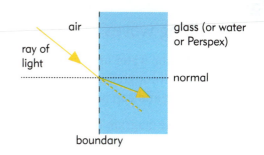

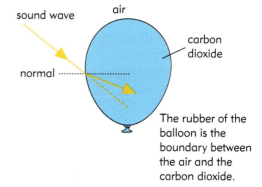

The rubber of the balloon is the boundary between the air and the carbon dioxide.

12 What <u>two</u> things happen to light waves as they move from air into glass?

13 Sound is refracted when it passes from air into carbon dioxide. Why does this happen?

What are waves?

We can't see light waves or sound waves.
But we <u>can</u> see water waves and waves travelling along a rope, so it is worth looking carefully at these waves.

This helps us to understand more about light waves and sound waves.

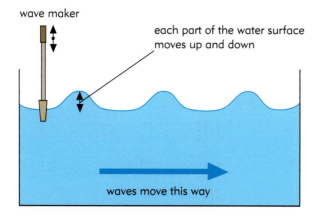

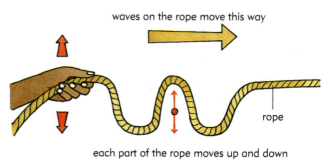

1 (a) What can waves travel along?
  (b) Do the particles in the water or rope move up and down or do they move along with the wave?

Describing waves

We say that the water and the rope are **disturbed** as waves travel through them. The disturbances can be different sizes and can be different distances apart.

2 Look at the diagram of a wave.
(a) What do we call the size of a disturbance?
(b) What do we call the distance between one disturbance and the next?

3 Look at the diagram. Write down:
(a) the amplitude,
(b) the wavelength
of the wave in cm.

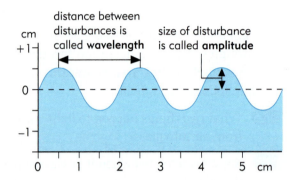

Differences between waves

The diagrams show three different sets of water waves: A, B and C.

4 Look at the diagrams of waves A, B and C.
(a) Write down the amplitude of each wave, in cm.
(b) Write down the wavelength of each wave, in cm.

5 (a) Which wave has the largest amplitude?
(b) Which wave has the shortest wavelength?

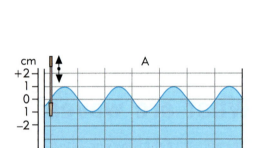

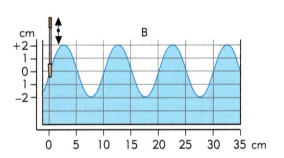

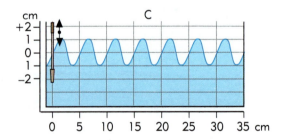

Another difference between waves

You can make a lot of waves each second or just a few waves each second.

If there are 10 waves each second, we say that the **frequency** is 10 **hertz** (Hz, for short).

The diagrams show what happens when you start to make water waves on calm water.

6 (a) How many complete waves are there after 1 second?
   (b) How many complete waves are there after 2 seconds?
   (c) How many complete waves are made during each second?
   (d) What is the frequency of the water waves?

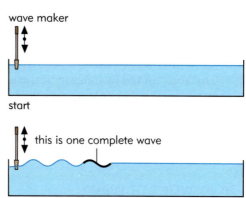

wave maker

start

this is one complete wave

1 second from start

2 seconds from start

Frequency and wavelength

The diagram shows what happens if you now make waves with double the frequency. The water is the same depth as before.

7 What happens to the wavelength when you double the frequency of the waves?

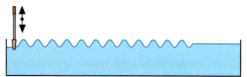

This is what you see after 2 seconds, if you make twice as many waves each second.

What do waves do?

Waves transfer **energy** from one place to another.

8 Look at the diagram.
   (a) What does the wave maker do?
   (b) Which way does the water move?
   (c) Where does the energy move to?

Waves transfer energy through the water from the wave maker to the foam rubber. But the water itself does <u>not</u> move from one end of the tank to the other.

9 Which way <u>does</u> the water move as the waves travel across it?

10 What happens to the energy carried by the waves when it is absorbed by the foam?

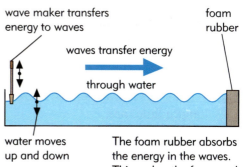

wave maker transfers energy to waves

foam rubber

waves transfer energy through water

water moves up and down

The foam rubber absorbs the energy in the waves. This makes the foam and the water slightly warmer.

The diagram shows how we can use the energy from water waves.

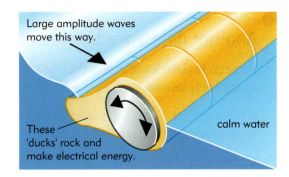

Large amplitude waves move this way.

These 'ducks' rock and make electrical energy.

calm water

11 (a) What useful form of energy do the 'ducks' produce from the energy carried by waves?

(b) You can tell by looking at the water that the 'ducks' have transferred most of the energy from the waves. Explain how you can tell.

Two types of wave

The diagram shows waves moving along a rope. Each wave is a disturbance that travels along the rope, but the rope itself does <u>not</u> travel along.

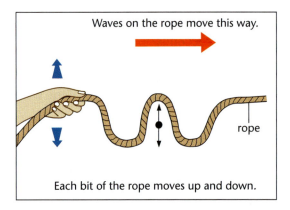

Waves on the rope move this way.

rope

Each bit of the rope moves up and down.

12 (a) In which direction are the waves travelling along the rope in the diagram?

(b) Describe how each bit of the rope moves as the waves pass along it.

Because each bit of the rope vibrates <u>at right angles</u> to the direction that the waves are travelling, we say that the waves are **transverse** waves. ('Transverse' means 'across'.)

The diagram shows waves moving along a spring. Each wave is a disturbance that travels along the spring, but the spring itself does <u>not</u> travel along.

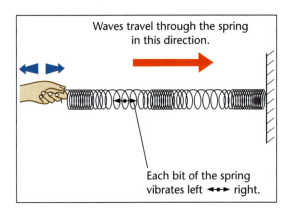

Waves travel through the spring in this direction.

Each bit of the spring vibrates left ←→ right.

13 (a) In which direction are the waves travelling along the spring in the diagram?

(b) Describe how each bit of the spring moves as the waves pass through it.

As the waves travel along the spring, each bit of the spring vibrates. These vibrations are <u>along the same direction</u> as the waves are travelling. So we say that the waves are **longitudinal** waves.

These diagrams show water waves and sound waves.

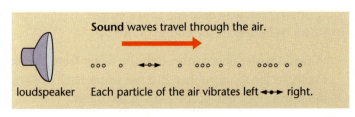

Sound waves travel through the air.

loudspeaker   Each particle of the air vibrates left ←→ right.

14 What types of wave (transverse or longitudinal) are:
(a) water waves?
(b) sound waves?

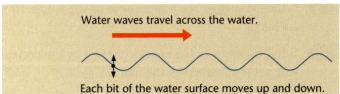

Water waves travel across the water.

Each bit of the water surface moves up and down.

A formula for waves

The wave maker shown in the diagrams has a frequency of 10 hertz (Hz). This means that it makes 10 waves every second. If you look at the diagrams, you will see that by the end of one second the first wave has travelled a distance equal to 10 wavelengths. So the **speed** of the waves is 10 wavelengths per second.

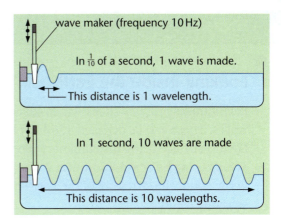

If the wave maker is changed to a frequency of 4 Hz, the wave-speed is now 4 wavelengths per second.

In fact, for <u>any</u> frequency:

$$\begin{array}{c}\text{wave-speed} \\ \text{(metres per second, m/s)}\end{array} = \begin{array}{c}\text{frequency} \\ \text{(hertz, Hz)}\end{array} \times \begin{array}{c}\text{wavelength} \\ \text{(metres, m)}\end{array}$$

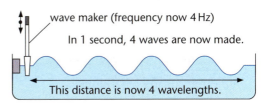

15 A sound with a frequency of 440 Hz has a wavelength in air of 0.75 m.
Calculate the speed of sound in air.
[Start by writing down the formula. Show all your working.]

> **Example**
> Some water waves have a wavelength of 5 cm and a frequency of 4 Hz.
> What is the wave-speed?
> wave-speed = frequency × wavelength
> = 4 × 5
> = <u>20 cm/s</u> (= 0.2 m/s)

Waves that go through the Earth

When there is an earthquake, vibrations travel through the Earth's crust and then through the inside of the Earth. They are called **seismic** waves. We can detect seismic waves far away from the earthquake that caused them. We use a **seismograph** to do this.

16 (a) Earthquakes often occur in remote areas where nobody lives. How do scientists know about these earthquakes?
(b) What have scientists learned from seismic waves about the inside of the Earth?

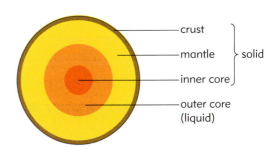

By studying seismic waves, scientists know about the different layers inside the Earth.

Looking inside the Earth

Earthquakes produce shock waves called **seismic** waves. These can be detected at other places on the Earth's surface using instruments called **seismographs** or **seismometers**. Most of what we know about the structure of the Earth comes from studying seismic waves.

There are two main types of seismic wave:

- primary waves or **P-waves** and
- secondary waves or **S-waves**.

There is also a third type of shock wave called surface waves. These are the waves that damage buildings during earthquakes.

The diagram provides information about P-waves and S-waves and what they tell us about the structure of the Earth.

17 (a) What type of wave are:
    (i) P-waves?
    (ii) S-waves?
  (b) Which travel faster, P-waves or S-waves?
  (c) Which of the two types of seismic wave cannot travel through liquid?
    Give a reason for your answer.

18 Copy and complete the table.

Part of Earth	Thickness (km)	Solid or liquid?
crust		
mantle		
outer core		
inner core		

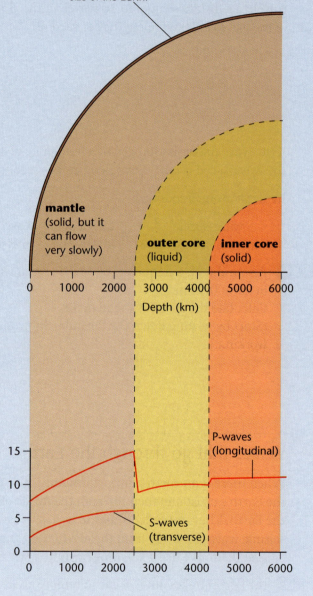

The Earth's **crust** is solid rock, 10–40 km thick. This is very thin compared with the size of the Earth.

mantle (solid, but it can flow very slowly)

outer core (liquid)

inner core (solid)

Depth (km)

Speed of waves (km/s)

P-waves (longitudinal)

S-waves (transverse)

How S-waves travel through the Earth

As the mantle gradually becomes denser, the speed of S-waves gradually increases. This means that they are gradually refracted, as shown on the diagram.

19 (a) Make a copy of the diagram.
    (b) Add a caption to your diagram explaining the curved path of the S-waves.

20 Why do S-waves produce a seismograph shadow zone?

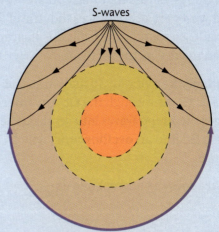

S-waves

seismograph <u>shadow zone</u>

How P-waves travel through the Earth

P-waves are refracted as they travel through the Earth's mantle in just the same way as S-waves, but the P-waves also travel through the Earth's core.

Whenever there are sudden changes in the speed of the P-waves, refraction produces a sudden change in their direction.

21 (a) Make a copy of the diagram.
    (b) Where do the sudden changes in the speed and direction of the P-waves mainly occur?
    (c) Which P-wave changes its speed as it passes through the Earth but does not change its direction?

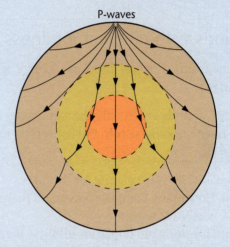

P-waves

Note: There is only a small change in speed at the boundary between the outer core and inner core. This means that there is only a small change in direction. (It is too small to show on this diagram.)

22 The small diagram shows a P-wave that takes a slightly different path from any of the P-waves shown on the earlier diagram.
    (a) Make a copy of this diagram.
    (b) State what is different about the path of this P-wave.
    (c) Add to your diagram the P-wave that sets out vertically downwards.

23 Do P-waves produce a shadow zone? Explain your answer.

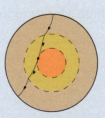

You get a seismograph record of P-waves <u>everywhere</u> on the Earth's surface.

Looking at sound waves

When things vibrate they make sound waves.
Sound waves are disturbances that travel through the air.
You can't see sound waves, but the diagram shows how
you can make a picture of them.

A **loudspeaker** vibrates to make a sound. The sound
travels as waves through the air. The **microphone**
changes the sound waves into an electrical signal. The
signal is fed into an **oscilloscope**, which changes it into
a picture.

1 What does the microphone do?

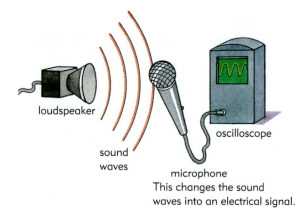

loudspeaker

sound
waves

oscilloscope

microphone
This changes the sound
waves into an electrical signal.

The sound waves are longitudinal but the
oscilloscope shows them as transverse waves.

Sounds with different loudness

The diagrams show oscilloscope pictures of two sounds,
A and B.

The two sounds are exactly the same note, so we say that
they have the same **pitch**. But sound B is louder than
sound A.

2 Look at the wave picture for sound A.
  (a) How many complete waves are there on the
      oscilloscope picture?
  (b) How long did it take for all these waves to hit the
      microphone?
  (c) How many complete waves would there be in a
      whole second?
  (d) What is the frequency of the sound waves?

The **loudness** of a sound depends on the amplitude of the
sound waves.

3 Why does sound B have a larger amplitude than
sound A?

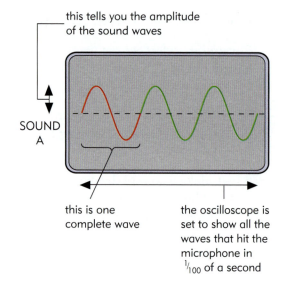

this tells you the amplitude
of the sound waves

SOUND
A

this is one
complete wave

the oscilloscope is
set to show all the
waves that hit the
microphone in
$1/100$ of a second

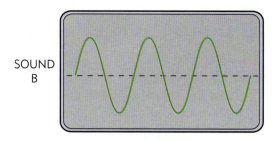

SOUND
B

Sound B is louder than sound A.

Sounds with different pitch

The diagram shows the oscilloscope picture of sound C.
Sound C is the same loudness as sound A, but sound C is a
higher note than sound A.

We say that it has a higher **pitch**.

4 (a) How many complete waves are there on the
oscilloscope picture of sound C?
(b) How long did it take for all these waves
to be made?
(c) How many complete waves would there be in a
whole second?
(d) What is the frequency of sound C?

5 Sound C has a higher frequency than sound A.
How does this change its pitch?

The higher the frequency of a sound, the higher its
pitch is.

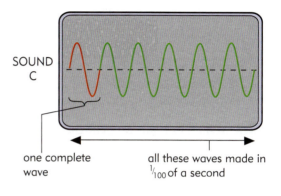

SOUND C

one complete wave · all these waves made in $\frac{1}{100}$ of a second

Another difference between sounds

Sounds A, B and C are 'pure' sounds.
They have only one frequency.
Most sounds are mixtures of different frequencies, but
they usually have one main frequency.
This frequency gives the sound its pitch.

Look at the pictures of sounds D, E and F.

6 Which is the loudest sound?

7 (a) Which sound has the highest pitch?
(b) What is the frequency of this sound?

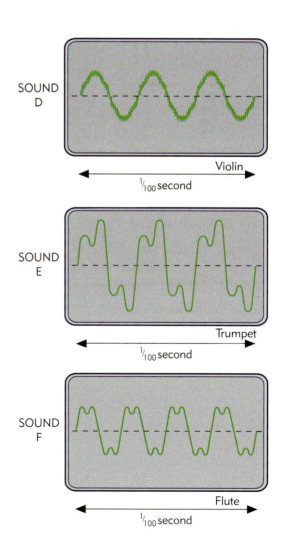

SOUND D

Violin · $\frac{1}{100}$ second

SOUND E

Trumpet · $\frac{1}{100}$ second

SOUND F

Flute · $\frac{1}{100}$ second

'Sound' you can't hear

If the frequency of a 'sound' is too high or too low you can't hear it.

The diagram shows what frequencies of sound humans can hear. It shows what some other animals can hear as well.

8 Which animal can hear the highest frequency?

9 What range of frequencies can a child hear?

10 As you get older, the range of frequencies you can hear changes.

How does it change?

11 Farmers sometimes control their sheepdogs with a 'silent' whistle.

How can a dog whistle be silent?

Sound with a very high frequency that is too high for humans to hear is called **ultrasound**.

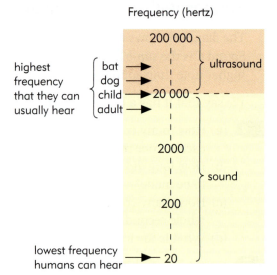

Frequency (hertz)

- highest frequency that they can usually hear: bat, dog, child → 20 000, adult
- 200 000 — ultrasound
- 2000, 200 — sound
- lowest frequency humans can hear → 20

Making ultrasounds

The photograph shows how you can make sound of any frequency, including ultrasound.

Inside a **signal generator**, an electrical circuit can produce electrical vibrations of any frequency. The signals are sent to a loudspeaker, which vibrates at the same frequency. Sound waves and ultrasound waves can be produced.

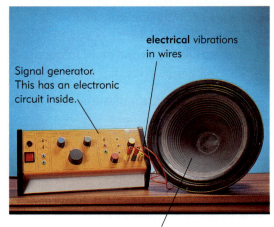

Signal generator. This has an electronic circuit inside.

electrical vibrations in wires

Vibrator (e.g. loudspeaker). This makes sound or ultrasound waves of the same frequency as the electrical signal.

Cleaning things with ultrasound

You can use ultrasound to clean a delicate wind-up watch without having to take it apart.

The diagram shows how you can do this.

12 (a) What is used to make molecules in the cleansing fluid vibrate?
(b) How do the vibrations affect the dirt in the watch?

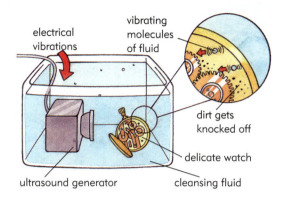

Ultrasound scans

Doctors use ultrasound to 'see' how a baby is developing inside its mother's womb. They make an ultrasound **scan**.

The doctor places an ultrasound generator on the mother's abdomen. The ultrasound travels through the mother's and the baby's bodies. When ultrasound travels from one substance to another, some ultrasound is reflected as an **echo**. An ultrasound detector collects the echoes. It changes them into a picture on the screen.

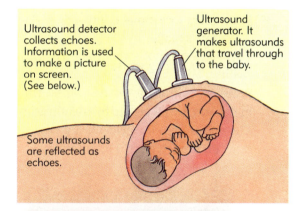

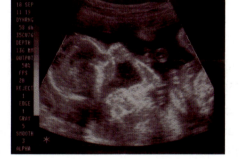

13 Look at the diagram.
(a) What does the ultrasound generator do?
(b) What does the ultrasound detector do?

14 Write down <u>one</u> other use of ultrasound scanning.

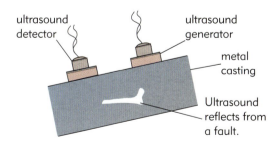

Using ultrasound to detect flaws in metal casings

To be strong, a metal casting must be solid metal all the way through. There should be no gaps or cracks in it.

Sometimes there are gaps or cracks inside the metal, so you can't see them. You can use ultrasound waves to detect flaws of this kind.

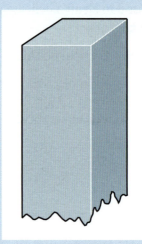

This metal casting needs to be strong. So there should be no gaps or cracks inside it.

The diagrams show what happens with a good casting. The ultrasound source makes short bursts (pulses) of ultrasound waves. These travel through the metal and are picked up by the ultrasound detector. The pulses from the detector are processed to make a trace on the screen of a monitor.

15 (a) Make a copy of the picture that you get on the screen from the good casting.
 (b) Explain why the screen shows <u>two</u> pulses.
 (c) Why is the left-hand pulse bigger than the right-hand pulse?

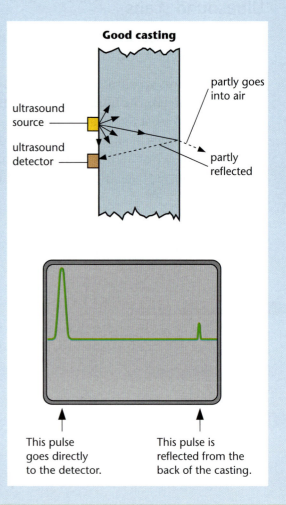

Good casting

ultrasound source

ultrasound detector

partly goes into air

partly reflected

This pulse goes directly to the detector.

This pulse is reflected from the back of the casting.

Higher

The diagrams show what happens if there is a crack or gap inside the metal. The ultrasound pulses are also reflected from the front and back of the <u>gap</u>. This means that the detector picks up <u>three</u> different reflections of each pulse. So the screen now shows <u>four</u> pulses.

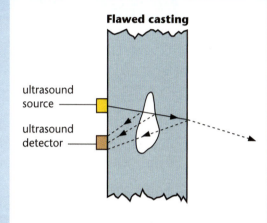

Flawed casting

ultrasound source

ultrasound detector

16 (a) Make a copy of the picture you get on the screen with the flawed casting. Label the pulses A, B, C and D, as on the diagram.
   (b) Underneath your diagram say what each of the pulses A–D represents.
   (c) Explain, as fully as you can, why the pulses are progressively smaller.

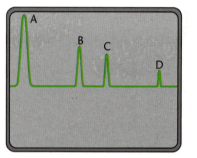

17 This diagram shows the ultrasound trace taken a few centimetres lower down on the casting than the one giving the traces A–D. Describe, in as much detail as you can, what this trace tells you. Give reasons for your answer.

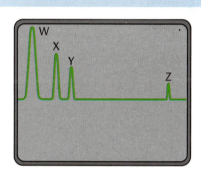

18 You can also use ultrasound to measure thickness. The reflection from the back of a steel plate occurs 0.1 millisecond after the pulse is made. How thick is the steel plate?

[The speed of ultrasound in steel is 1500 m/s. The diagram exaggerates the angles. The ultrasound travels almost horizontally through the steel and is reflected straight back.]

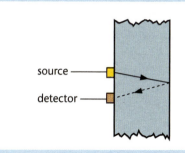

source

detector

Electromagnetic waves

Remember from Key Stage 3

- Sunlight looks white but it is made up of many different colours.

- A prism can be used to split sunlight into a spectrum of many colours.

- The colours in the spectrum gradually change in this order: red, orange, yellow, green, blue, violet.

- A prism bends (refracts) light.

- Different colours of light refract by different amounts so they change direction by different amounts.

- This effect is called dispersion.

- Light travels as waves and different colours of light have different wavelengths.

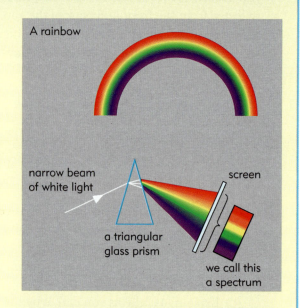

A rainbow

narrow beam of white light

a triangular glass prism

screen

we call this a spectrum

1 Look at the diagram.
(a) Which colour is refracted (changes direction) least?
(b) Which colour is refracted (changes direction) most?

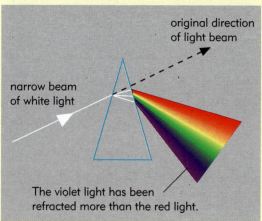

original direction of light beam

narrow beam of white light

The violet light has been refracted more than the red light.

2 Look at the diagram. Which colour has the longer wavelength?

3 How many violet waves are there:
(a) in a thousandth of a millimetre?
(b) in a whole millimetre?

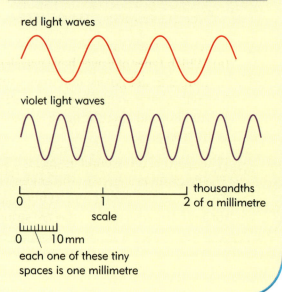

red light waves

violet light waves

0   1   2   thousandths of a millimetre

scale

0   10 mm

each one of these tiny spaces is one millimetre

Waves beyond the ends of the rainbow

White light is a mixture of many different colours. Different colours of light have different wavelengths.

We know that there are 'sounds' that we can't hear. This is because their frequency is too high or too low. Our ears can only hear certain frequencies.

There is also 'light' that we can't see.

This is because its wavelength is too long or too short. Our eyes can only see certain wavelengths.

The diagram shows how we can tell that these waves are there, even though we can't see them.

4 (a) How do we know that there are waves outside the red end of the spectrum?
　(b) What do we call these waves?

5 (a) How do we know there are waves outside the violet end of the spectrum?
　(b) What do we call these waves?

Infrared and ultraviolet waves are also called infrared and ultraviolet <u>radiation</u>.

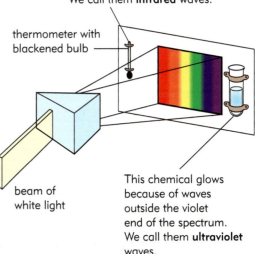

The thermometer gets hot because of waves outside the red end of the spectrum. We call them **infrared** waves.

thermometer with blackened bulb

beam of white light

This chemical glows because of waves outside the violet end of the spectrum. We call them **ultraviolet** waves.

There are waves with longer wavelengths than infrared waves.

There are also waves with shorter wavelengths than ultraviolet waves.

All of these waves are parts of a bigger spectrum, which we call the **electromagnetic spectrum**.

6 Look at the diagram.
　(a) Which types of waves have wavelengths that are longer than light?
　(b) Which types of waves have wavelengths that are shorter than light?

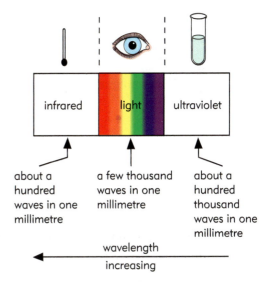

infrared | light | ultraviolet

about a hundred waves in one millimetre

a few thousand waves in one millimetre

about a hundred thousand waves in one millimetre

wavelength increasing

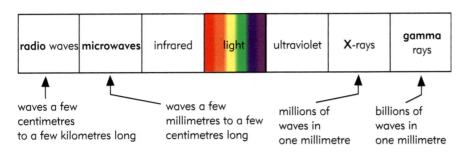

radio waves | **microwaves** | infrared | light | ultraviolet | **X-rays** | **gamma** rays

waves a few centimetres to a few kilometres long

waves a few millimetres to a few centimetres long

millions of waves in one millimetre

billions of waves in one millimetre

How fast do electromagnetic waves travel?

All the different kinds of electromagnetic waves travel at the same speed through space. So the waves with the shortest wavelength also have the highest frequency.

All electromagnetic waves are **transverse** waves. [See page 476 if you can't remember what this means.]

7 Look at the diagram.
  (a) Which type of wave has the lowest frequency?
  (b) Which type of wave has the highest frequency?

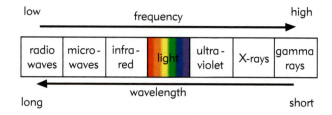

low   frequency   high

| radio waves | micro-waves | infra-red | light | ultra-violet | X-rays | gamma rays |

long   wavelength   short

8 Use information from the box.
  (a) How fast do all electromagnetic waves travel?
  (b) Do radio waves and X-rays travel at the same speed?

There are many different types of electromagnetic waves.

9 Write down the names of <u>seven</u> different types of electromagnetic waves.

10 Write down <u>two</u> things that are the same about all the types of electromagnetic waves.

> In space <u>all</u> electromagnetic waves travel at the same speed.
>
> This speed is 300 million metres per second.
>
> Electromagnetic waves are also called electromagnetic radiation.

What happens when electromagnetic waves hit things?

All electromagnetic waves do not need a substance to travel through. So they can travel easily through empty space (a **vacuum**).

But electromagnetic waves often bump into matter in the form of solids, liquids or gases.

The diagrams show what can then happen.

11 What <u>three</u> things can happen when electromagnetic waves hit matter?

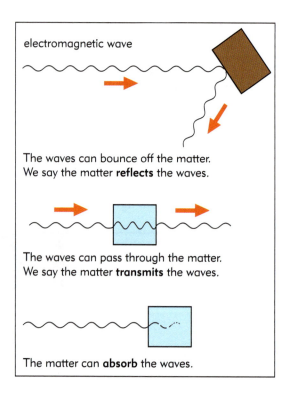

electromagnetic wave

The waves can bounce off the matter. We say the matter **reflects** the waves.

The waves can pass through the matter. We say the matter **transmits** the waves.

The matter can **absorb** the waves.

Absorbing electromagnetic radiation

Sometimes when radiation hits a solid, liquid or gas, some of the radiation is **transmitted** and some is **absorbed**.

This is what happens when light hits polythene.

The diagram shows how you can find out how much light polythene lets through.

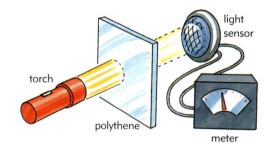

12 (a) What thickness of polythene absorbs half the light?
(b) What percentage of light passes through 4 mm of polythene?
(c) How does the absorption of light change when the polythene is thicker?

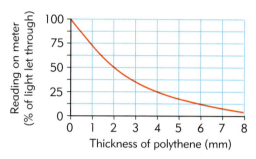

What happens when electromagnetic waves are absorbed?

Like all waves, electromagnetic waves carry energy from one place to another. We say that the waves **transfer** energy.

When waves are absorbed, this energy is transferred to the material that absorbs them. The diagram shows what happens then.

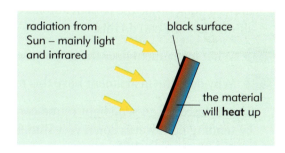

13 A black surface absorbs light and infrared radiation. What happens to the energy carried by this radiation?

14 Radio waves are absorbed by an aerial. What happens to the energy carried by these waves?

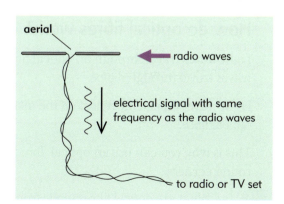

489

Surfaces that partly reflect

Light passes quite easily through **transparent** substances. But when light passes from one transparent substance to another, some light is always **reflected**.

15 Look at the top diagram.
What percentage of light does the glass reflect?

16 The piece of glass transmits only about 89 per cent of the light that falls on it. Explain, as fully as you can, what happens to the other 11 per cent.

Different types of electromagnetic radiation are transmitted, reflected or absorbed by different materials.

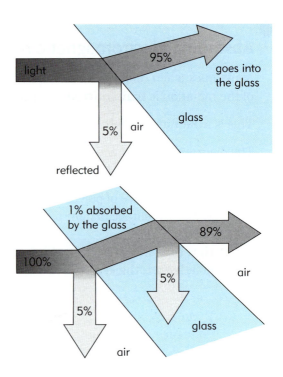

Changing the direction of waves

A doctor thinks a patient has an stomach ulcer, so she needs to look inside the patient's stomach. The diagram shows how she can do this.

1 What instrument does the doctor use?

2 Why does the instrument have this name?

The endoscope can send light round corners. It does this by sending light waves down very thin fibres made from glass, called **optical fibres**.

Using an endoscope ('endo' means 'inside', 'scope' means 'looking for').

How do optical fibres work?

Light waves travel through an optical fibre just like sound waves travel through a pipe.

Light waves repeatedly reflect off the inside surfaces of the optical fibre.

This is why you can use an optical fibre to send light round corners.

3 (a) What are optical fibres made from?
(b) How thick is an optical fibre?

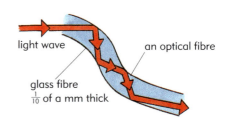

Light waves travelling through an optical fibre.

Why doesn't light get out of an optical fibre?

The diagrams show what happens to some beams of light when they meet the boundary between glass and air.

4 (a) What percentage of light passes out of the glass and is refracted?

(b) What percentage of light is reflected back into the glass?

The diagrams show what happens as a ray of light makes a bigger and bigger angle with a normal.

5 Look at the diagrams.

(a) What happens to a ray of light which reaches the boundary at an angle of 42°?

(b) What happens to the light which reaches the boundary at a bigger angle?

Total internal reflection happens when light reflects off the inside surface of the glass. The angle at which you start to get total internal reflection is called the **critical angle**.

6 What is the critical angle for glass?

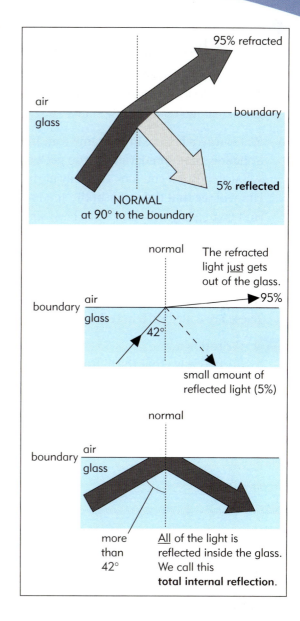

95% refracted

air

glass

boundary

NORMAL at 90° to the boundary

5% **reflected**

normal — The refracted light just gets out of the glass.

boundary — air — glass

42°

95%

small amount of reflected light (5%)

normal

boundary — air — glass

more than 42° — All of the light is reflected inside the glass. We call this **total internal reflection**.

How an endoscope works

The diagram shows how an endoscope lets a doctor see the inside of your stomach.

7 Look at the diagram.

(a) How does light travel from the lamp to the patient's stomach?

(b) What happens to some of the light that shines onto the inside of the stomach?

(c) Why is there a second bundle of optical fibres?

How an endoscope works.

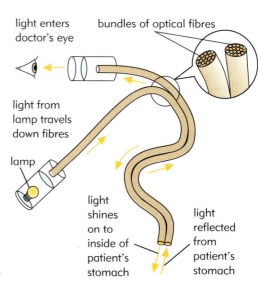

light enters doctor's eye

bundles of optical fibres

light from lamp travels down fibres

lamp

light shines on to inside of patient's stomach

light reflected from patient's stomach

Waves bending round corners

Light travels in straight lines. So when light waves go past the edge of an opaque object you get a shadow.

Water waves behave in the same way. Instead of a shadow that is in darkness, you get a shadow of still water. But if you look carefully at the picture, you will see that the water in the shadow area isn't <u>completely</u> still. There has been some bending of the waves around the edge of each barrier. This is called <u>diffraction</u>.

Waves can bend round a barrier because the edge of the barrier acts as a new source of waves. These are called **diffracted waves**.

8 Write down:
  (a) <u>two</u> ways that the diffracted waves are the <u>same</u> as the original waves;
  (b) <u>two</u> ways that the diffracted waves are <u>different</u> from the original waves.

The edge of each barrier acts as a new source of waves. These diffracted waves are much smaller in height than the original waves. We say that they have a smaller <u>amplitude</u>. But they have the same frequency and the same wavelength as the original waves.

Diffraction of sound waves

Sound waves can be diffracted. This is one of the reasons why you can often hear sounds around the corners of buildings, where you would expect a sound shadow.

Long-wavelength sound waves are more strongly diffracted than short-wavelength sound waves.

This is why the person shown in the diagram can hear the tuba clearly but can hardly hear the flute at all.

9 (a) Does the flute or the tuba have a higher pitch and a shorter wavelength?
  (b) Is the note from the flute or from the tuba more strongly affected?

The fact that sound can be diffracted is further evidence that it travels as waves.

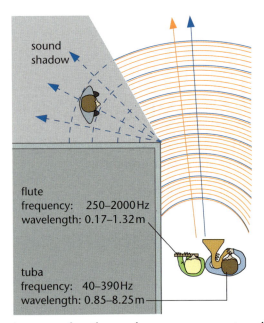

sound shadow

flute
frequency: 250–2000 Hz
wavelength: 0.17–1.32 m

tuba
frequency: 40–390 Hz
wavelength: 0.85–8.25 m

Long wavelength sound waves are more strongly diffracted by the edge of the building. So you can hear the tuba clearly from round the corner.

Diffraction of radio waves

Radio waves can also be diffracted. The diagram shows one of the ways that this can happen.

10 Explain how people can often receive radio waves even though they are in the 'radio shadow' of a hill.

As with sound waves, radio waves with longer wavelengths are more strongly diffracted.

The fact that radio signals can be diffracted is further evidence that they actually do travel as waves. You get diffraction with all types of electromagnetic waves, including light. But as the wavelength of the waves becomes shorter it becomes more difficult to see the effects of the diffraction.

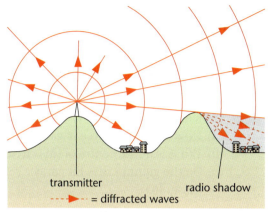

transmitter   radio shadow
- - - ▸ = diffracted waves

Radio waves are diffracted by the top of the hill. This means that the town receives a signal even though it is in the shadow of the hill.

Using different types of radiation

You can kill living cells by giving them a high dose of gamma radiation.

Smaller doses can damage cells. Damage to the cells of your body may cause cancer.

Killing bacteria with gamma rays

We sometimes want to kill harmful bacteria. We can do this using gamma radiation.

1 Write down <u>two</u> uses of gamma rays to kill harmful bacteria.

2 Using gamma rays, you can kill the bacteria on things inside completely sealed packets.
(a) Why is this possible?
(b) Why is this very useful?

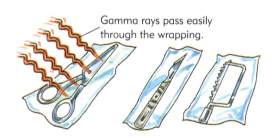

Gamma rays pass easily through the wrapping.

A surgeon's instruments must have no bacteria on them, so we sterilise them with gamma rays.

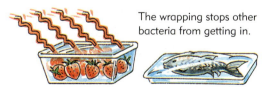

The wrapping stops other bacteria from getting in.

Bacteria make food go bad. If we kill the bacteria, the food stays fresh for longer.

Killing cancer cells with gamma radiation

Doctors can use gamma rays to kill cancer cells inside a person's body, but they must be careful not to damage healthy cells. The diagram shows how they can do this.

3 Look at the diagram.
   (a) How does the source of gamma rays move?
   (b) What is at the centre of this circle?
   (c) Do the gamma rays hit cancer cells or healthy cells most often?

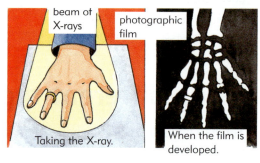

Source of gamma radiation moves in a circle.

Cancer cells at centre of circle. The radiation hits these cells all the time.

Using X-rays safely

X-rays can pass easily through some substances but not through others.

To use X-rays safely and in a useful way, we need to know what substances they will, or won't, pass through.

The diagrams show how you can use X-rays to make a shadow picture of the bones inside a person's hand.

X-rays pass easily through skin and flesh but are absorbed by bone and metal. Photographic film absorbs any X-rays that fall on it. These parts of the film then go black when the film is developed.

4 The X-ray photograph in the diagram shows a broken finger. Which finger is this?

5 Doctors can use X-rays to see whether your lungs are healthy. How do they know if there is diseased tissue in your lungs?

X-rays can damage the cells of your body. Because metals absorb X-rays, they can be used to protect you.

6 Look at the photograph.
   Then write down <u>two</u> other ways of reducing the risk of damaging the cells in people's bodies with X-rays.

beam of X-rays photographic film

Taking the X-ray.

When the film is developed.

The X-rays do not pass through the areas that show up as white. You can't see the skin or flesh because X-rays pass through these areas easily, and turn the photographic film black.

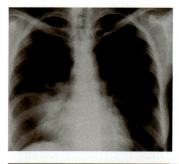

A chest X-ray. Diseased tissue absorbs X-rays more than healthy tissue.

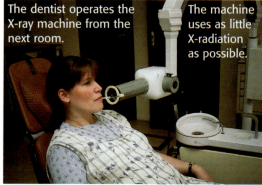

The dentist operates the X-ray machine from the next room.

The machine uses as little X-radiation as possible.

Taking an X-ray photograph at the dentist's.

Using ultraviolet (UV) radiation

The Sun sends out lots of ultraviolet (UV) radiation, some of which falls on the Earth.
Most of this is absorbed by the Earth's atmosphere but some of it gets through.

7 You get a lot of ultraviolet radiation if you go skiing in the mountains.
Write down <u>two</u> reasons for this.

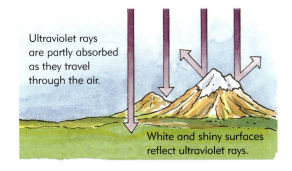

Ultraviolet rays are partly absorbed as they travel through the air.

White and shiny surfaces reflect ultraviolet rays.

The diagrams show some of the effects ultraviolet rays can have on your body.

8 Write down:
(a) <u>two</u> reasons why people might want to let ultraviolet rays on to their skin;
(b) <u>two</u> ways in which ultraviolet radiation can harm your body.

Ultraviolet rays can damage your eyes. So you have to wear dark glasses. Sunbeds can give people with pale skin a **tan**.

Ultraviolet rays absorbed by your skin are used to make vitamin D, but they can also cause skin **cancer**.

Changing ultraviolet radiation into light

Some substances can absorb the energy from ultraviolet radiation and use it to produce light. We say that these substances are **fluorescent**.

The diagrams show some uses for fluorescent substances.

9 Write down <u>two</u> uses for fluorescent substances.

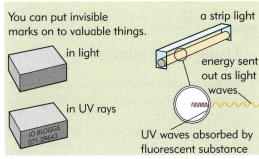

You can put invisible marks on to valuable things.

a strip light

in light

in UV rays

JO BLOGGS
071 29643

energy sent out as light waves

UV waves absorbed by fluorescent substance

Using fluorescent substances.

Using infrared (IR) radiation

Hot things send out infrared (IR) radiation, and when things absorb infrared radiation they get hot.
So infrared rays are often called **heat rays**.

Infrared radiation can be used for cooking, for example in toasters and grills.

10 An electric toaster has a shiny surface between the heating elements and the outer case.
Write down <u>two</u> reasons for this.

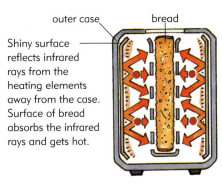

outer case   bread

Shiny surface reflects infrared rays from the heating elements away from the case. Surface of bread absorbs the infrared rays and gets hot.

How a toaster works.

You can often change channels on a television set, or switch on a video player, using a remote control. The diagram shows how this works.

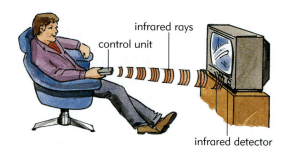

infrared rays
control unit
infrared detector

11 (a) How is your instruction carried to the television set or video player?

(b) Why must you point the remote control at the television set or video player?

Long-distance telephone messages used to be sent as electrical signals through copper wires. They are now mainly carried by infrared rays inside optical fibres.

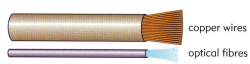

copper wires
optical fibres

12 Write down <u>two</u> advantages of using the optical fibres.

The cable of optical fibres has a much **smaller** diameter but it can still carry the same number of telephone calls.

With copper wires the signal gets weak so you need to boost the signal every 4 or 5 km. With optical fibres there is less **weakening** of the signal.

Using microwaves

To make good use of microwaves, we need to know what will reflect them, transmit them or absorb them.

The diagram shows how a microwave oven works.

13 (a) Why is the case of the microwave oven made of metal?

(b) What happens to water molecules when they absorb microwaves?

(c) Why should you use food containers made from plastic, pottery or glass in a microwave oven?

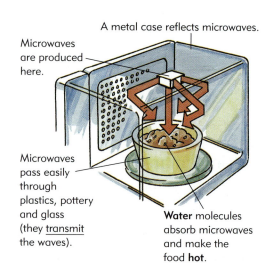

A metal case reflects microwaves.

Microwaves are produced here.

Microwaves pass easily through plastics, pottery and glass (they <u>transmit</u> the waves).

Water molecules absorb microwaves and make the food **hot**.

The diagram shows how television satellites use microwaves with certain wavelengths.

14 (a) What wavelengths of microwaves are used for satellite television?

(b) Why are these wavelengths used?

15 How does a metal satellite dish help the aerial collect a strong signal?

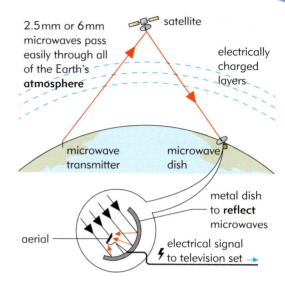

Using radio waves

When we use any kind of waves, we need to know what will reflect them, what will transmit them and what will absorb them.

Radio waves are used to carry radio and television signals. The diagrams show why they are suitable for this job.

16 What substances will radio waves pass through easily?

17 What happens when radio waves are absorbed by an aerial?

18 Why can't you send a radio message to or from a submarine?

The Earth's atmosphere has electrically charged layers. One of the electrically charged layers reflects radio waves with long wavelengths.

The diagram shows how we can use these reflections.

19 Why is it useful to be able to reflect long-wavelength radio waves?

20 What is the wavelength of these radio waves?

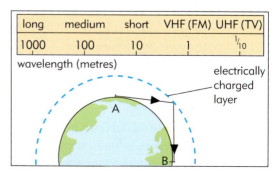

Sending radio waves round the curved surface of the Earth.

Why is everything going digital?

In recent years, the companies who provide telephone services and who transmit TV programmes have been changing over to **digital** signals.

All these use digital information.

> **1** Write down the names of <u>two</u> other common digital devices, besides telephones and television sets.

To understand what digital signals are, and why they are used, it is easiest to start by looking at the type of signal that digital signals are replacing. These are called **analogue** signals.

What are analogue signals?

Before we can send sound to distant places, we first have to convert it to an electrical signal. We can do this using a microphone.

Speech and music both consist of sound waves. The amplitudes and frequencies of these sound waves are constantly changing. These sound waves may be any frequency (within a certain range) and any amplitude (up to a certain maximum). In other words, both the frequency and amplitude of the sound waves vary continuously (without any gaps or jumps).

The electrical signal produced from sound waves by a microphone is very much like the sound waves themselves. The electrical signal can also include any frequency (in a range) and can have any amplitude (up to a maximum). Because the electrical signal varies continuously, just like the sound signal that produced it, we say that it is an <u>analogue</u> signal.

> **2** Though an analogue signal from a sound is very much like the original sound, it also different in some ways. Write down <u>two</u> differences.
>
> **3** (a) What device is used to change an electrical signal back into sound waves again at the place where the signal arrives?
>
> (b) Analogue signals can be sent long distances using microwaves or radio waves.
> What <u>two</u> additional appliances do you need to be able to send and receive signals in this way?

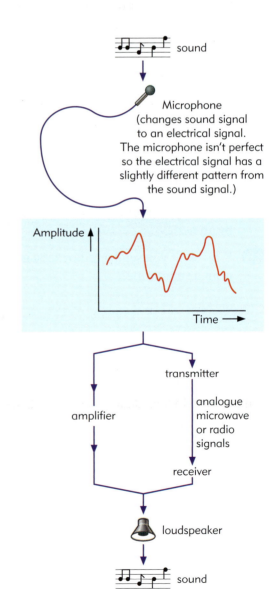

What are digital signals?

To produce a digital signal from the sound of speech or music, a microphone is first used to produce an electrical analogue signal.

But instead of leaving the signal like this, the frequencies and amplitudes of the signal are measured thousands of times each second. This information is then changed into a series of electrical **pulses**, each lasting a tiny fraction of a second.

Each of the pulses can vary in only one way: it is either 'on' (1) or 'off' (0), it is either present or it is absent. This is why it is called a <u>digital</u> signal.

4 Write down the sequence of 'on's and 'off's in the digital signal shown in the diagram.

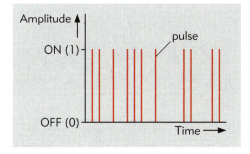

A digital signal is nothing like the sound that produced it. It is a code for the frequencies and amplitudes of the sound.

Because they do not vary continuously like sounds themselves, digital signals are <u>not</u> just like them. They are only <u>approximations</u> to the sounds. However, if we measure the sound often enough and in enough detail when we make the digital signal, our ears can't tell the difference. In fact, digital signals are better in some ways than analogue signals. The diagrams show why.

5 Write down <u>two</u> ways that digital signals are better than analogue signals.

Analogue

Amplitude decreases. Pattern also changes. So it sounds different. Some quality is lost.

Digital

Amplitude decreases. But the pattern stays the same. So quality is kept. You can also send more information in the same time.

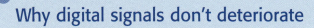

Higher

Why digital signals don't deteriorate

When signals travel long distances:

- they gradually become weaker;
- they may pick up interference which shows itself as 'noise'.

To overcome the problem of the signal gradually weakening, it may be amplified every so often along its journey.

How analogue signals are affected

The diagrams show how the above factors cause a deterioration in the quality of an analogue signal.

Graph A shows the amplitudes of the different frequencies of sound in an analogue signal at one particular time.

As the signal travels, different frequencies may be weakened to different extents. This changes the signal, i.e. its quality deteriorates.

Interference means that the signal picks up 'noise'.

This noise adds itself to the signal so that it is even less like the original. The quality of the signal has deteriorated further.

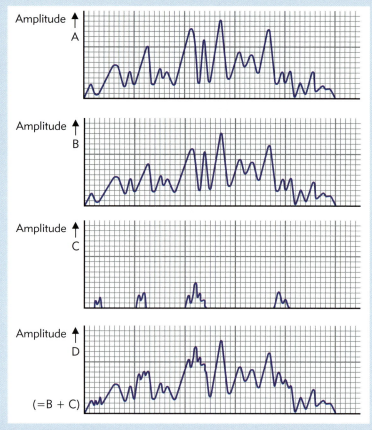

Whenever the signal is amplified the defects are also amplified. So there is no improvement in the quality of the signal.

6 (a) Describe the <u>two</u> ways in which the quality of analogue signals can deteriorate.

(b) Amplifying analogue signals every so often does <u>not</u> improve their quality. Why not?

Higher

How digital signals are affected

The diagrams show how the factors listed at the top of page 500 affect a digital signal.

Graph P shows a small fragment of a digital signal.

All the pulses in a digital signal are identical, so any weakening of a digital signal is likely to affect all the pulses equally.

Interference means that the signal picks up 'noise'.

This noise adds itself to the signal. However, when the signal is amplified, or when it is decoded, pulses that are above a certain amplitude are treated as 'on' and pulses below that amplitude are treated as 'off'.

This means that the signal is restored to its original quality. So there is no deterioration.

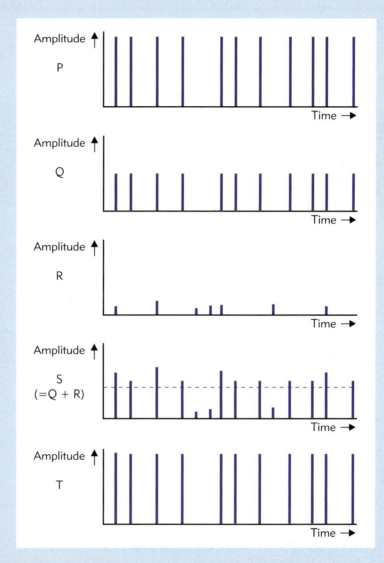

7 Explain why the quality of a digital signal is unaffected:
   (a) by a reduction in its amplitude as it travels;
   (b) by 'noise'.

8 Analogue and digital TV signals can both be received via satellite. Very heavy rain clouds can lower the picture quality from analogue signals but can cause the picture from digital signals to break up altogether.
   How can these different effects be explained?

Radiation that harms your body

When electromagnetic waves are absorbed they release energy. This energy can damage cells or even kill them. Different types of electromagnetic radiation damage cells in different ways.

Radiation that 'cooks' cells

Your cells become hot if they absorb infrared radiation or microwaves.

This heat can damage or kill the cells.

1 Look at the diagrams.
   (a) Why are microwaves more dangerous than infrared radiation?
   (b) Why can't you accidentally damage your cells with the microwaves from a microwave oven?

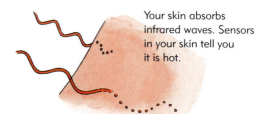

Your skin absorbs infrared waves. Sensors in your skin tell you it is hot.

Microwaves go deeper into your body. There are no temperature sensors there so you don't know the cells are getting hot.

There's nothing to stop you putting your hand under a grill...

...but if you open the door of a microwave oven it switches off.

Radiation that causes skin cancer

Ultraviolet radiation from the Sun is mainly absorbed by your skin.

This can damage the molecules inside skin cells.

The cells can then start to multiply very quickly and also spread to other parts of the body. This is called **cancer** and may cause death.

2 What type of electromagnetic radiation usually causes skin cancer?

3 People with dark skins are less likely to get skin cancer. Why is this?

4 How can you protect yourself against skin cancer? Explain your answer.

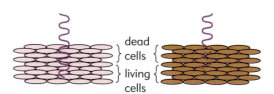

If you have pale skin, ultraviolet rays can get through to the living cells.

If you have dark skin, more ultraviolet rays are absorbed by the layers of dead cells.

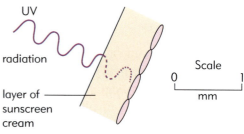

How to protect your skin against cancer. Sunscreen cream absorbs the ultraviolet rays.

Radiation that can cause cancer inside people's bodies

X-rays can pass fairly easily through the soft parts of your body.

Gamma radiation can pass quite easily through any part of your body.

Both types of radiation can cause cancer.

5 Look at the diagram.
(a) Why can X-rays and gamma rays cause cancer?
(b) Where in a person's body can these types of radiation cause cancer?

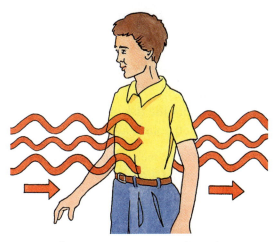

X-rays and gamma rays can pass through your body. But some are absorbed by your cells. This can cause cancer anywhere in your body.

Can any other sorts of radiation cause cancer?

Substances that give out gamma radiation are called **radioactive** substances. The diagram shows two other types of radiation which radioactive substances can give out.

6 What <u>three</u> types of radiation can radioactive substances give out?

All three types of radiation from radioactive substances can damage molecules in cells.

So they can all cause cancer.

Scientists <u>know</u> that the types of radiation mentioned on these pages can harm the cells in your body. Some scientists also <u>suspect</u> that other forms of radiation may harm the cells in your body (see pages 506–507).

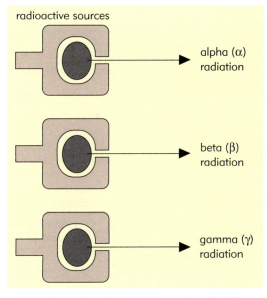

We say that radioactive sources <u>emit</u> radiation. A radioactive source can emit more than one type of radiation.

How much harmful radiation do you get?

X-rays and the radiation from radioactive substances can definitely cause cancer anywhere in our bodies.
So we should avoid these types of radiation if we can.

What amount of radiation is safe?

The amount of radiation your body gets is called your radiation <u>dose</u>. The graph shows how the risk of cancer depends on the size of your radiation dose.

7 How does the risk of cancer change with the amount of radiation your body receives?

Unfortunately, we're surrounded by harmful radiation. So we can't avoid it all.

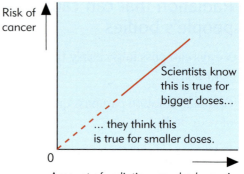

Risk of cancer / Amount of radiation your body receives

Scientists know this is true for bigger doses...
... they think this is true for smaller doses.

Radiation that we can't avoid

We are bombarded with radiation from space, called **cosmic radiation**. There are also radioactive substances all around us and even inside the cells our bodies are made from. So our bodies receive radiation all the time, which we call **background radiation**.

8 The boxes on this page and the maps on the next page show the main sources of background radiation.

Copy the table. Then fill in the figures for each source of the background radiation that your body gets.

How much background radiation my body gets each year	
Source of background radiation	**Annual radiation dose (units)**
cosmic rays	
buildings	
food and drink	
the ground	
the air	
TOTAL	

Cosmic rays

These come from the Sun and from space.

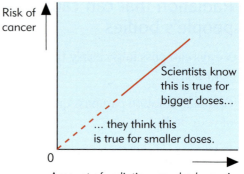

Earth

We all get 250 units of cosmic radiation each year.
You get 1 unit more for every 30 m you live above sea level.

Buildings

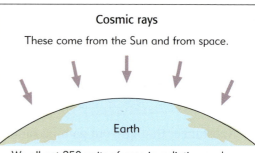

There are radioactive substances in stone, brick and concrete. This gives you about 350 units of radiation a year.

Food and drink

Radioactive substances dissolve. They get in plants, animals that eat the plants and into water. This gives you about 300 units of radiation each year.

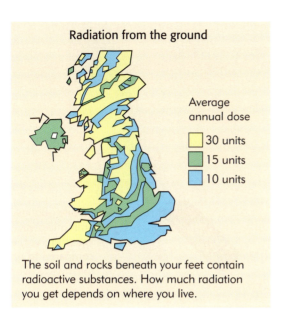

Radiation from the ground

Average annual dose

- 30 units
- 15 units
- 10 units

The soil and rocks beneath your feet contain radioactive substances. How much radiation you get depends on where you live.

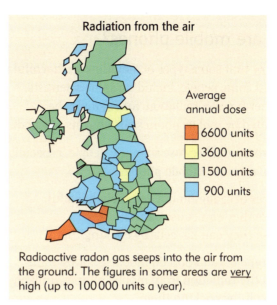

Radiation from the air

Average annual dose

- 6600 units
- 3600 units
- 1500 units
- 900 units

Radioactive radon gas seeps into the air from the ground. The figures in some areas are <u>very</u> high (up to 100 000 units a year).

Radiation from nuclear power stations

If you live within a mile of a nuclear power station, you should add an extra 5 units a year to your radiation dose.

This is much less than many people think.

Radiation that depends on what you do

The table shows the main sources of any extra radiation which people may receive.

Other sources of radiation	Dose
dental X-ray	20 units each time
chest/leg/arm X-ray	50 units each time
flying	4 units per hour

9 (a) Write down any of the sources that you think have affected you during the past year.
  (b) Add your radiation dose from these sources to your total from background radiation.

Scientists think that the average annual radiation dose in Britain is reasonably safe.

10 (a) What is the average annual radiation dose in Britain?
  (b) How does your own annual dose compare with the average?

The average radiation dose in Britain is about 2500 units each year.

11 Some people get a bigger radiation dose because of their job. Write down <u>two</u> jobs that give people a bigger radiation dose.

How safe are mobile phones?

Scientists <u>know</u> that some types of radiation are harmful to our bodies. But some people, including some scientists, <u>think</u> that other types of radiation, for example the radiation emitted by mobile phones, also harms our bodies.

This is very worrying, because so many people, especially young people, now use mobile phones.

12 Which types of radiation are definitely harmful:
(a) because they can cause cancer?
(b) because they can kill cells by cooking them?

13 (a) What type of radiation do mobile phones emit?
(b) Why is the radiation from mobile phones unlikely to kill cells by cooking them?

Mobile phones emit microwaves. These are <u>very</u> weak compared to a microwave oven.

Do mobile phones cause cancer?

Several surveys have been done, in Sweden and in the USA as well as in the UK, to see if there is a link between mobile phones and brain cancer (see Box).

14 Write down <u>two</u> reasons why these surveys do not show a definite link between mobile phones and brain cancer.

Mobile phones and brain cancer

Some surveys show a very slight increase in brain cancer amongst people who use mobile phones.

However:
- not all surveys show this;
- the number of cases is very small;
[If you toss a coin just three times there's quite a good chance you'll get heads every time.]

- there was a decrease of other types of cancer amongst mobile phone users. [So do mobile phones <u>protect</u> people against other forms of cancer?]

Can microwaves cause cancer?

Many scientists do not believe that microwaves can damage cells in a way that causes cancer.

The diagram shows how other scientists have tried to prove them wrong.

15 (a) What did these other scientists do?
(b) What did their experiments show?

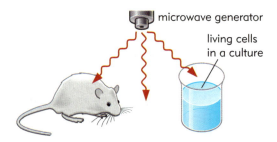

microwave generator

living cells in a culture

Some experimenters have found slight increases in cancer. Others have found no difference or even a decrease in cancer.

Mobile phones and the media

Reports in newspapers and on TV are sometimes made very dramatic so that they grab people's attention. This means that they are not always very fair or balanced. In the newspaper article, the reported facts are all correct but the headline is only an opinion.

16 (a) Why is the headline only an opinion?
(b) What would have to be done to show a definite link between mobile phone masts and brain cancer?

Man killed by microwaves

Just one year after a mobile phone mast was put up near his house, John Smith (47) developed a brain tumour. Six months later he was dead.

His widow, Mary Smith (45) is campaigning to get all mobile phone masts removed from places close to where people live and work.

Should you use a mobile phone?

Many of the things that we do have some risk.
For example, every time you cross the street there is a risk that you might get knocked down.

So we always have to balance the **benefits** of whatever we want to do against the **risks**. This is what each of us has to do with mobile phones.

17 Given that there <u>might</u> be a health risk with mobile phones, which of the following do you think is the best policy?

A Never use a mobile phone.
B Use a mobile phone only in real emergencies.
C Use a mobile phone whenever it's useful, but for no longer than you need to.
D Use a mobile phone whenever you feel like it and for as long as you like.

Are power lines a health hazard?

Some people, including some scientists, believe that living close to high-voltage power lines can cause **leukaemia** (a blood cancer). Others disagree.

As with mobile phones, there is no definite evidence either way about power lines.

18 What do you think should be the government's policy about power lines?

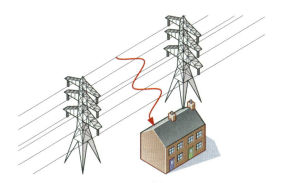

Power lines give out very low frequency, very long wavelength radiation.

Investigating radioactive substances

Substances that give out alpha, beta or gamma radiation are called <u>radioactive</u> substances. We say that these substances **emit** radiation.

Radioactive substances emit radiation all the time. There is nothing you can do to stop this.

1. Look at the diagrams.
   Then write down what happens to the amount of radiation a radioactive substance emits:
   (a) when you heat it up or cool it down;
   (b) when you dissolve it;
   (c) when you break it up into small pieces.

You can't make a radioactive substance emit radiation faster or slower.

This makes it different from other kinds of radiation, which you <u>can</u> control.

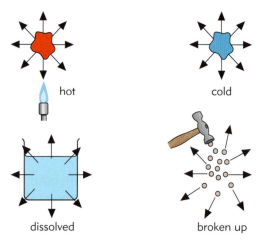

hot   cold

dissolved   broken up

Nothing you can do to a radioactive substance changes how much radiation it emits.

An object emits light if you make it hot enough.

A cold object emits less infrared radiation than a hotter object.

2. (a) What can you do to a substance to make it emit light?
   (b) How can you make a substance emit less infrared radiation?

Detecting radiation.

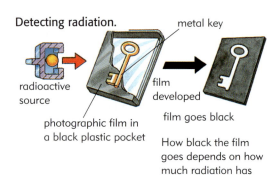

radioactive source

photographic film in a black plastic pocket

metal key

film developed

film goes black

How black the film goes depends on how much radiation has fallen on it.

How do we know that radioactive substances emit radiation?

You can't see, hear or feel the radiation from radioactive substances, but you can tell that it is there.
The diagrams show how you can do this.

3. Write down <u>two</u> ways of detecting the radiation from a radioactive substance.

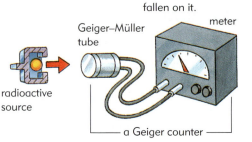

Geiger–Müller tube

meter

radioactive source

a Geiger counter

The faster the source emits radiation, the bigger the reading on the meter. You always get <u>some</u> reading because of background radiation.

4   Which would be the best way:
    (a) of measuring how <u>fast</u> a radioactive source is emitting radiation?
    (b) of telling how <u>much</u> radiation a person has been exposed to during a whole week?

5   Even when it isn't near a radioactive source, a Geiger counter gives a small reading. Why is this?

How do we know that there are three types of radiation?

Radioactive substances emit three types of radiation.

6   What are the <u>three</u> types of radiation called?

We know radioactive sources emit three different types of radiation because they are absorbed by different materials.

7   Look at the diagrams.
    Write down what is needed to absorb the radiation emitted:
    (a) from source A;
    (b) from source B;
    (c) from source C.

Average background reading = 10 or 11		
Source A	Source B	Source C
reading = 75	reading = 80	reading = 65
reading = 11 — a few cm of air	reading = 79 — very thin paper	reading = 64 — 3 mm of metal
reading = 11 — very thin paper	reading = 10 — 3 mm of metal	reading = 16 — 2 cm of lead
A emits **alpha** (α) radiation.	B emits **beta** (β) radiation.	C emits **gamma** (γ) radiation.

How does a Geiger counter work?

The diagram shows what happens inside a Geiger–Müller (G–M) tube.

8   (a) How does the radiation affect gas molecules in the G–M tube?
    (b) What is the effect of using stronger radiation?
    (c) How does this alter the reading?

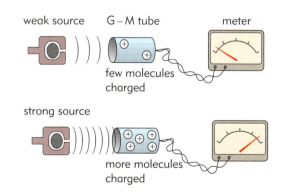

weak source   G–M tube   meter

few molecules charged

strong source

more molecules charged

When radiation hits gas molecules in the G–M tube they become electrically charged.

Alpha, beta, gamma: which is the most dangerous?

Those atoms of elements that are radioactive are called radio-isotopes or **radionuclides**.

It is dangerous to expose your body to the radiation from radionuclides. If the radiation is absorbed by living cells, it can kill them or make them cancerous.

There are three types of radiation from radionuclides: **alpha** (α), **beta** (β) and **gamma** (γ). Which type is the most dangerous depends on whether the source of the radiation is inside or outside of your body.

The diagram on the right shows what happens to the radiation if a radionuclide is inside a cell.

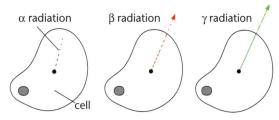

The diagram below shows how easily the different types of radiation can get through the dead outer layers of your skin to the living cells below.

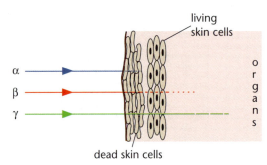
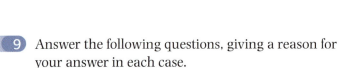

9 Answer the following questions, giving a reason for your answer in each case.
(a) Which type of radiation is <u>least</u> harmful, and which is <u>most</u> harmful when the radionuclide that emits it is inside one of your body's living cells?
(b) Which type of radiation is <u>least</u> harmful and which is <u>most</u> harmful when the source of the radiation is outside your body?

Higher

What are alpha, beta and gamma radiation?

The diagrams show some of the differences between alpha, beta and gamma radiation.

An alpha particle is a helium nucleus.

A beta particle is a fast-moving electron. It is emitted from the <u>nucleus</u> of an atom.

Symbol	Name	Mass	Charge
+	proton	1	+1
●	neutron	1	0
e⁻	electron	almost 0	−1

10 What is:
  (a) an alpha (α) particle?
  (b) a beta (β) particle?
  (c) a gamma (γ) ray?

11 What are the mass and charge of:
  (a) an alpha particle?
  (b) a beta particle?
  (c) a gamma ray?

The electromagnetic spectrum

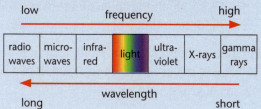

low ——— frequency ——→ high

| radio waves | micro-waves | infra-red | light | ultra-violet | X-rays | gamma rays |

←——— wavelength ———

long   short

Another difference between alpha, beta and gamma radiation

Alpha and beta radiation consist of electrically charged particles. Because these charged particles are moving they are, in effect, an electric current. So if alpha or beta particles pass through a magnetic field, a force acts on them, just as it does on a wire that has a current flowing through it (see page 419).

The diagram shows what happens to alpha, beta and gamma radiation when they pass through a magnetic field.

 12 (a) Describe what, if anything, happens to each type of radiation as it passes through a magnetic field.
  (b) Explain the differences between what happens to the three types of radiation.

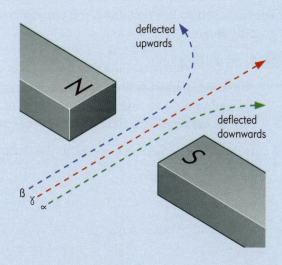

deflected upwards

deflected downwards

Working with radiation

Looking at radioactive decay

You can never tell when a particular radioactive atom will emit radiation and decay. It is a **random** process. But a sample of radioactive material usually contains billions of radioactive atoms so, on average, it emits radiation at a fairly steady rate.

Over a period of time, the sample of radioactive material gradually becomes less radioactive. The time it takes for the radiation to fall to half its original level is called the **half-life**. During one half-life, half of the atoms of the radionuclide will have decayed.

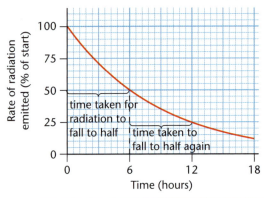

1 (a) What is the half-life of the radionuclide shown on the graph?
(b) Plot a similar graph using the data in the table. Find the half-life of this radionuclide.

Different radionuclides have different half-lives. Half-lives can vary from a tiny fraction of a second to billions of years.

Rate of radiation emitted (% of start)	100	80	60	40	20
Time (hours)	0	3	7	13	23

Using radioactive substances to tell dates

As the radioactive atoms in a substance decay, the substance emits less radiation. You can use this idea to date things.

2 Look at the graph.

Some archaeologists find a piece of wood in an ancient tomb. The carbon in the wood is only 60 per cent as radioactive as the carbon in some new wood.

How old is the piece of wood from the tomb?.

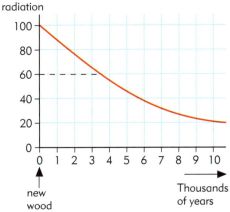

Decay graph for radioactive carbon

Using radioactive substances safely

Radioactive substances are very useful. We use them to kill cancer cells or harmful bacteria. We also use them as fuels in nuclear power stations.

But the radiation from radioactive materials is very dangerous. So we must do everything we can to protect our bodies from this radiation.

3   Why is the radiation from radioactive substances dangerous?

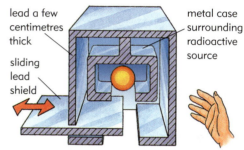

Shielding a small radioactive source (for killing bacteria or cancer cells).

Shielding the radiation from radioactive sources

We need to stop the radiation from radioactive sources from reaching our bodies. To do this we must **shield** the radioactive sources with substances that absorb the radiation.

4   (a) What are small, not very strong, radioactive sources used for?
    (b) How are these radioactive sources shielded?

5   (a) Where would you find large and very powerful radioactive sources?
    (b) How are these radioactive sources shielded?

Thick layers of metal and concrete are needed so that almost all of the gamma radiation is absorbed.

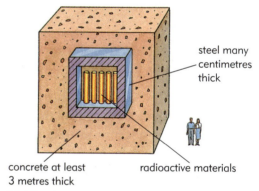

Shielding large, very powerful, radioactive sources in nuclear power stations.

Keeping radioactive substances contained

Radioactive substances are especially dangerous if they get inside our bodies. The radiation that can damage cells is then emitted very near to the cells, or even inside them.

6   Radioactive substances from nuclear power stations sometimes escape into the environment.
    Write down <u>three</u> ways in which this can happen.

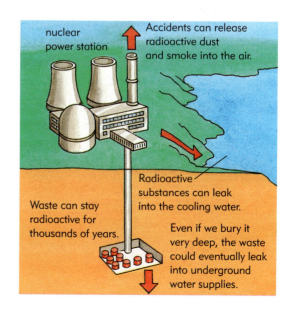

Checking radiation doses

People who work with radioactive materials often wear **film badges**. They can then check how much radiation their bodies have received each week.

7 Look at the diagrams.
What does the developed film tell you?

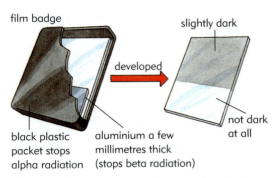

film badge

slightly dark

developed

not dark at all

black plastic packet stops alpha radiation

aluminium a few millimetres thick (stops beta radiation)

When the film is developed, it goes dark in the places where radiation has been absorbed. The film goes darkest where the most radiation has fallen.

Another use for radioactive substances

Radioactive substances can help us to make things into thin sheets.

For example, aluminium cooking foil is very thin. So it only partly absorbs beta radiation.

The thicker the foil is, the more beta radiation it absorbs. We can use this idea to control the thickness of aluminium foil when we are making it. The diagrams show how we can do this.

8 The rollers squeeze the aluminium sheet into thin foil. Beta rays are aimed through the foil at a G–M tube. A control box connected to the G–M tube controls how hard the rollers squeeze.
   (a) How does the amount of radiation reaching the G–M tube change as the foil gets thicker?
   (b) If the foil is too thick, how should the pressure on the rollers change?

9 Explain why alpha radiation and gamma radiation are not suitable for controlling the thickness of aluminium foil.

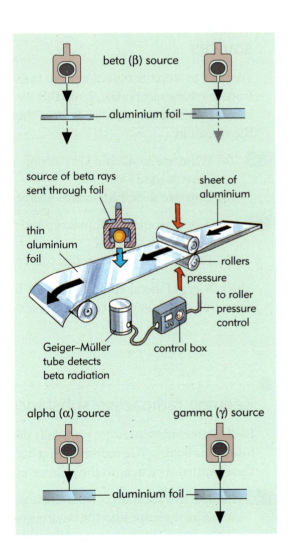

beta (β) source

aluminium foil

source of beta rays sent through foil

sheet of aluminium

thin aluminium foil

rollers

pressure

to roller pressure control

Geiger–Müller tube detects beta radiation

control box

alpha (α) source

gamma (γ) source

aluminium foil

Dating rocks

Uranium atoms are radioactive. They have a very long half-life. They decay to produce a series of radionuclides, each of which has a relatively short half-life. When the last of this series of radionuclides decays, it produces stable atoms of **lead**.

Some igneous rocks contained uranium atoms when they were first formed. We can date these rocks by comparing the numbers of uranium atoms and lead atoms that are now in the rock. The example explains how we can do this.

10 An igneous rock contains one atom of uranium-238 for every three atoms of lead-206 (formed from the decay of the uranium).
    (a) What percentage of the uranium atoms is still there?
    (b) How old is the rock? [Use the graph.]

Other rocks contained atoms of a **potassium** radio-isotope when they were first formed. These atoms decay to form a stable isotope of a gas called **argon**. Sometimes this argon gets trapped inside the rock. We can then date the rock by comparing the numbers of potassium-40 and argon-40 atoms that it now contains.

11 A rock has three times as many argon-40 atoms as potassium-40 atoms.
    (a) What percentage of potassium atoms is still there?
    (b) In one half-life, half the potassium atoms change into argon atoms. After two half-lives, what proportion of potassium atoms remains?
    (c) How old is the rock?

12 (a) Draw a graph showing the decay of potassium-40 over <u>three</u> half-lives.
    (b) Use the graph to estimate the age of a rock that contains two potassium-40 atoms for every three argon-40 atoms.

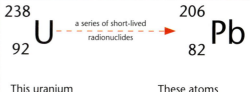

This uranium radionuclide has a half-life of 4.5 billion years.

These atoms of lead are stable.

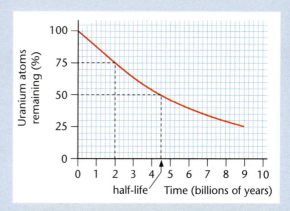

Example
A sample of rock contains 3 atoms of uranium-238 for every 1 atom of lead-206. In other words, 75% of the uranium-238 atoms are still there. From the graph, the rock is about 2 billion years old.

This potassium radionuclide has a half-life of 1.3 billion years.

These atoms of argon are stable.

Using radioactive tracers

Small amounts of radionuclides can be used to find out what happens to substances inside the bodies of plants and animals. The picture shows an example. Radionuclides that are used in this way are called **tracers**.

13 (a) What would be a suitable half-life for the iodine radionuclide used as a tracer – seconds, minutes, hours, days or weeks?
  (b) What type of radiation should the iodine radionuclide emit?
  Give <u>two</u> reasons for your answer.

14 Scientists suspect that chemical pollution in a river is coming from a waste dump.
  (a) How could they use a radionuclide to find out?
  (b) What properties should the radionuclide have to make this test as effective and as safe as possible?

A doctor injects her patient with a radionuclide of iodine.

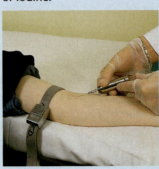

After a few hours, she can measure the uptake of iodine by the patient's thyroid gland. She does this with a radiation detector <u>outside</u> the patient's body.

Remember

Alpha radiation is very easily absorbed. Inside your body it can cause serious damage to your cells. Beta radiation can pass quite easily through the soft parts of your body. Gamma radiation can pass through your body very easily indeed.

Nuclear fission

In nuclear reactors, atoms with a very large nucleus – for example, uranium atoms – are bombarded with neutrons. The diagram shows what happens.

15 Describe <u>in words</u> the nuclear fission reaction shown in the diagram.
[Your answer should include why the reaction is called nuclear fission.]

16 The fission of one gram of nuclear fuel can release about a million times more energy than burning one gram of an ordinary fuel such as coal or gas. Explain why.

Did you know?
The fission of an atom releases a lot more energy than when two atoms make a chemical bond.

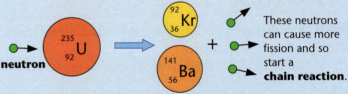

The new atoms produced by fission are radioactive.

The fission (splitting) of a uranium atom.

What are atoms made of?

Radioactive substances, like all other substances, are made from **atoms**. They emit radiation because of changes inside their atoms.

To understand these changes, you need to know how atoms themselves are made.

What's inside an atom?

In some substances all the atoms are the same; these substances are called <u>elements</u>.

Gold is an element. The diagram shows what each atom of gold is made of.

1 Each gold atom contains three different kinds of particles.
(a) What are the <u>three</u> different kinds of particles called?
(b) Which of these particles have an electrical charge?

2 Gold has no electrical charge. What does this tell you about the charge of the protons and the charge of the electrons in the atom?

Every atom has the same number of protons as electrons.

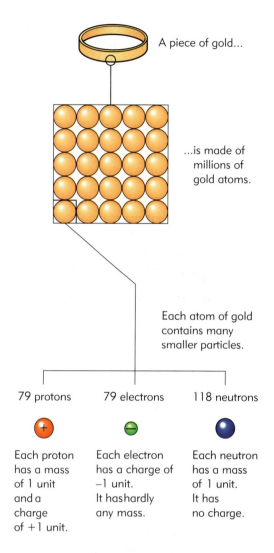

A piece of gold...

...is made of millions of gold atoms.

Each atom of gold contains many smaller particles.

79 protons	79 electrons	118 neutrons
Each proton has a mass of 1 unit and a charge of +1 unit.	Each electron has a charge of −1 unit. It has hardly any mass.	Each neutron has a mass of 1 unit. It has no charge.

The 'Christmas pudding' model of an atom

The particles inside atoms are very, very small.

You can't see them even with a very powerful microscope.

But scientists once thought that atoms were probably made up a bit like a Christmas pudding.

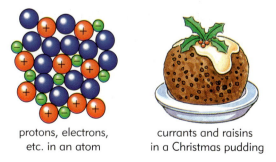

protons, electrons, etc. in an atom

currants and raisins in a Christmas pudding

'Christmas pudding' model of an atom.

When scientists tested their idea about atoms, they found out that they were wrong.

You can make gold into <u>very</u> thin sheets, just a few atoms thick. This is called **gold leaf**. Ernest Rutherford and his students tried firing some very fast particles at a sheet of gold leaf.

The diagrams showed what the scientists expected to happen and what actually happened.

3 (a) What did the scientists expect to happen?
   (b) What actually happened?

This test meant that scientists had to change their mind about how atoms were made. They did so very quickly. This was partly because Rutherford was a very eminent scientist but also because there were no well established theories that depended on the earlier model of the atom.

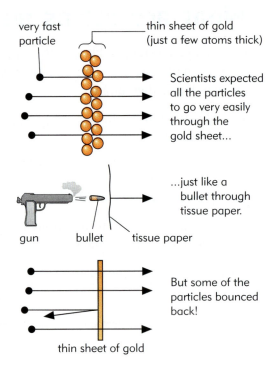

very fast particle — thin sheet of gold (just a few atoms thick)

Scientists expected all the particles to go very easily through the gold sheet...

...just like a bullet through tissue paper.

gun   bullet   tissue paper

But some of the particles bounced back!

thin sheet of gold

A new model of an atom

Scientists now think that atoms are mainly empty space. The protons and neutrons are in a very small **nucleus**. Electrons move about in the space around the nucleus.

4 Why did a <u>few</u> of the fast moving particles bounce back from the sheet of gold leaf?

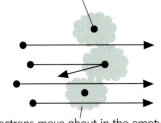

nucleus of atom (protons and neutrons)

electrons move about in the empty space around the nucleus

New model of an atom.

Why are some atoms radioactive?

In most atoms the nucleus doesn't change, so the numbers of protons and neutrons stay the same. We say that these atoms have a **stable** nucleus.

But some atoms have an **unstable** nucleus. Sooner or later, an unstable nucleus will emit radiation. The nucleus changes when it does this, and we say that the nucleus **decays**.

The diagrams show two carbon atoms.

One carbon atom has a stable nucleus; the other carbon atom has an unstable nucleus.

5 Copy the table. Then complete it.

	Number of protons	Number of neutrons	Number of nucleons
stable carbon atom	6	6	12
unstable carbon atom			

6 (a) What is the <u>same</u> about both carbon atoms?
(b) What is <u>different</u> about the two carbon atoms?

All atoms of the same element have the same number of **protons**.
Atoms of the same element can have different numbers of **neutrons**.
Atoms of the same element with different numbers of neutrons are called **isotopes**.

Nucleus of a stable carbon atom.   Nucleus of an unstable carbon atom.

Both atoms have six electrons in the space around the nucleus.

Particles that we find in the nucleus are called nucleons.

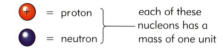 = proton ⎱ each of these nucleons has a
= neutron ⎰ mass of one unit

We can show a stable carbon atom like this:

this is the mass or → 12 **C** ← this stands for carbon
nucleon number

this is the proton → 6
number

We call this a carbon-12 atom.

We can show an unstable carbon atom like this:

14 **C**
6

We call this a carbon-14 atom.

What happens when an unstable nucleus splits up?

When an unstable atom decays:

- it changes into an atom of a different element with a different number of protons in its nucleus;
- it emits radiation.

The diagram shows what happens to the nucleus of a carbon-14 atom when it decays.

7 (a) What does a carbon-14 atom emit when it decays?
(b) What does its nucleus change into?

8 (a) What is the <u>same</u> about a carbon-14 nucleus and a nitrogen-14 nucleus?
(b) How is a nitrogen-14 nucleus <u>different</u> from a carbon-14 nucleus?

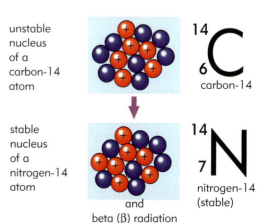

unstable nucleus of a carbon-14 atom

14 **C**
6
carbon-14

stable nucleus of a nitrogen-14 atom

14 **N**
7
nitrogen-14 (stable)

and beta (β) radiation

Nuclear equations

When a radioactive atom decays, an atom of a different element is produced. The examples explain how you can show what happens in the form of a **nuclear equation**.

This line of numbers shows the total numbers of protons and neutrons. These <u>balance</u>: 226 = 222 + 4.

This line of numbers shows the numbers of protons. These <u>balance</u>: 88 = 86 + 2.

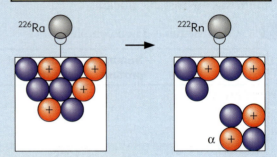

Example 1

A radioactive atom of radium emits an alpha particle from its nucleus and changes into an atom of radon. We can write down what happens like this:

$$^{226}_{88}\text{Ra} \rightarrow ^{222}_{86}\text{Rn} + ^{4}_{2}\text{He} \ (\alpha \text{ particle})$$

Both these lines balance.

The bottom line tells you that a beta particle is emitted when a neutron decays into a proton.

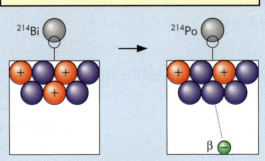

Example 2

An atom of bismuth emits a beta particle from its nucleus and changes into an atom of polonium. We can write down what happens like this:

$$^{214}_{83}\text{Bi} \rightarrow ^{214}_{84}\text{Po} + ^{0}_{-1}\text{e} \ (\beta \text{ particle})$$

9 Suggest <u>two</u> different ways of finding out which type(s) of radiation a radioactive source emits. [You may need to refer back to page 509 for part of your answer.]

10 Copy down the following nuclear equations. Under each one, write down <u>in words</u> what has happened.

(a) $^{222}_{86}\text{Rn} \rightarrow ^{218}_{84}\text{Po} + ^{4}_{2}\text{He} \ (\alpha)$

(b) $^{228}_{88}\text{Ra} \rightarrow ^{228}_{89}\text{Ac} + ^{0}_{-1}\text{e} \ (\beta)$ [Ac = actinium]

Index

515260